AN INTRODUCTION TO FINITE ELEMENT, BOUNDARY ELEMENT, AND MESHLESS METHODS

With Applications to Heat Transfer and Fluid Flow

Darrell W. Pepper
University of Nevada Las Vegas
Alain J. Kassab
University of Central Florida
Eduardo A. Divo
Embry-Riddle Aeronautical University

ASME PRESS

Library of Congress Cataloging-in-Publication Data

Pepper, D. W. (Darrell W.)
An introduction to finite element, boundary element, and meshless methods with applications to heat transfer and fluid flow/Darrell W. Pepper, University of Nevada Las Vegas, Alain J. Kassab, University of Central Florida, Eduardo A. Divo, Embry-Riddle Aeronautical University.
pages cm
Includes bibliographical references and index.
ISBN 978-0-7918-6033-5
1. Fluid dynamics—Mathematical models. 2. Heat—Transmission—Mathematical models. 3. Finite element method. 4. Boundary element methods. 5. Meshfree methods (Numerical analysis) I. Kassab, A. (Alain J.) II. Divo, E. III. Title.

QA911.P39 2014
532'.05015182—dc23
2014009054

DEDICATION

To the students and masters of these elegant numerical methods, as well as future numerical methods yet to come.

TABLE OF CONTENTS

PREFACE

This book stems from our experiences in teaching numerical methods to both engineering students and experienced, practicing engineers in industry. The emphasis in this book deals with finite element, boundary element, and meshless methods. Much of the material comes from courses we have conducted over many years at our institutions, including AIAA home study and ASME short courses presented over several decades, as well as from the suggestions and recommendations of our colleagues and students. There are numerous books on applied numerical methods, many of them being finite element and boundary element textbooks available in the literature today. However, there are very few books dealing with meshless methods, especially those showing how nearly all of these numerical schemes originate from the fundamental principles of the method of weighted residuals. We find that when students once master the concepts of the finite element method (and meshing), it's not long before they begin to look at more advanced numerical techniques and applications, especially the boundary element and meshless methods (since a mesh is not required). Our intent in this book is to provide a simple explanation of these three powerful numerical schemes, and to show how they all fall under the umbrella of the more universal method of weighted residuals approach.

The book is divided into three sections, beginning with the finite element method, then progressing through the boundary element method, and finally ending with the meshless method. Each section serves as a stand-alone description, but it is apparent to see how each conveniently leads to the other techniques. We recommend that the reader begin with the finite element method, as this serves as the primary basis for defining the method of weighted residuals.

We begin by introducing the basic fundamentals of the finite element method using simple examples. Particular attention is given to the development of the discrete set of algebraic equations, beginning with simple one-dimensional problems that can be solved by inspection, and continuing to two- and three-dimensional elements. Once these principles are grasped, we then introduce the concept of boundary elements, and the relative ease with which one reduces the dimensionality of a problem (a great relief when solving large problems, or problems with infinite domain boundaries). The boundary element technique is a natural extension of the finite element method, and becomes greatly appreciated by users. While the method has some limitations regarding the wide range of applications afforded by the finite element technique, it is still a very popular and useful method. It is finding use in crack growth and related applications dealing with structural mechanics, and couples nicely with finite element meshes.

The more recent introduction of meshless methods is rapidly becoming a method now being used by practitioners of both finite element and boundary element methods. The method is simple to grasp, and simple to implement. The power of the method is becoming more appreciated with time. The meshless method has been shown to yield solutions with accuracies comparable to finite element methods employing an extensive number of elements, yet requiring no mesh (or connectivity of nodes). While there is much left to discover with regards to some of the formulation and parameters used in the development of the meshless method, it is a method with much promise and wide spread applications. We have used it for structural analysis, fluid flow, heat transfer, and various biomedical applications.

We provide computer files in both MathCad and MATLAB that are used to illustrate the setup and subsequent solutions of these example problems. These computer codes are not elegant nor optimized for efficiency, but do provide the reader with the logic and steps necessary to obtain solutions. The code listings are available from the www.fbm.centecorp.com website, along with example data files.

There are many commercially available finite element codes available in the market, and a few that are free via the web. We tend to use COMSOL because of its ease of use, and it multiphysics capabilities. COMSOL is a very versatile finite element code that handles a wide variety of applications, including fluid flow, heat transfer, solid mechanics, and electrodynamics. This package runs on PCs.

Because many finite element and boundary element books are written for the structurally oriented engineer, those nonstructural engineers and students more interested in the fluid-thermal fields must sift through undesired concepts and applications before finding a relevant problem area. We have found that students quickly grasp the basic concepts of heat transfer and can easily follow the principles of heat flow and one degree of freedom (temperature). A simple generic approach is utilized in this book that is focused on the transport and diffusion of heat (scalar transport); we then illustrate how one can extend these basic approaches to wider applications, with emphasis on the nonlinear equations for fluid motion.

We wish to thank our colleagues and former students who have greatly contributed to the material presented in this book. We began some years ago by offering several free short courses stemming from the information within this book to our colleagues in the ASME Heat Transfer Division. We gaged their reactions and interests, and have incorporated their suggestions in arranging the presentation of information and material. We especially wish to thank Erik Pepper and Mrs. Julie Longo for their efforts in editing the manuscript and graphical images in this book, and to our ASME Press Editor, Mary Grace Stefanchik, for her helpful comments and editorial assistance; we also wish to thank our former students and colleagues for their patience in reading and suggestions for revising the manuscript.

Darrell W. Pepper
Alain Kassab
Eduardo Divo
September 9, 2014

Overview

THE METHOD OF WEIGHTED RESIDUALS (MWR)

This book is focused on three numerical methods utilized for analysis of field problems in heat transfer and fluid flow: the Finite Element Method (FEM), the Boundary Element Method (BEM), and the Meshless Method (MM). The three numerical methods discussed in this book, and for that matter, most of the other commonly utilized numerical methods, including Finite Difference Method (FDM) [1] and Finite Volume Method (FVM) [2,3], can be formulated in the single over-arching framework of the Method of Weighted Residuals (MWR) [4].

In order to develop the MWR formulation, let us consider a typical steady-state heat transfer problem where the temperature, $T(x,y)$, is governed by the heat conduction equation and subjected to either first kind (prescribed temperature) or second kind (prescribed temperature gradient) boundary conditions,

G.E.: $\nabla^2 T(x,y) + u_G = 0 \qquad \vec{r} \in \Omega$

B.C.'s: $T(x_s,y_s) = T_s \qquad \vec{r}_s \in \Gamma_T$

$\left.\dfrac{\partial T}{\partial n}\right|_{(x_s,y_s)} = q_s \qquad \vec{r}_s \in \Gamma_q$

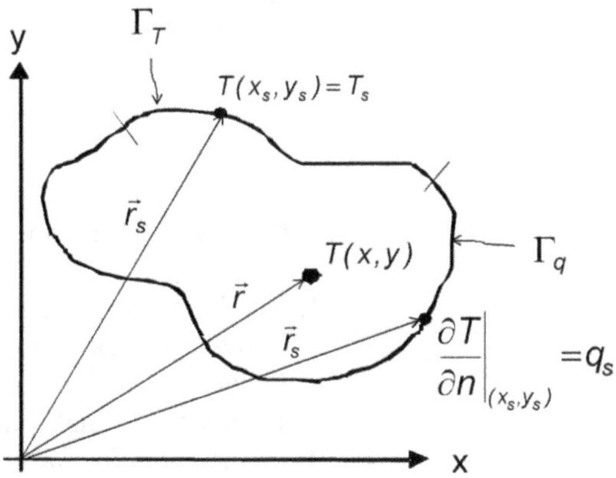

where $\vec{r}_s$ is the position vector to a point (x_s,y_s) on the boundary Γ binding a domain Ω. As a note, we choose this problem as an illustrative example, and the procedure we now outline can apply to any other governing scalar or vector linear or non-linear equation subject to any other type of boundary condition not listed above.

The basic premise of MWR is to approximate the temperature by a set of trial functions, $\phi_j(x, y)$, as

$$\tilde{T}(x,y) = \sum_{j=1}^{N} \alpha_j \phi_j(x,y) \tag{1}$$

We are free to choose to have localized or global support, with the only obvious requirement that the trial functions must be linearly independent. The expansion coefficients, α_j, may have physical meaning, such as representing nodal temperatures in FDM and FVM, or may be arbitrary.

Introducing Eq. (1) into the governing equation leads to a domain residual, $R_\Omega(x, y)$,

$$R_\Omega(x,y) = \nabla^2 \tilde{T}(x,y) + u_G \quad (x,y) \in \Omega \tag{2}$$

Introducing Eq. (1) into the boundary conditions leads to boundary residuals. In particular, this leads to a residual, $R_{\Gamma_T}(x,y)$, on the Γ_T portion of the boundary where a first kind boundary condition is imposed

$$R_{\Gamma_T}(x,y) = \tilde{T}(x_s,y_s) - T_s \qquad (x_s,y_s) \in \Gamma_T \tag{3}$$

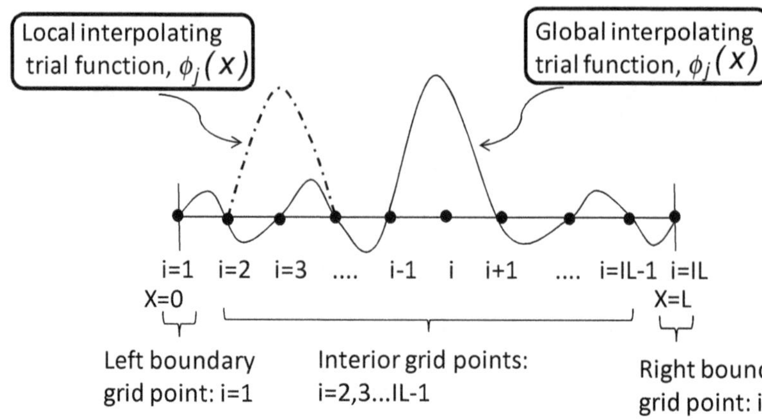

Figure 1. Illustration of 1-D local and global trial functions, $\phi_j(x)$.

and to a residual, $R_{\Gamma_q}(x,y)$, on the Γ_q portion of the boundary where a second kind boundary condition is imposed

$$R_{\Gamma_q}(x,y) = \frac{\partial \tilde{T}(x_s,y_s)}{\partial n} - q_s \qquad (x_s,y_s) \in \Gamma_q \tag{4}$$

Depending on our *choice of trial functions* any of these residuals may be zero, and that choice broadly differentiates numerical methods from a MWR perspective as being:

1. *Interior methods:* trial functions satisfy the boundary conditions, and this leads to a domain residual only.
2. *Boundary methods:* trial functions satisfy the governing equation, and this leads to boundary residuals only.
3. *Mixed methods:* trial functions satisfy neither the governing equation nor the boundary conditions, and this leads to both a domain and boundary residuals.

The FDM, FVM, and FEM are mixed methods with trial functions that have local support. The BEM is a boundary method, and the MM is a mixed method with trial functions that have, depending on the technique, either global or local support as referenced Fig. 1, with the latter the most widely used in practice.

The next task in MWR is to determine the unknown expansion coefficients by minimizing the residual. To this end, weighting functions are introduced: (a) a weighting function, $W_\Omega(x,y)$, for the domain residual $R_\Omega(x,y)$; (b) a weighting function, $w_{\Gamma_T}(x,y)$, for the boundary residual, $R_{\Gamma_T}(x,y)$, on portion Γ_T of the boundary; (c) a weighting function, $w_{\Gamma_q}(x,y)$, for the boundary residual, $R_{\Gamma_q}(x,y)$, on portion Γ_q of the boundary. A weighted residual statement is then formulated to solve for the expansion coefficients,

$$\oiint_\Omega R_\Omega(x,y)w_{\Omega,j}(x,y)d\Omega + \int_{\Gamma_T} R_{\Gamma_T}(x,y)w_{\Gamma_T,j}(x,y)d\Gamma + \int_{\Gamma_q} R_{\Gamma_q}(x,y)w_{\Gamma_q,j}(x,y)d\Gamma = 0 \quad j = 1,2...N \tag{5}$$

What further differentiates MWR techniques from each other is the *choice of the weighting functions* that leads to the following common minimization techniques:

1. **Collocation Method:** A set of collocation points, $\vec{r}_i$, is distributed on the domain and the boundary and the choice for the weighting function is the Dirac delta function, $\delta(\vec{r}-\vec{r}_i)$, acting at each one of these points,

$$w_j(\vec{r}) = \delta(\vec{r}-\vec{r}_j) \tag{6}$$

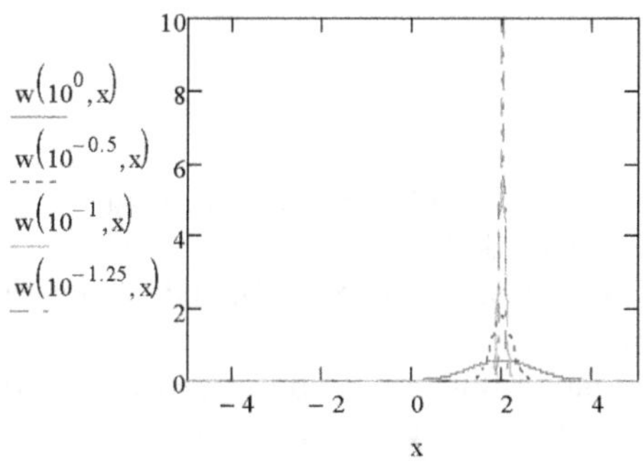

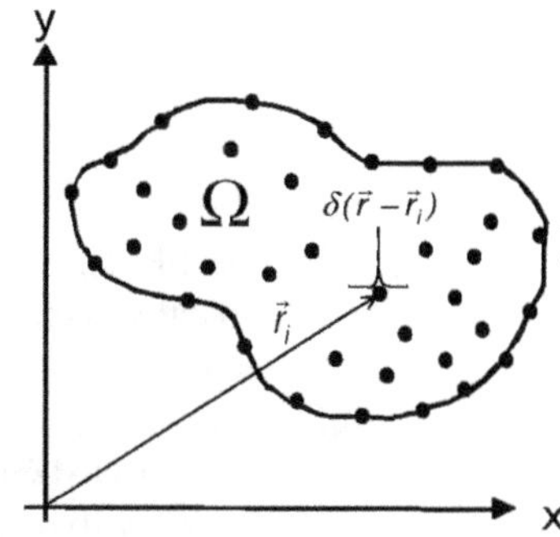

Figure 2. Plot of 1-D delta sequence acting at $x_i = 2$ and tending to Dirac delta function as $k \rightarrow 0$, and a Dirac delta function acts at a point (x_i, y_i) in a 2D domain.

The Dirac delta function is defined by its action on other functions, namely

$$\int_{-\infty}^{+\infty} \delta(x-x_i) f(x) dx = f(x_i) \quad \text{and} \quad \int_{-\infty}^{+\infty} \delta(x-x_i) dx = 1 \tag{7}$$

The Dirac delta function can be approximated numerically by any number of so-called delta sequences [6], for instance the following sequence obeys the property of a delta function in the given limit,

$$\delta(x-x_i) = \lim_{k \rightarrow 0} \left[\frac{e^{-\left(\frac{x-x_i}{k^2}\right)^2}}{\sqrt{\pi k}} \right] \tag{8}$$

as seen in Fig. 2. Multidimensional delta functions can be constructed as products of 1D delta functions. In Cartesian coordinates for instance: $\delta(x, y; x_i, y_i) = \delta(x-x_i)\delta(y-y_i)$. Collocation MWR is used to solve the governing equations in strong form and is the method employed to formulate the FDM and strong-form meshless methods. The FDM is a collocation MWR with local shape functions, typically taken as polynomials, the collocation points, $\vec{r}_i$, are called the mesh/grid points and are produced automatically by mesh generation procedures, the expansion coefficients, α_j, are the FDM nodal temperatures.

2. **Subdomain Method:** The domain Ω is subdivided into N-subdomains Ω_j, and the weighting function is chosen to be

$$\begin{aligned} w_j(\vec{r}) &= 1 \quad \text{if} \quad \vec{r} \in \Omega_j \\ &= 0 \quad \text{if} \quad \vec{r} \notin \Omega_j \end{aligned} \tag{9}$$

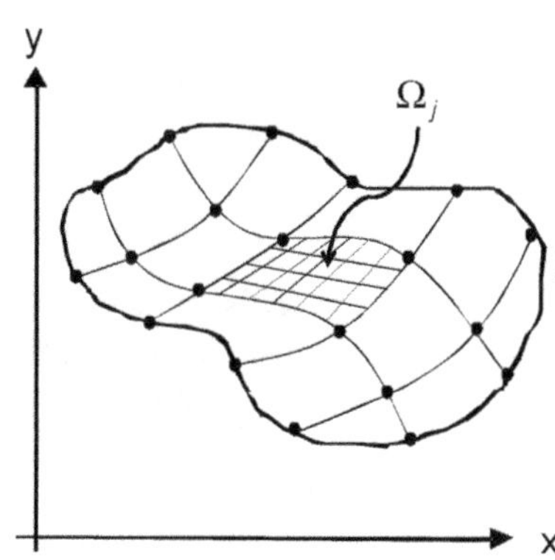

The FVM is a subdomain MWR with local shape functions, typically taken as polynomials, the subdomains are called finite volumes and are generated automatically by mesh generation techniques, and the expansion coefficients, α_j, are the FVM nodal temperatures.

3. **Galerkin Method:** The weighting function is chosen to be the expansion function itself, that is

$$w_j(\vec{r}) = \phi_j(\vec{r}) \tag{10}$$

The FEM is most often formulated using Galerkin MWR, using local shape functions, typically taken as polynomials, that are defined over a set of N-subdomains Ω_j called finite elements, and the expansion coefficients, α_j, are the FEM nodal temperatures.

4. **Least-Squares:** The weighting function is chosen to be the partial of the residual with respect to the expansion coefficients, α_j, that is

$$w_j(\vec{r}) = \frac{\partial R(\vec{r})}{\partial \alpha_j} \tag{11}$$

For example, supposing that we are considering a domain residual, we have

$$\oiint_{\Omega} R_{\Omega}(\vec{r}) \frac{\partial R_{\Omega}(\vec{r})}{\partial \alpha_j} d\Omega = \frac{1}{2} \frac{\partial}{\partial \alpha_j} \oiint_{\Omega} R_{\Omega}^2(\vec{r}) d\Omega \tag{12}$$

which is obviously a least-square minimization with respect to the expansion coefficients, α_j. There are some FEM formulations and meshless method formulations that utilize the concept of least-squares. There is another MWR formulation that minimizes using moments of the residual and the reader is referred to [4] for details on that method. The method of moments MWR finds applications as a numerical method in electromagnetics.

MWR EXAMPLE PROBLEM: FDM, FVM, FEM, BEM AND MM

Let us consider a simple 1D problem where the temperature is governed in a region $X \in [0, L]$ by the following non-homogeneous differential equation and first kind boundary conditions,

G.E.: $\dfrac{d^2T(x)}{dx^2} + T(x) + x = 0 \qquad x \in [0, L]$

B.C.'s: $T(0) = T_o$
$T(L) = T_L$

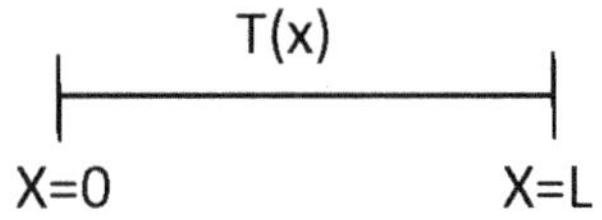

The exact solution to this problem is readily obtained as,

$$T(x) = T_o \cos(x) + \left(\frac{T_L + L - T_o \cos(L)}{\sin(L)} \right) \sin(x) - x \tag{13}$$

with the exact derivative of the temperature given by

$$q(x) = -T_o \sin(x) + \left(\frac{T_L + L - T_o \cos(L)}{\sin(L)} \right) \cos(x) - 1 \tag{14}$$

This temperature profile is illustrated in Fig. 3 for values of $T_o = 15$ and $T_L = 25$. We shall use this problem to illustrate the five numerical methods, FDM, FVM, FEM, BEM, and Localized Collocation Meshless Method (LCMM) formulated by the MWR principle corresponding to the particular method. The final result of the approximation process is an algebraic set of equations that are the discrete analog of the governing equation and boundary conditions that is solved by an appropriate numerical procedure.

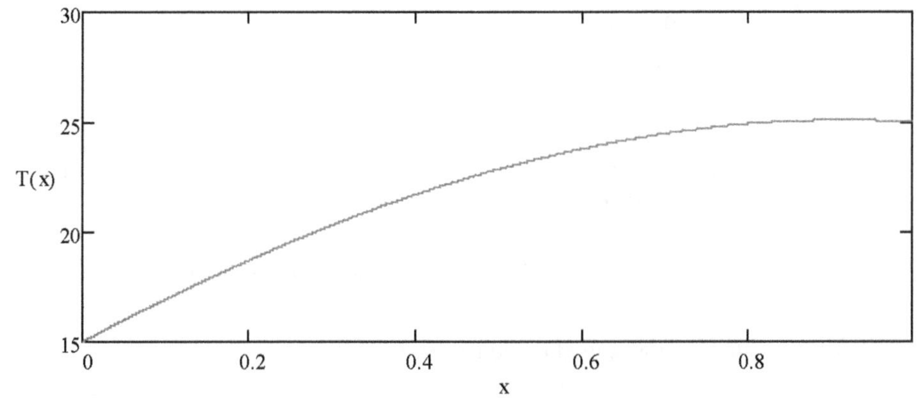

Figure 3. Temperature distribution for the MWR example problem with $T_o = 15$ and $T_L = 25$.

Finite Difference Method (FDM) – Collocation MWR with Local Polynomial Trial Functions

In the FDM [1], we lay out a set of i = 1,2...IL grid points to discretise the domain x ∈ $[0, L]$. This is usually accomplished by a grid generator. We identify the interior grid points, i = 2,3...IL – 1 and boundary grid points, i = 1 and i = IL. Here the grid spacing, $\Delta x = L/(IL-1)$, is uniform, although in general this is not the case as grid adaption is used to resolve regions of high gradients. The solution is sought at discrete locations, x_i, and denoted as $T(x_i) = T_i$, or the FDM nodal values of the temperature. Using collocation MWR, and placing the Dirac delta function at any interior node, x_i, we integrate the residual over the domain

$$\int_0^1 \left(\frac{d^2\tilde{T}}{dx^2} + \tilde{T} + x \right) \delta(x - x_i)\, dx = 0 \quad \text{for} \quad i = 2,3 \ldots IL-1 \tag{15}$$

and there results the residual equation at the grid point x_i,

$$\left. \left(\frac{d^2\tilde{T}}{dx^2} + \tilde{T} + x \right) \right|_{x_i} = 0 \qquad \text{for} \quad i = 2,3 \ldots IL-1 \tag{16}$$

Using a local quadratic polynomial approximation for, $\tilde{T}(x)$, over grid points i – 1, i and i + 1, with the origin x = 0 located at the grid point x_i,

$$\tilde{T}(x) = \alpha_1 + \alpha_2 x + \alpha_3 x^2 \tag{17}$$

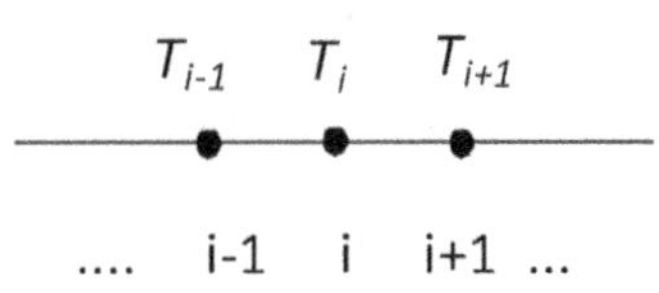

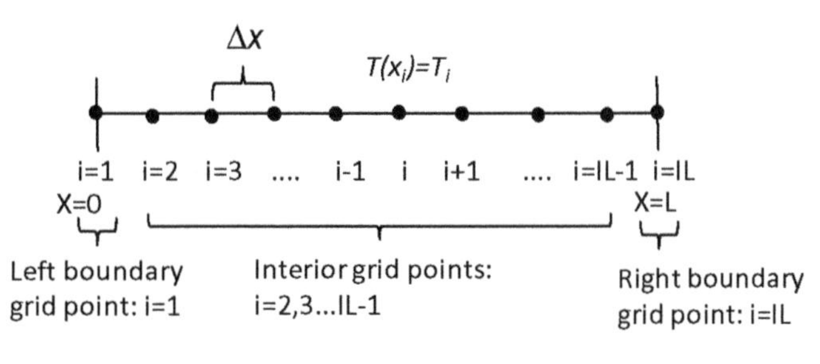

Figure 4. Discretization of the 1D domain used in the FDM.

one finds that

$$\tilde{T}(x) = T_i + \left(\frac{T_{i+1} - T_{i-1}}{2\Delta x} \right) x + \left(\frac{T_{i+1} - 2T_i + T_{i-1}}{2\Delta x^2} \right) x^2 \tag{18}$$

and upon introducing the above local approximation for the temperature into Eq. (16), we arrive at the interior FDM algebraic equation,

$$\frac{T_{i-1} - 2T_i + T_{i+1}}{\Delta x^2} + T_i + x_i = 0 \quad \text{for} \quad i = 2,3...IL-1 \tag{19}$$

that is re-arranged in the tri-diagonal form

$$\left(\frac{1}{\Delta x^2} \right) T_{i-1} + \left(1 - \frac{2}{\Delta x^2} \right) T_i + \left(\frac{1}{\Delta x^2} \right) T_{i+1} = -x_i \tag{20}$$

Defining the FDM coefficients, $a_i = \left(\frac{1}{\Delta x^2} \right), b_i = \left(1 - \frac{2}{\Delta x^2} \right), c_i = \left(\frac{1}{\Delta x^2} \right), d_i = -x_i$, and applying the first kind boundary conditions at $x = 0$ and $x = L$, the following set of tri-diagonal FDM equations is readily assembled and efficiently solved by the Thomas Algorithm,

$$\begin{bmatrix} 1 & 0 & 0 & 0 & \cdots & 0 \\ a_2 & b_2 & c_2 & 0 & \cdots & 0 \\ 0 & a_3 & b_3 & c_3 & \cdots & 0 \\ \vdots & \vdots & \ddots & \ddots & \ddots & \vdots \\ 0 & \cdots & 0 & a_{IL-1} & b_{IL-1} & c_{IL-1} \\ 0 & 0 & 0 & 0 & 0 & 1 \end{bmatrix} \begin{Bmatrix} T_1 \\ T_2 \\ T_3 \\ \vdots \\ T_{IL-1} \\ T_{IL} \end{Bmatrix} = \begin{Bmatrix} T_o \\ d_2 \\ d_3 \\ \vdots \\ d_{IL-1} \\ T_L \end{Bmatrix} \tag{21}$$

Using IL = 6 grid points, the MATHCAD spreadsheet calculation for the FDM is provided below:

Finite Difference Method:

$IL := 6$ ...no of grid points

$\Delta x := \dfrac{L}{IL - 1}$ $\quad \Delta x = 0.2$ .... mesh spacing

$x_i := (i - 1)\cdot \Delta x$ ... x-location of ith grid point

$a := \dfrac{1}{\Delta x^2}$ $\quad b := \left(1 - \dfrac{2}{\Delta x^2}\right)$ $\quad c := \dfrac{1}{\Delta x^2}$ ...FDM coefficients

$A_{i,j} := 0$

$d_i := 0$ ...initialize problem

Load FDM equations

...load left boundary condition

$i := 1$

$A_{i,i} := 1$

$d_i := To$

$i := 2, 3 .. IL - 1$ ...load interior FDE's

$A_{i,i-1} := a$

$A_{i,i} := b$

$A_{i,i+1} := c$

$d_i := -x_i$

$i := IL$

$A_{i,i} := 1$ ...load right hand side temperature boundary conditions FDE

$d_i := TL$

$i := 1, 2 .. IL$

$$A = \begin{pmatrix} 1 & 0 & 0 & 0 & 0 & 0 \\ 25 & -49 & 25 & 0 & 0 & 0 \\ 0 & 25 & -49 & 25 & 0 & 0 \\ 0 & 0 & 25 & -49 & 25 & 0 \\ 0 & 0 & 0 & 25 & -49 & 25 \\ 0 & 0 & 0 & 0 & 0 & 1 \end{pmatrix} \qquad d = \begin{pmatrix} 15 \\ -0.2 \\ -0.4 \\ -0.6 \\ -0.8 \\ 25 \end{pmatrix}$$

...echo FDM matrix and RHS

$T_{FDM} := A^{-1}\cdot d$ ...solve

$err_{FDM_i} := \dfrac{\left|T_{FDM_i} - T(x_i)\right|}{\left|T(x_i)\right|}$ ....compute relative error

$T_{FDM_i} =$	$T(x_i) =$	$err_{FDM_i} =$
15	15	0
18.73251	18.72608	$3.437\cdot 10^{-4}$
21.70772	21.69763	$4.653\cdot 10^{-4}$
23.79863	23.78822	$4.375\cdot 10^{-4}$
24.91359	24.90653	$2.833\cdot 10^{-4}$
25	25	0

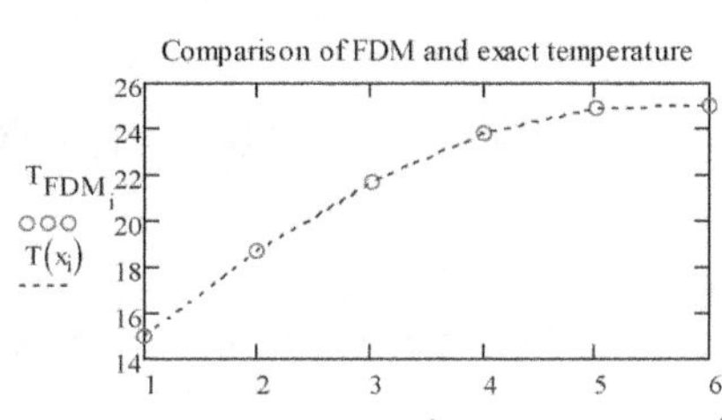

Finite Volume Method – Subdomain MWR with Local Polynomial Trial Functions

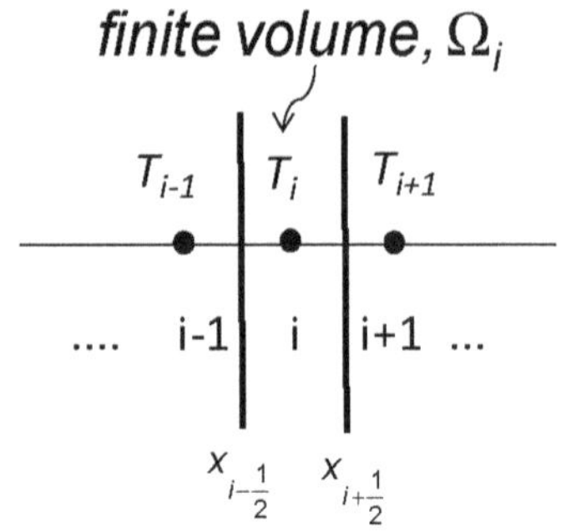

In the finite volume method (FVM) [2, 3], the same discretization as in Fig. 4 may be used, except that now a subdomain (finite volume) $\Omega_i \in \left[x_{i-\frac{1}{2}}, x_{i+\frac{1}{2}}\right]$ surrounding each grid point x_i is defined to extend from, $x_{i-\frac{1}{2}}$, the half way mark between i – 1 and i, and $x_{i+\frac{1}{2}}$ located at the half-way mark between i and i + 1. In this case the subdomain MWR is applied and

$$\int_0^1 \left(\frac{d^2\tilde{T}}{dx^2} + \tilde{T} + x \right) w_i(x)\,dx = 0 \qquad \text{for} \quad i = 2,3 \ldots IL-1 \tag{22}$$

leads to

$$\int_{x_{i-\frac{1}{2}}}^{x_{i+\frac{1}{2}}} \left(\frac{d^2\tilde{T}}{dx^2} + \tilde{T} + x \right) dx = 0 \qquad \text{for} \quad i = 2,3 \ldots IL-1 \tag{23}$$

since $w_i(x) = 0$ outside the subdomain Ω_i. Integrating the second derivative leads to

$$\left.\frac{d\tilde{T}}{dx}\right|_{x_{i+\frac{1}{2}}} - \left.\frac{d\tilde{T}}{dx}\right|_{x_{i-\frac{1}{2}}} + \int_{x_{i-\frac{1}{2}}}^{x_{i+\frac{1}{2}}} \left(\tilde{T} + x\right) dx = 0 \qquad \text{for} \quad i = 2,3 \ldots IL-1 \tag{24}$$

Noting that the first two terms are related to the flux in and out of the subdomain (finite volume) Ω_i, and this expression integrates the source term over the finite volume, unlike FDM that collocates and samples the generation term at the grid point, the FVM expresses a conservation principle on the grid. This is a distinction that becomes very important in non-linear and multi-dimensional problems. We are now left with introducing the approximation for $\tilde{T}(x)$ to arrive at the FVM algebraic analog. In FVM, various local interpolations are utilized. We shall use local linear interpolation between grid points to evaluate the $T(x)$ at the finite volume faces, so that,

$$\tilde{T}(x) = \alpha_1 + \alpha_2 x = \begin{cases} \dfrac{T_{i+1} + T_i}{2} + \left(\dfrac{T_{i+1} - T_i}{\Delta x} \right) x & \text{for} \quad x \in [x_i, x_{i+1}] \\[2ex] \dfrac{T_i + T_{i-1}}{2} + \left(\dfrac{T_i - T_{i-1}}{\Delta x} \right) x & \text{for} \quad x \in [x_{i-1}, x_i] \end{cases} \tag{25}$$

Resulting in the following expressions for the derivatives in Eq. (24),

$$\left.\frac{d\tilde{T}}{dx}\right|_{x_{i\pm\frac{1}{2}}} = \begin{cases} \left.\dfrac{d\tilde{T}}{dx}\right|_{x_{i+\frac{1}{2}}} = \left(\dfrac{T_{i+1} - T_i}{\Delta x} \right) \\[2ex] \left.\dfrac{d\tilde{T}}{dx}\right|_{x_{i-\frac{1}{2}}} = \left(\dfrac{T_i - T_{i-1}}{\Delta x} \right) \end{cases} \tag{26}$$

For consistency, using the trapezoidal rule that integrates a linear interpolation, and the interpolation we developed in Eq. (25), the integral of $\tilde{T}(x)$ over the finite volume is evaluated as

$$\int_{x_{i-\frac{1}{2}}}^{x_{i+\frac{1}{2}}} \tilde{T}(x)dx = \left(\frac{\Delta x}{8}\right)T_{i-1} + \left(\frac{3\Delta x}{4}\right)T_i + \left(\frac{\Delta x}{8}\right)T_{i+1} \tag{27}$$

Integrating the source term analytically over of finite volume and putting it all together, Eq. (24) becomes

$$\left(\frac{T_{i+1}-T_i}{\Delta x}\right) - \left(\frac{T_i - T_{i-1}}{\Delta x}\right) + \left[\left(\frac{\Delta x}{8}\right)T_{i-1} + \left(\frac{3\Delta x}{4}\right)T_i + \left(\frac{\Delta x}{8}\right)T_{i+1}\right] = -\frac{1}{2}\left(x^2_{i+\frac{1}{2}} - x^2_{i-\frac{1}{2}}\right) \tag{28}$$

Diving by Δx, we arrive at the FVM algebraic analog,

$$\left(\frac{1}{\Delta x^2} + \frac{1}{8}\right)T_{i-1} + \left(\frac{3}{4} - \frac{2}{\Delta x^2}\right)T_i + \left(\frac{1}{\Delta x^2} + \frac{1}{8}\right)T_{i+1} = -\frac{1}{2}\left(x_{i+\frac{1}{2}} - x_{i-\frac{1}{2}}\right) \tag{29}$$

Defining the FVM coefficients, $a_i = \left(\frac{1}{\Delta x^2} + \frac{1}{8}\right), b_i = \left(\frac{3}{4} - \frac{2}{\Delta x^2}\right), c_i = \left(\frac{1}{\Delta x^2} + \frac{1}{8}\right), d_i = -\frac{1}{2}\left(x_{i+\frac{1}{2}} + x_{i-\frac{1}{2}}\right)$, and applying the first kind boundary conditions at x = 0 and x = L, we arrive at the same tri-diagonal form as in Eq. (21), except with different coefficients. Again using IL = 6 for consistency, the MATHCAD spreadsheet for the FVM implementation and its solution is provided:

Finite Volume Method:

$$a := \left(\frac{1}{8} + \frac{1}{\Delta x^2}\right) \quad b := \left(\frac{3}{4} - \frac{2}{\Delta x^2}\right) \quad c := \left(\frac{1}{8} + \frac{1}{\Delta x^2}\right)$$ ...FVM coefficients

$A_{i,j} := 0$

$d_i := 0$ ...initialize problem

Load FVM equations

$i := 1$ ...load left boundary condition

$A_{i,i} := 1$

$d_i := To$

$i := 2, 3 .. IL - 1$ ...load interior FDE's

$A_{i,i-1} := a$

$A_{i,i} := b$

$A_{i,i+1} := c$

$d_i := -(i-1)\cdot\Delta x$

$i := IL$

$A_{i,i} := 1$ ...load right hand side temperature boundary conditions FDE

$d_i := TL$

$i := 1, 2 .. IL$

$$A = \begin{pmatrix} 1 & 0 & 0 & 0 & 0 & 0 \\ 25.125 & -49.25 & 25.125 & 0 & 0 & 0 \\ 0 & 25.125 & -49.25 & 25.125 & 0 & 0 \\ 0 & 0 & 25.125 & -49.25 & 25.125 & 0 \\ 0 & 0 & 0 & 25.125 & -49.25 & 25.125 \\ 0 & 0 & 0 & 0 & 0 & 1 \end{pmatrix} \qquad d = \begin{pmatrix} 15 \\ -0.2 \\ -0.4 \\ -0.6 \\ -0.8 \\ 25 \end{pmatrix}$$

$T_{FVM} := A^{-1}\cdot d$ ...solve

$$err_{FVM_i} := \frac{\left|T_{FVM_i} - T(x_i)\right|}{\left|T(x_i)\right|}$$ ....compute relative error

$T_{FVM_i} =$	$T(x_i) =$	$err_{FVM_i} =$
15	15	0
18.7229	18.72608	$1.699\cdot10^{-4}$
21.69264	21.69763	$2.299\cdot10^{-4}$
23.78308	23.78822	$2.162\cdot10^{-4}$
24.90304	24.90653	$1.4\cdot10^{-4}$
25	25	0

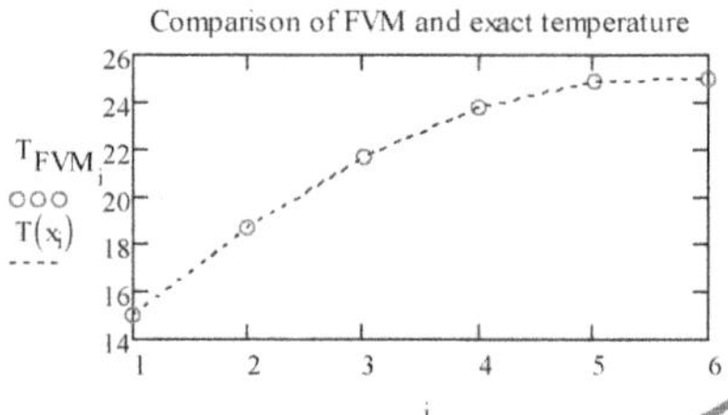

Finite Element Method – Galerkin MWR with Local Polynomial Trial Functions

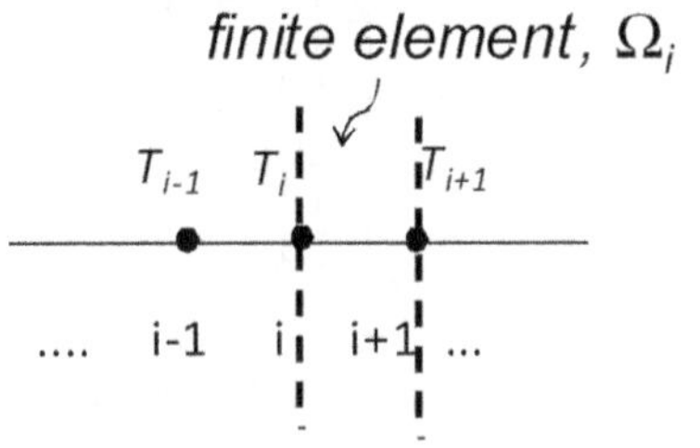

In the FEM [5], the same discretization as in Fig. 4 may be used, interpreting each region Ω_i $[x_i, x_{i+1}]$ as a finite element "i". Applying Galerkin minimization

$$\int_0^1 \left(\frac{d^2\tilde{T}}{dx^2} + \tilde{T} + x \right) w_k(x)dx = 0 \tag{30}$$

and given that $w_i(x) = \phi_i(x)$ and, that in FEM, the trial functions $\phi_i(x)$ are local and are zero outside the finite element Ω_i, then

$$\int_{x_i}^{x_{i+1}} \left(\frac{d^2\tilde{T}}{dx^2} + \tilde{T} + x \right) \phi_k(x)dx = 0 \tag{31}$$

In FEM parlance, this is called the *strong form* in that the approximation $\tilde{T}(x)$ is required to be twice differentiable, hence, linear trial functions may not be used. Technically, the approximation, $\tilde{T}(x)$, is required to be C^2 continuous in strong form. Consequently, integration by parts is used on the highest order derivative (and in 2D and 3D this is equivalent to applying Green's first identity), and there results the so-called *weak form* statement

$$\left(\phi_k(x) \frac{d\tilde{T}}{dx} \right) \Bigg|_{x_i}^{x_{i+1}} - \int_{x_i}^{x_{i+1}} \left(\frac{d\tilde{T}}{dx} \frac{d\phi_k}{dx} \right) dx + \int_{x_i}^{x_{i+1}} \left(\tilde{T} + x \right) \phi_k(x) dx = 0 \tag{32}$$

As only first order derivatives appear, the weak statement admits linear interpolation. Consequently, using linear interpolating functions $\phi_1^{(i)}(x)$ and $\phi_2^{(i)}(x)$, we have within a finite element "i"

$$\tilde{T}(x) = T_i \phi_1^{(i)}(x) + T_{i+1} \phi_2^{(i)}(x) \tag{33}$$

and, the interpolating functions are defined as the linear Lagrange interpolating functions within each element such that

$$\phi_1^{(i)} = \begin{cases} 1 & \text{if} \quad x = x_i \\ 0 & \text{if} \quad x = x_{i+1} \end{cases} \qquad \phi_2^{(i)} = \begin{cases} 1 & \text{if} \quad x = x_{i+1} \\ 0 & \text{if} \quad x = x_i \end{cases} \tag{34}$$

and, moreover, these functions are defined to be zero outside the finite element "i". It is straightforward to find that

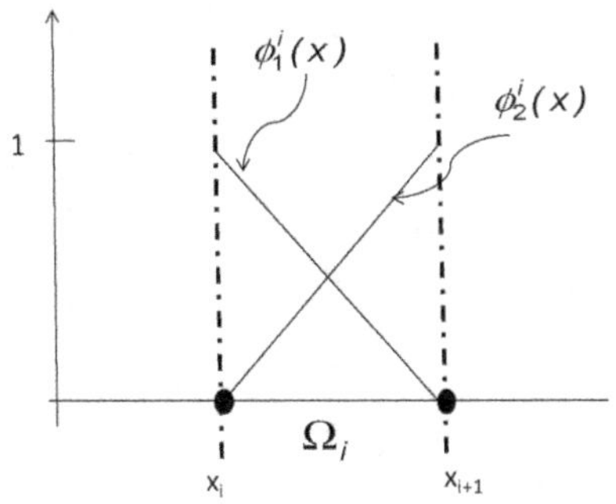

$$\phi_1^{(i)} = \begin{cases} \left(\dfrac{x - x_{i+1}}{x_i - x_{i+1}} \right) & \text{if} \quad x \in [x_i, x_{i+1}] \\ 0 & \text{if} \quad \notin [x_i, x_{i+1}] \end{cases} \qquad \phi_2^{(i)} = \begin{cases} \left(\dfrac{x - x_i}{x_{i+1} - x_i} \right) & \text{if} \quad x \in [x_i, x_{i+1}] \\ 0 & \text{if} \quad \notin [x_i, x_{i+1}] \end{cases} \tag{35}$$

By setting k = 1,2 in each element "i", we arrive at the FEM equation for each finite element

$$\begin{bmatrix} \int_{x_i}^{x_{i+1}} \left(\frac{d\phi_1^i}{dx}\frac{d\phi_1^i}{dx} - \phi_1^i\phi_1^i \right) dx & \int_{x_i}^{x_{i+1}} \left(\frac{d\phi_1^i}{dx}\frac{d\phi_2^i}{dx} - \phi_1^i\phi_2^i \right) dx \\ \int_{x_i}^{x_{i+1}} \left(\frac{d\phi_2^i}{dx}\frac{d\phi_1^i}{dx} - \phi_2^i\phi_1^i \right) dx & \int_{x_i}^{x_{i+1}} \left(\frac{d\phi_2^i}{dx}\frac{d\phi_2^i}{dx} - \phi_2^i\phi_2^i \right) dx \end{bmatrix} \begin{Bmatrix} T_i \\ T_{i+1} \end{Bmatrix} = \begin{Bmatrix} \int_{x_i}^{x_{i+1}} \left(x\phi_1^i \right) dx \\ \int_{x_i}^{x_{i+1}} \left(x\phi_2^i \right) dx \end{Bmatrix} + \begin{Bmatrix} -\left.\frac{dT}{dx}\right|_{x_i} \\ \left.\frac{dT}{dx}\right|_{x_{i+1}} \end{Bmatrix} \tag{36}$$

or in matrix form the algebraic finite element equation is

$$\begin{bmatrix} K_{11}^i & K_{12}^i \\ K_{21}^i & K_{22}^i \end{bmatrix} \begin{Bmatrix} T_i \\ T_{i+1} \end{Bmatrix} = \begin{Bmatrix} F_1^i \\ F_2^i \end{Bmatrix} \tag{37}$$

where we have defined,

$$K_{mn}^i = \int_{x_i}^{x_{i+1}} \left(\frac{d\phi_m^i}{dx}\frac{d\phi_n^i}{dx} - \phi_m^i\phi_n^i \right) dx \tag{38}$$

$$F_1^i = \int_{x_i}^{x_{i+1}} \left(x\phi_1^i \right) dx - \left.\frac{dT}{dx}\right|_{x_i} = f_1^i - q_i \quad \text{and} \quad F_2^i = \int_{x_i}^{x_{i+1}} \left(x\phi_2^i \right) dx + \left.\frac{dT}{dx}\right|_{x_{i+1}} = f_2^i + q_{i+1} \tag{39}$$

Given that at each node there is an influence from the interpolating function $\phi_1^{(i)}$ from element "i" and from the interpolating function $\phi_2^{(i-1)}$ from element "i – 1", a nodal equation can be assembled assuming that the derivatives are continuous at common nodes as

$$K_{21}^i T_i + (K_{22}^i + K_{11}^{i+1}) T_{i+1} + K_{12}^{i+1} T_{i+2} = f_2^i + f_1^{i+1} \quad \text{for} \quad i = 2,3 \ldots IL-1 \tag{40}$$

For the first node, the flux at the 1st node does not cancel out and

$$K_{11}^1 T_1 + K_{12}^2 T_2 = f_1^1 - q_1 \quad \text{for} \quad i = 1 \tag{41}$$

And similarly for node IL the flux at node IL does not cancel out and

$$K_{21}^{IL} T_{IL-1} + K_{22}^{IL} T_{IL} = f_2^{IL} + q_{IL} \quad \text{for} \quad i = IL \tag{42}$$

This process is automatically accomplished via the loading of the local matrix equation into a global matrix equation using the connectivity matrix as described in detail later in the FEM section of this book. The result is a tri-diagonal matrix set of equations in the form

$$\begin{bmatrix} a_1 & b_1 & 0 & 0 & \cdots & 0 \\ a_2 & b_2 & c_2 & 0 & \cdots & 0 \\ 0 & a_3 & b_3 & c_3 & \cdots & 0 \\ \vdots & \vdots & \ddots & \ddots & \ddots & \vdots \\ 0 & \cdots & 0 & a_{IL-1} & b_{IL-1} & c_{IL-1} \\ 0 & 0 & 0 & 0 & a_{IL} & b_{IL} \end{bmatrix} \begin{Bmatrix} T_1 \\ T_2 \\ T_3 \\ \vdots \\ T_{IL-1} \\ T_{IL} \end{Bmatrix} = \begin{Bmatrix} -q_1 \\ 0 \\ 0 \\ \vdots \\ 0 \\ q_L \end{Bmatrix} + \begin{Bmatrix} f_1^1 \\ f_2^1 + f_1^2 \\ f_2^2 + f_1^3 \\ \vdots \\ f_2^{IL-1} + f_1^{IL} \\ f_2^{IL} \end{Bmatrix} \tag{43}$$

The final step in the FEM is to re-arrange the above according to the imposed boundary conditions, moving unknowns to the left and knowns to the right when necessary. For example, since we imposed the temperatures at nodes i = 1 and i = IL, then q_1 and q_{IL} are unknown and consequently the first and last unknowns are switched with corresponding column switches.

$$\begin{bmatrix} 1 & b_1 & 0 & 0 & \cdots & 0 \\ 0 & b_2 & c_2 & 0 & \cdots & 0 \\ 0 & a_3 & b_3 & c_3 & \cdots & 0 \\ \vdots & \vdots & \ddots & \ddots & \ddots & \vdots \\ 0 & \cdots & 0 & a_{IL-1} & b_{IL-1} & 0 \\ 0 & 0 & 0 & 0 & 0 & -1 \end{bmatrix} \begin{Bmatrix} q_1 \\ T_2 \\ T_3 \\ \vdots \\ T_{IL-1} \\ q_{IL} \end{Bmatrix} = \begin{Bmatrix} -a_1 T_1 \\ -a_2 T_1 \\ 0 \\ \vdots \\ 0 \\ 0 \end{Bmatrix} + \begin{Bmatrix} 0 \\ 0 \\ 0 \\ \vdots \\ -c_{IL-1} T_{IL} \\ -b_{IL} T_{IL} \end{Bmatrix} + \begin{Bmatrix} f_1^1 \\ f_2^1 + f_1^2 \\ f_2^2 + f_1^3 \\ \vdots \\ f_2^{IL-1} + f_1^{IL} \\ f_2^{IL} \end{Bmatrix} \tag{44}$$

This tri-diagonal matrix set of equations is then solved. A MATHCAD spreadsheet implementing the FEM solution of the example problem is provided:

Finite Element Method

$\phi 1(i,x) := \dfrac{xx(i+1) - x}{xx(i+1) - xx(i)}$ $\qquad d\phi 1\_dx := \dfrac{-1}{\Delta x}$ $\qquad$ ...define linear finite element shape functions

$\phi 2(i,x) := \dfrac{xx(i) - x}{xx(i) - xx(i+1)}$ $\qquad d\phi 2\_dx := \dfrac{1}{\Delta x}$

Assemble FEM Equations:

$A_{i,j} := 0$ $\qquad$ ...initialize problem

$d_i := 0$

Generate individual element equations and load matrices

$K11(i) := \displaystyle\int_{(i-1)\cdot\Delta x}^{i\cdot\Delta x} d\phi 1\_dx \cdot d\phi 1\_dx - \phi 1(i,x)\cdot\phi 1(i,x)\, dx$

$K12(i) := \displaystyle\int_{(i-1)\cdot\Delta x}^{i\cdot\Delta x} d\phi 1\_dx \cdot d\phi 2\_dx - \phi 1(i,x)\cdot\phi 2(i,x)\, dx$

$K21(i) := K12(i)$

$K22(i) := \displaystyle\int_{(i-1)\cdot\Delta x}^{i\cdot\Delta x} d\phi 2\_dx \cdot d\phi 2\_dx - \phi 2(i,x)\cdot\phi 2(i,x)\, dx$

$f1(i) := \displaystyle\int_{(i-1)\cdot\Delta x}^{i\cdot\Delta x} \phi 1(i,x)\cdot x\, dx$

$f2(i) := \displaystyle\int_{(i-1)\cdot\Delta x}^{i\cdot\Delta x} \phi 2(i,x)\cdot x\, dx$

Assemble stiffness matrix and RHS force (load) vector

$i := 1$ $\qquad$ .. load left boundary equation

$A_{i,i} := K11(i)$

$A_{i,i+1} := K12(i)$

$d_i := f1(i)$

$i := 2, 3 .. IL - 1$ $\qquad$ ...load interior FEM equations

$A_{i,i-1} := K21(i-1)$

$A_{i,i} := K22(i-1) + K11(i)$

$A_{i,i+1} := K12(i)$

$d_i := f2(i-1) + f1(i)$

$i := IL$ $\qquad$ ...load right boundary FEM equation

$A_{i,i-1} := K21(i-1)$

$A_{i,i} := K22(i)$

$d_i := f2(i)$

assembled FEM matrix equations [K]{x}={f} before imposing BC's:

$$A = \begin{pmatrix} 4.933 & -5.033 & 0 & 0 & 0 & 0 \\ -5.033 & 9.867 & -5.033 & 0 & 0 & 0 \\ 0 & -5.033 & 9.867 & -5.033 & 0 & 0 \\ 0 & 0 & -5.033 & 9.867 & -5.033 & 0 \\ 0 & 0 & 0 & -5.033 & 9.867 & -5.033 \\ 0 & 0 & 0 & 0 & -5.033 & 4.933 \end{pmatrix} \qquad d = \begin{pmatrix} 6.667\times 10^{-3} \\ 0.04 \\ 0.08 \\ 0.12 \\ 0.16 \\ 0.113 \end{pmatrix}$$

Rearrange to impose BC's

$U1 := To$

$U2 := TL$

$dd := d - \left(A^{\langle IL \rangle} \cdot U2 + A^{\langle 1 \rangle} \cdot U1\right)$

$$A^{\langle 1 \rangle} = \begin{pmatrix} 4.933 \\ -5.033 \\ 0 \\ 0 \\ 0 \\ 0 \end{pmatrix} \qquad A^{\langle IL \rangle} = \begin{pmatrix} 0 \\ 0 \\ 0 \\ 0 \\ -5.033 \\ 4.933 \end{pmatrix}$$

...modify force vector by adding negative of first column times temperature at left wall and negative of last column times temperature to account for temperature boundary conditions

$A1\_new_i := 0 \qquad A1\_new_1 := 1$

$AIL\_new_i := 0 \qquad AIL\_new_{IL} := -1$

$A^{\langle 1 \rangle} := A1\_new$

$A^{\langle IL \rangle} := AIL\_new$

$$A^{\langle 1 \rangle} = \begin{pmatrix} 1 \\ 0 \\ 0 \\ 0 \\ 0 \\ 0 \end{pmatrix} \qquad A^{\langle IL \rangle} = \begin{pmatrix} 0 \\ 0 \\ 0 \\ 0 \\ 0 \\ -1 \end{pmatrix}$$

...modify first and last columns of stifness matrix to account for temperature boundary conditions

Modified FEM equations read:

$$A = \begin{pmatrix} 1 & -5.033 & 0 & 0 & 0 & 0 \\ 0 & 9.867 & -5.033 & 0 & 0 & 0 \\ 0 & -5.033 & 9.867 & -5.033 & 0 & 0 \\ 0 & 0 & -5.033 & 9.867 & -5.033 & 0 \\ 0 & 0 & 0 & -5.033 & 9.867 & 0 \\ 0 & 0 & 0 & 0 & -5.033 & -1 \end{pmatrix} \qquad dd = \begin{pmatrix} -73.993 \\ 75.54 \\ 0.08 \\ 0.12 \\ 125.993 \\ -123.22 \end{pmatrix}$$

$T_{FEM} := A^{-1} \cdot dd$ ...solve

$i := 1, 2..IL \qquad U_i := T(x_i)$

$U_1 := Q(x_1)$

$U_{IL} := Q(x_{IL})$

...exact solution at grid points accounting for the fact that FEM is solving for q(x) at x=0 and x=L

$$errFEM_i := \frac{\left|T_{FEM_i} - U_i\right|}{\left|U_i\right|}$$

$T_{FEM_i} =$	$U_i =$	$errFEM_i =$
20.22922	20.26688	$1.858\cdot 10^{-3}$
18.71971	18.72608	$3.398\cdot 10^{-4}$
21.68765	21.69763	$4.6\cdot 10^{-4}$
23.77793	23.78822	$4.325\cdot 10^{-4}$
24.89955	24.90653	$2.801\cdot 10^{-4}$
-2.10775	-2.13152	0.011

<u>Note:</u> first and last entries in solution vector are the heat fluxes at the two boundaries as the temperature is specified there

It is noted the integrals required in the FEM are carried out automatically in practice using Gauss-type quadratures, and in the MATHCAD spreadsheets, they are computed using Romberg integration.

Boundary Element Method – Collocation MWR of Boundary Integral Equation

The BEM formulation [7–9] is closely related to the Green's function method [6], and it begins by introducing a weight function, w(x), much like in the MWR and integrating over the domain $x \in [0, L]$ we find

$$\int_0^L \left(\frac{d^2T}{dx^2} + T + x \right) w(x) dx = 0 \tag{45}$$

The choice of the test function will come later and it will be chosen judiciously to allow solution of the problem. Integrating the integral of the second derivative multiplied by the weight function by parts twice (recall that in FEM the integration by parts was done only once), we find

$$\int_0^1 \left(\frac{d^2w}{dx^2} + w \right) T(x) dx + w \frac{dT}{dx} \bigg|_0^L - T \frac{dw}{dx} \bigg|_0^L + \int_0^1 w(x) x dx = 0 \tag{46}$$

Notice that the integration by parts has shifted the differential operator from $T(x)$ to $w(x)$ and has added four terms involving boundary conditions to the equation. It is at this point that we make a choice for $w(x)$, specifically, we require that $w(x)$ solves

$$\frac{d^2w}{dx^2} + w = -\delta(x - x_i) \quad \text{for} \quad -\infty < x < +\infty \tag{47}$$

where $\delta(x - x_i)$ is the Dirac delta function acting at the point, x_i. This equation can be solved for $w(x)$ using Fourier transforms, as shown in the Appendix, and the solution is

$$w^*(x, x_i) = -\frac{1}{2} \sin(| x - x_i |) \tag{48}$$

This is called the Green's free space solution for the differential equation we are considering, and, being able to find it is a critical component of any BEM formulation. With this choice the integral involving the temperature becomes

$$\int_0^L \left(\frac{d^2w^*}{dx^2} + w^*(x, x_i) \right) T(x) dx = \int_0^L -\delta(x - x_i) T(x) dx = \begin{cases} -T(x_i) & \text{if} \quad x_i \in (0, L) \\ -\frac{1}{2} T(x_i) & \text{if} \quad x_i = 0 \quad \text{or} \quad L \end{cases} \tag{49}$$

where the factor of one half at the two boundary points, $x_i = 0$ or $x_i = L$, is due to the fact that the integral over the domain only sees half of the Dirac delta function, and, specifically, the sifting property becomes

$$\lim_{\varepsilon \to 0} \int_\varepsilon^L -\delta(x - x_i) T(x) dx = \lim_{\varepsilon \to 0} \int_0^{L-\varepsilon} -\delta(x - x_i) T(x) dx = -\frac{1}{2} T(x_i) \tag{50}$$

at the boundary points. This can readily be verified by numerically carrying out the integral using the delta sequence in Eq. (8). Consequently, we can write the following expression for our problem

$$c_i T(x_i) + T \frac{dw^*}{dx} \bigg|_0^L = w^*(x, x_i) \frac{dT}{dx} \bigg|_0^L + \int_0^1 w^*(x, x_i) x dx \tag{51}$$

where,

$$c_i = \begin{cases} 1 & \text{if} \quad x \in (0, L) \\ \dfrac{1}{2} & \text{if} \quad x = 0 \quad \text{or} \quad x = L \end{cases} \tag{52}$$

For our problem, we know T(0) and T(L), but we do not know $\left.\dfrac{dT}{dx}\right|_{x=0}$ or $\left.\dfrac{dT}{dx}\right|_{x=L}$. We then formulate a boundary value problem by applying Eq. (51) while taking $x_i = 0$ and L. This leads to two equations in two unknowns

$$\begin{aligned} \frac{1}{2}T(0) + T\left.\frac{dw^*}{dx}\right|_0^L &= w^*(x,0)\left.\frac{dT}{dx}\right|_0^L + \int_0^1 w^*(x,0)x\,dx \\ \frac{1}{2}T(L) + T\left.\frac{dw^*}{dx}\right|_0^L &= w^*(x,L)\left.\frac{dT}{dx}\right|_0^L + \int_0^1 w^*(x,L)x\,dx \end{aligned} \tag{53}$$

The derivative of the Green's free space solution is

$$\frac{dw^*}{dx} = -\frac{1}{2}\cos(|x - x_i|)\,\text{sgn}(x - x_i) \tag{54}$$

where sgn(x) denotes the signum function. In 1D, the BEM leads to a 2 point-boundary value problem expressed in Eq. (53) which can be readily cast into matrix form as

$$\begin{bmatrix} \dfrac{1}{2} - \left.\dfrac{dw^*}{dx}\right|_{x=0,x_i=0} & \left.\dfrac{dw^*}{dx}\right|_{x=L,x_i=0} \\ -\left.\dfrac{dw^*}{dx}\right|_{x=0,x_i=L} & \dfrac{1}{2} - \left.\dfrac{dw^*}{dx}\right|_{x=L,x_i=L} \end{bmatrix} \begin{Bmatrix} T_o \\ T_L \end{Bmatrix} = \begin{bmatrix} -w^*\big|_{x=0,x_i=0} & w^*\big|_{x=L,x_i=0} \\ -w^*\big|_{x=0,x_i=L} & w^*\big|_{x=L,x_i=L} \end{bmatrix} \begin{Bmatrix} q_o \\ q_L \end{Bmatrix} + \begin{Bmatrix} \int_0^1 w^*(x,0)x\,dx \\ \int_0^1 w^*(x,L)x\,dx \end{Bmatrix} \tag{55}$$

or anticipating further BEM development in the sequel section on BEM, we can write the 1D BEM equations in the standard BEM form

$$[H]\{T\} = [G]\{q\} + \{b\} \tag{56}$$

Just as in FEM, at this stage the algebraic equations are re-arranged in standard linear form [A]{x} = {b}, gathering all the unknowns in the vector {x}. In our case, where first kind conditions are imposed on both ends, the BEM equations can be solved for the unknown boundary fluxes directly as.

$$\{q\} = [G]^{-1}[H]\{T\} - [G]^{-1}\{b\} \tag{57}$$

Once the fluxes are determined, then Eq. (51) can be used to find T(x) at any point $x \in [0, L]$. A MATHCAD spreadsheet implementing the BEM is provided:

Boundary Element Method

$ws(y, x_i) := \frac{-1}{2} \cdot \sin(|y - x_i|)$ **Fundamental solution**

$dws(y, x_i) := \frac{d}{dy} ws(y, x_i) \rightarrow -\frac{\cos(|y - x_i|) \cdot \text{signum}(y - x_i, 0)}{2}$ **Derivative of the fundamental solution**

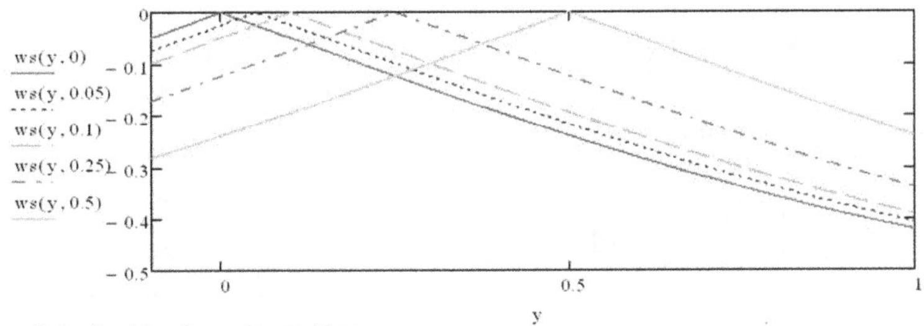

Solve the boundary value problem:

1. build up influence matrices [H] and [G]

$$H := \begin{pmatrix} \frac{1}{2} - dws(0, 0) & dws(L, 0) \\ -dws(0, L) & \frac{1}{2} - dws(L, L) \end{pmatrix} \qquad H = \begin{pmatrix} 0.5 & -0.27 \\ -0.27 & 0.5 \end{pmatrix}$$

$$G := \begin{pmatrix} -ws(0, 0) & ws(L, 0) \\ -ws(0, L) & ws(L, L) \end{pmatrix} \qquad G = \begin{pmatrix} 0 & -0.421 \\ 0.421 & 0 \end{pmatrix}$$

2. impose first kind boundary conditions and re-arrange to generate right hand side vector:

$$b := H \cdot \begin{pmatrix} T(0) \\ T(L) \end{pmatrix} - \begin{bmatrix} \int_0^L ws(y, 0) \cdot y \, dy \\ \int_0^L ws(y, L) \cdot y \, dy \end{bmatrix} = \begin{pmatrix} 0.897 \\ 8.527 \end{pmatrix}$$

2. Solve the equations for the normal derivatives at x=0 and x=L

$$\begin{pmatrix} qo \\ qL \end{pmatrix} := G^{-1} \cdot b \qquad \begin{pmatrix} qo \\ qL \end{pmatrix} = \begin{pmatrix} 20.267 \\ -2.132 \end{pmatrix}$$

Compare with exact boundary derivatives:

$Q(0) = 20.267$

$Q(L) = -2.132$

Compute solution at any interior point and compare with exact solution:

$$T_{bem}(x) := [-(T(L) \cdot dws(L, x) - T(0) \cdot dws(0, x)) + (ws(L, x) \cdot qL - ws(0, x) \cdot qo)] + \int_0^L ws(y, x) \cdot y \, dy$$

$T(0.75) = 24.722$ $\quad T_{bem}(0.75) = 24.722$ $\quad T(0.75) - T_{bem}(0.75) = 1.316 \times 10^{-4}$

$T(0.5) = 22.86$ $\quad T_{bem}(0.5) = 22.86$ $\quad T(0.5) - T_{bem}(0.5) = 3.553 \times 10^{-15}$

$T(0.25) = 19.545$ $\quad T_{bem}(0.25) = 19.545$ $\quad T(0.25) - T_{bem}(0.25) = 0$

$T_{BEM_i} := T_{bem}(x_i)$ $\qquad T_{BEM_1} := T(0)$ $\qquad T_{BEM_{IL}} := T(L)$

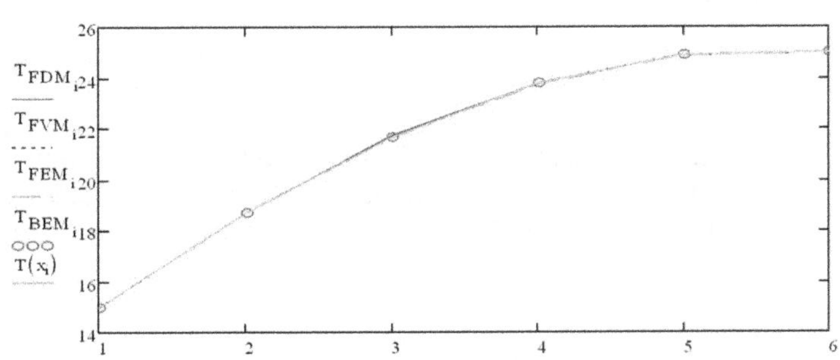

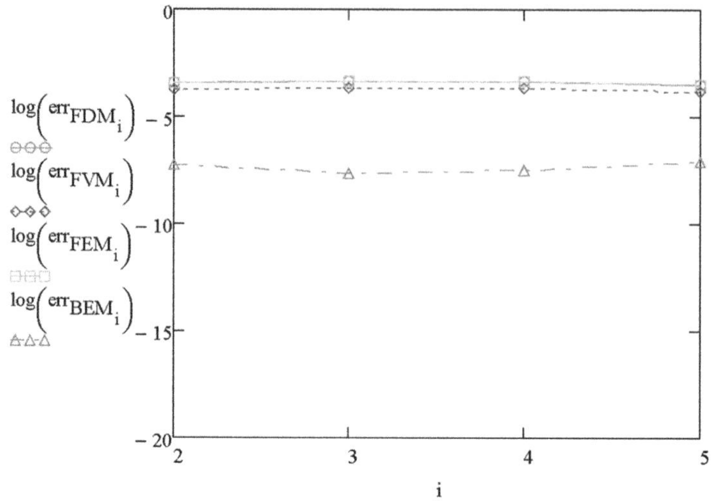

i =	err_{BEM_i} =
1	0
2	$5.166 \cdot 10^{-8}$
3	$2.091 \cdot 10^{-8}$
4	$2.924 \cdot 10^{-8}$
5	$7.915 \cdot 10^{-8}$
6	0

Figure 5. Comparison of relative error between FDM, FVM, FEM, and BEM at internal points in sample problem with IL = 6.

A comparison of the error in solving the problem by the methods covered so far is provided in Fig. 5 and reveals that BEM provides by far the most accurate solution, with FVM providing the most accurate solution, between FDM, FVM, and FEM. The accuracy of the BEM is due to the fact that the boundary value problem that is solved is an exact expression and that the temperature is evaluated at the interior using an exact expression. In 2D and 3D, the BEM requires discretization of the corresponding boundary integral equation. The limitation of BEM is that one must be able to find the fundamental solution and that is not always possible. As such, although very accurate, the BEM cannot be applied to as wide a range of problems as FDM, FVM and FEM.

Meshless Method – Collocation MWR with Global Radial-Basis Function (RBF) Trial Functions

Meshless methods find their roots in the spectral and pseudo-spectral techniques, [10,11], where global Chebycheff and Legendre polynomial expansion on regular point distributions are used in MWR along with domain decomposition. More recent use of global and local radial basis function (RBF) expansions on arbitrary point distributions [12,13] has led to a class of numerical methods called meshless methods [14–19] where a wide variety of governing equations have been successfully solved in strong or weak form. In solving our example problem, we will use the Hardy Multiquadric family of RBF's [13] defined by,

$$\phi_j(x,y) = \left[1 + \frac{r^2(x,y,x_j,y_j)}{S}\right]^{n-\frac{3}{2}} \tag{58}$$

where, $r(x, y, x_j, y_j)$ is the radial (Euclidean) distance from the expansion point (x_j, y_j) to any point (x, y), S is a shape parameter that controls the flatness of the RBF and is set by the user and n is an integer. With n = 1, we retrieve the inverse multiquadric

$$\phi_j(x,y) = \frac{1}{\sqrt{1 + \dfrac{r^2(x,y,x_j,y_j)}{S}}} \tag{59}$$

that we will use to solve our problem. Specifically, we will use a global expansion for the 1D temperature as

$$\tilde{T}(x)=\sum_{j=1}^{N}\alpha_j\phi_j(x,x_j) \tag{60}$$

Notice that in our case, the coefficients of the expansion have no physical meaning in contrast to FDM, FVM, and FEM. A plot of this RBF interpolation function is provided in Fig. 6, with a shape factor S = 10^2 and with r normalized with respect to a point spacing Δx taken to be constant in order to use the same IL = 6 points discretization utilized to solve the problem with FDM, FVM, and FEM. What is striking is that the RBF interpolator is nearly flat. This is the characteristic of such interpolants that high levels (spectral) of accuracy are obtained by controlling the shape parameter to provide a nearly flat profile for functions that do not feature discontinuities [20].

We will solve our problem in strong form, use the same point distribution as in FDM, FVM, and FEM, and, as such, we require the second derivative of the temperature

$$\frac{d^2\tilde{T}(x)}{dx^2}=\sum_{j=1}^{N}\alpha_j\frac{d^2\phi_j(x,x_j)}{dx^2} \tag{61}$$

Introducing the RBF expansion for the temperature Eq. (60) and its second derivative, Eq. (61), into the governing equation, and collocating at the interior points,

$$\sum_{j=1}^{N}\alpha_j\frac{d^2\phi_j(x_i,x_j)}{dx^2}+\sum_{j=1}^{N}\alpha_j\phi_j(x_i,x_j)=-x_i \quad \text{for} \quad i=2,3...IL-1 \tag{62}$$

and in addition at the boundaries we collocate the RBF expansion to impose the boundary conditions

$$\begin{aligned}&\sum_{j=1}^{N}\alpha_j\phi_j(x_1,x_j)=T_o \quad \text{for} \quad i=1\\&\sum_{j=1}^{N}\alpha_j\phi_j(x_{IL},x_j)=T_L \quad \text{for} \quad i=IL\end{aligned} \tag{63}$$

Defining the operator

$$L_j(x_i,x_j)=\frac{d^2\phi_j(x,x_j)}{dx^2}+\phi_j(x_i,x_j) \tag{64}$$

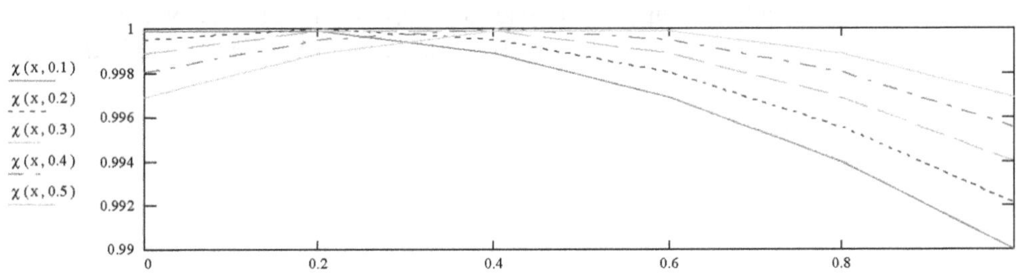

Figure 6. Plot of the inverse Hardy Multiquadric RBF at several sample locations on [0,1] with S=10^2.

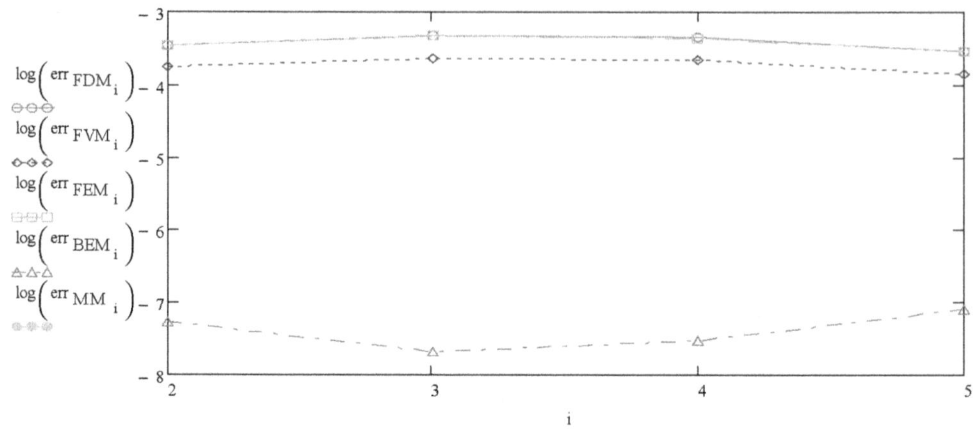

i =	err_{FDM_i} =	err_{FVM_i} =	err_{FEM_i} =	err_{BEM_i} =	err_{MM_i} =
2	$3.437\cdot10^{-4}$	$1.699\cdot10^{-4}$	$3.398\cdot10^{-4}$	$5.166\cdot10^{-8}$	$-9.493\cdot10^{-5}$
3	$4.653\cdot10^{-4}$	$2.299\cdot10^{-4}$	$4.6\cdot10^{-4}$	$2.091\cdot10^{-8}$	$-9.39\cdot10^{-5}$
4	$4.375\cdot10^{-4}$	$2.162\cdot10^{-4}$	$4.325\cdot10^{-4}$	$2.924\cdot10^{-8}$	$-9.405\cdot10^{-5}$
5	$2.833\cdot10^{-4}$	$1.4\cdot10^{-4}$	$2.801\cdot10^{-4}$	$7.915\cdot10^{-8}$	$-9.538\cdot10^{-5}$

Figure 7. Comparison of the error in computing the temperature at the interior points i = 2,3...IL – 1.

The meshless method discrete analog to our problem is assembled in a fully populated matrix set of equations,

$$\begin{bmatrix} \phi_1(x_1,x_1) & \phi_2(x_1,x_2) & \phi_3(x_1,x_3) & \phi_4(x_1,x_4) & \cdots & \phi_{IL}(x_1,x_{IL}) \\ L_1(x_2,x_1) & L_2(x_2,x_2) & L_3(x_2,x_3) & L_4(x_2,x_4) & \cdots & L_{IL}(x_1,x_{IL}) \\ L_1(x_3,x_1) & L_2(x_3,x_2) & L_3(x_3,x_3) & L_4(x_3,x_4) & \cdots & L_{IL}(x_2,x_{IL}) \\ \vdots & \vdots & \vdots & \vdots & \ddots & \vdots \\ L_1(x_{IL-1},x_1) & L_2(x_{IL-1},x_2) & L_3(x_{IL-1},x_3) & L_4(x_{IL-1},x_4) & \cdots & L_{IL}(x_{IL-1},x_{IL}) \\ \phi_1(x_{IL-1},x_1) & \phi_2(x_{IL-1},x_2) & \phi_3(x_{IL-1},x_3) & \phi_4(x_{IL-1},x_4) & \cdots & \phi_{IL}(x_{IL},x_{IL}) \end{bmatrix} \begin{Bmatrix} \alpha_1 \\ \alpha_2 \\ \alpha_3 \\ \vdots \\ \alpha_{IL-1} \\ \alpha_{IL} \end{Bmatrix} = \begin{Bmatrix} T_o \\ -x_2 \\ -x_3 \\ \vdots \\ -x_{IL-1} \\ T_L \end{Bmatrix} \quad (65)$$

or in compact form, we have:$[C]\{\alpha\} = \{d\}$. These equations are readily solved by direct methods to yield the expansion coefficients, α. Once these are found the temperature can be evaluated anywhere using the RBF expansion, Eq. (60). A MATHCAD spreadsheet of the meshless method we just outlined is provided.

The solutions obtained by all the 5 methods are now compared in a composite plot provided in Fig. 7 and accompanying set of tables. A MATHCAD spreadsheet used in all calculations in this introduction is provided on the accompanying website for this book: www.fbm.centecorp.com. Accompanying MAPLE and MATLAB files are also provided.

Meshless method solution using Multiquadric Radial Basis Functions (RBF):

Define multiquadric interpolant ? (x) and shape factor "S": $S := 10^3$ $n := 1$

$$\chi(y,x_i) := \left[1 + \frac{(y-x_i)^2}{S\cdot\Delta x^2}\right]^{n-\frac{3}{2}}$$

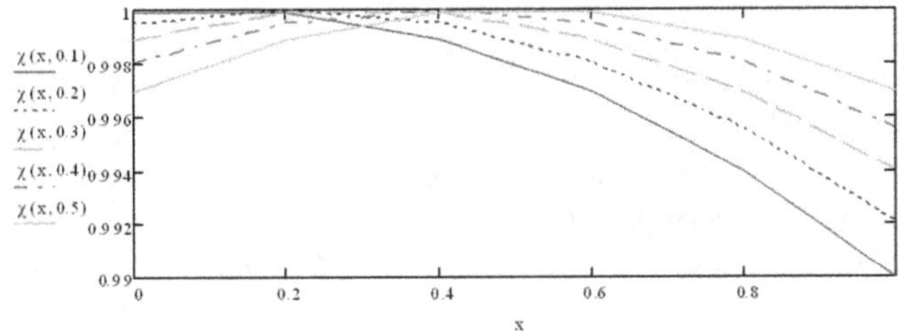

Second derivative of the RBF:

$$dd\chi(y,x_i) := \frac{d^2}{dy^2}\chi(y,x_i) \rightarrow \frac{3\cdot\left(\frac{x_i}{20}-\frac{y}{20}\right)^2}{4\cdot\left[\frac{(x_i-y)^2}{40}+1\right]^{\frac{5}{2}}} - \frac{1}{40\cdot\left[\frac{(x_i-y)^2}{40}+1\right]^{\frac{3}{2}}}$$

Define elements of interpolating matrix and RHS vector:

$$L\_int(z,y) := dd\chi(z,y) + \chi(z,y)$$

Build meshless matrix equations

$i := 2, 3.. IL - 1$ $k := 1, 2.. IL$

$C_{i,k} := L\_int(x_i, x_k)$ $b_i := -x_i$

$C_{1,k} := \chi(x_1, x_k)$ $b_1 := T(0)$

$C_{IL,k} := \chi(x_{IL}, x_k)$ $b_{IL} := T(L)$

Echo matrix:

$$C = \begin{pmatrix} 1 & 1 & 0.998 & 0.996 & 0.992 & 0.988 \\ 0.975 & 0.975 & 0.975 & 0.973 & 0.972 & 0.969 \\ 0.973 & 0.975 & 0.975 & 0.975 & 0.973 & 0.972 \\ 0.972 & 0.973 & 0.975 & 0.975 & 0.975 & 0.973 \\ 0.969 & 0.972 & 0.973 & 0.975 & 0.975 & 0.975 \\ 0.988 & 0.992 & 0.996 & 0.998 & 1 & 1 \end{pmatrix}$$

Conditioning number of the meshless coefficient matrix (using L2 norm):

$$\text{cond2}(C) = 6.145 \times 10^{10}$$

Solve for the coefficients:

$$\alpha := C^{-1}\cdot b$$

use meshless expansion to find the solution at interior points:

$$T_{meshless}(y) := \sum_{j=1}^{IL} \left(\alpha_j \cdot \chi(y, x_j)\right)$$

Compare with exact solution at meshless expansion points:

$$x = \begin{pmatrix} 0 \\ 0.2 \\ 0.4 \\ 0.6 \\ 0.8 \\ 1 \end{pmatrix} \quad T(x) = \begin{pmatrix} 15 \\ 18.726 \\ 21.698 \\ 23.788 \\ 24.907 \\ 25 \end{pmatrix} \quad T_{meshless}(x) = \begin{pmatrix} 15 \\ 18.726 \\ 21.698 \\ 23.788 \\ 24.907 \\ 25 \end{pmatrix} \quad T(x) - T_{meshless}(x) = \begin{pmatrix} -2.095\times10^{-7} \\ -9.493\times10^{-5} \\ -9.39\times10^{-5} \\ -9.405\times10^{-5} \\ -9.538\times10^{-5} \\ -1.993\times10^{-7} \end{pmatrix}$$

....anywhere else $xo := 0.34$

$T_{meshless}(xo) = 20.894$ $T(xo) = 20.894$ $T(xo) - T_{meshless}(xo) = -9.085\times10^{-5}$

REFERENCES

[1] Roache, P.J., *Fundamentals of Computational Fluid Dynamics*, Hermosa Press, Albuquerque, New Mexico, 1998.

[2] Anderson, D.A., Tannehill, J.C., and Pletcher, R.H., *Computational Fluid Mechanics and Heat Transfer*, second edition, McGraw Hill, 1997.

[3] Fletcher, C.A.J., *Computational Techniques for Fluid Dynamics*, Vol. I and II, Springer-Verlag, New York, 1991.

[4] Finlayson, B.A., *The Method of Weighted Residuals and Variational Principle*s, Academic Press, New York, 1972.

[5] Pepper, D. and Heinrich, J., *The Finite Element Method: Basic Concepts and Applications*, Taylor and Francis, New York, 1992.

[6] Greenberg, M.D., *Application of Green's Functions in Science and Engineering*, Prentice Hall, Englewood Cliffs, New Jersey, 1971.

[7] Brebbia, C.A., Telles, J.C.F., and Wrobel, L., *Boundary Element Techniques in Engineering: Theory & Application in Engineering*, Springer-Verlag, New York, 1984.

[8] Divo, E. and Kassab, A.J., *Boundary Element Method for Heat Conduction with Applications in Non-Homogeneous Media*, Wessex Institute of Technology (WIT) Press, Southampton, UK, and Boston, USA, 2003.

[9] Kassab, A.J., Wrobel, L.C., Bialecki, R., and Divo, E., "Boundary Elements in Heat Transfer," *Chapter 4 in Handbook of Numerical Heat Transfer*, Vol. 1, 2nd Edition, Minkowycz, W., Sparrow, E.M., and Murthy, J. Y. (eds.), John Wiley and Sons, pp. 125–166, 2006.

[10] Gottlieb, D. and Orzag, S.A., *Numerical Analysis of Spectral Methods: Theory and Applications*, Society for Industrial and Applied Mathematics, Bristol, England, 1977.

[11] Maday, Y. and Quateroni, A., "Spectral and Pseudo-Spectral Approximations of the Navier-Stokes Equations," *SIAM J. Numerical Analysis*, Vol. 19, No. 4, pp. 761–780, 1982.

[12] Buhmann, M.D., *Radial Basis Functions: Theory and Implementation*, Cambridge University Press, Cambridge, 2003.

[13] Hardy, R.L., "Multiquadric Equations of Topography and Other Irregular Surfaces," *Journal of Geophysical Research*, Vol. 176, pp. 1905–1915.

[14] Fasshauer, G., "RBF Collocation Methods as Pseudo-Spectral Methods," *Boundary Elements XVII*, Kassab, A., Brebbia, C.A., and Divo, E. (eds.), WIT Press, pp. 47–57, 2005.

[15] Kansa, E.J., "Multiquadrics - a Scattered Data Approximation Scheme with Applications to Computational Fluid Dynamics I - Surface Approximations and Partial Derivative Estimates," *Comp. Math. Appl.*, Vol. 19, pp. 127–145, 1990.

[16] Kansa, E.J., "Multiquadrics - a Scattered Data Approximation Scheme with Applications to Computational Fluid Dynamics II - Solutions to Parabolic, Hyperbolic and Elliptic Partial Differential Equations," *Comp. Math. Appl.*, Vol. 19, pp. 147–161, 1990.

[17] Vertnik, R. and Sarler, B., "Meshless Local Radial Basis Function Collocation Method for Convective-Diffusive Solid-Liquid Phase Change Problems," *International Journal of Numerical Methods for Heat and Fluid Flow*, Vol. 16, No. 5, pp. 617–640, 2006.

[18] Divo, E. and Kassab, A.J., "Localized Meshless Modeling of Natural Convective Viscous Flows," *Numerical Heat Transfer, Part B: Fundamentals*, Vol. 53, pp. 487–509, 2008.

[19] Divo, E.A. and Kassab, A.J., "An Efficient Localized RBF Meshless Method for Fluid Flow and Conjugate Heat Transfer," *ASME Journal of Heat Transfer*, Vol. 129, pp. 124–136, 2007.

[20] Cheng, A.H.-D., Golberg, M.A., Kansa, E.J., and Zammito, G., "Exponential Convergence and H-c Multiquadric Collocation Method for Partial Differential Equations," *Numerical Methods in P.D.E.*, Vol. 19, No. 5, pp. 571–594, 2003.

Appendix A DERIVATION OF THE 1D FUNDAMENTAL SOLUTION FOR $T'' + T = -\delta(x - x_i)$

We seek the solution of

$$\frac{d^2 w^*}{dx^2} + w^* = -\delta(x - x_i) \quad \text{for} \quad -\infty < x < +\infty \tag{66}$$

that can be solved by the use of the Fourier transform. The Fourier transform, $\Im[f(x)]$, of a function f(x) defined on the interval $-\infty < x < +\infty$ and its back-transform, $\Im^{-1}[\overline{f}(\lambda)]$, are

$$\Im[f(x)] = \overline{f}(\lambda) = \int_{-\infty}^{+\infty} f(x) e^{i\lambda x} dx \quad \text{and} \quad \Im^{-1}[\overline{f}(\lambda)] = f(x) = \frac{1}{2\pi} \int_{-\infty}^{+\infty} \overline{f}(\lambda) e^{-i\lambda x} dx \tag{67}$$

where $i = \sqrt{-1}$ is the imaginary number, and λ is the wavelength in Fourier space. Noting that the Fourier transform of the second derivative of f(x) is, $\Im[f''(x)] = (-i\lambda)^2 \overline{f}(\lambda)$, using the sifting property of the Dirac Delta function, and taking the transform of the above equation yields the algebraic equation

$$(-i\lambda)^2 \overline{w}^*(\lambda, x_i) + \overline{w}^*(\lambda, x_i) = e^{i\lambda x_i} \tag{68}$$

where, $\overline{w}^*(\lambda, x_i)$ is the Fourier transform of the fundamental solution. Solving for the Fourier transform of the fundamental solution, we find

$$\overline{w}(\lambda, x_i) = \frac{e^{i\lambda x_i}}{1 - \lambda^2} \tag{69}$$

Inverting back to real space using the inversion formula,

$$w^*(x, x_i) = \frac{1}{2\pi} \int_{-\infty}^{+\infty} \frac{e^{-i\lambda(x - x_i)}}{1 - \lambda^2} d\lambda \tag{70}$$

This integral along the real axis can be evaluated by means of contour integration. There are two cases that must be considered for inversion:

1. *For $(x - x_i) > 0$:* a semi-circular contour in the upper-half plane indented along the real axis at the two real poles located at $\lambda = \pm 1$ can be used along with the residue theorem to yield

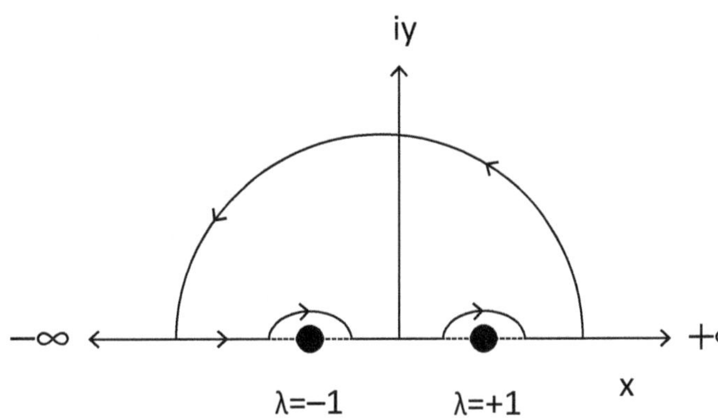

$$w^*(x, x_i) = -\frac{1}{2} \sin(x - x_i) \tag{71}$$

2. *For $(x - x_i) < 0$:* a semi-circular contour in the lower-half plane indented along the real axis at the two real poles located at $\lambda = \pm 1$ can be used along with the residue theorem to yield

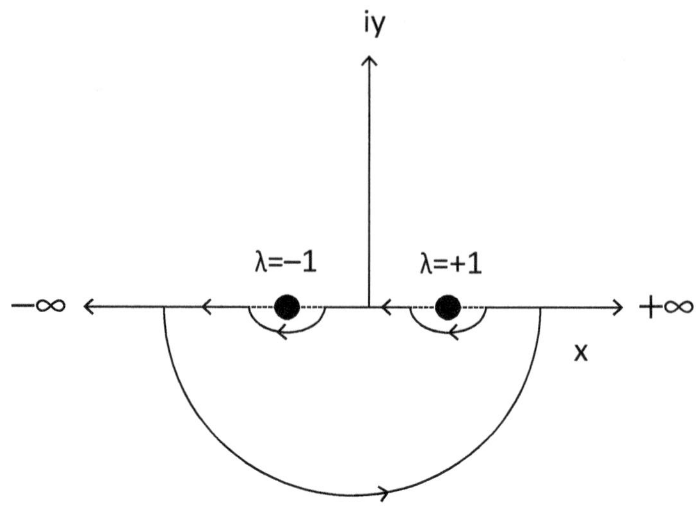

$$w^*(x,x_i) = \frac{1}{2}\sin(x - x_i) \tag{72}$$

Accounting for the fact that sin(x) is odd, that is that sin(−x) = −sin(x), the above two results are combined into the sought-after fundamental solution

$$w^*(x,x_i) = -\frac{1}{2}\sin(|x - x_i|) \tag{73}$$

APPENDIX B – MATLAB

The following list of 1-D examples closely follow the description of the various techniques used to solve the equation

$$\frac{\partial^2 T(x)}{\partial x^2} + T(x) + x = 0, \quad 0 \le x \le 1$$
$$T(0) = 15$$
$$T(L) = 25$$

The descriptions of the numerical methods using MathCad are very clear and easy to follow in the Introduction. Since MATLAB is a very popular programming language, we also want to include the use of MATLAB in setting up these schemes. MATLAB 13 is used to create the 1-D programs for the FDM, FVM, FEM, BEM, and MEM techniques. The references by Chapra and Canale [B1], Kattan [B2], Qin and Wang [B3], Coleman [B4], and Attaway [B5] lists various MATLAB code listings for solving ODE and PDE equations. The code listings included in this appendix incorporate the basic and common routines used in MATLAB to simplify the complexity and length of the codes. MAPLE versions are listed in Appendix C, and FORTRAN versions are available on the website www.fbm.centecorp.com.

1. Finite Difference Method (FDM)

```
%FINITE DIFFERENCE METHOD

%number of grid points
il=6;
%mesh spacing
deltax=1/(il-1);

T0=15;
TL=25;

%x location of the ith grid point
for i=1:il
 x(i)=(i-1)*deltax;
end

%FDM coefficients
a=1/(deltax^2);

b=1-(2/(deltax^2));

c=a;

%initialize problem

A=zeros(il,il);
```

```
d=zeros(il,1);

%load FDM equations
%load left boundary condition
A(1,1)=1;
d(1,1)=T0;

%load interior FDE's

for i=2:il-1
    A(i,i-1)=a;
    A(i,i)=b;
    A(i,i+1)=c;
d(i,1)=-x(i);
end
A(6,6)=1;
d(6,1)=TL;

%echo FDM matrix and RHS
disp(A);
disp(d);

%solve
TFDM=inv(A)*d

T(1)=15;
T(2)=18.72608;
T(3)=21.69763;
T(4)=23.78822;
T(5)=24.90653;
T(6)=25;

for i=1:il
    errFDM(i)=abs(TFDM(i)-T(i))/abs(T(i));
end
errFDM

plot(x,TFDM,'or');
hold on;
plot(x,T);

save('results.mat','TFDM')
```

1.0000	0	0	0	0	0
25.0000	–49.0000	25.0000	0	0	0
0	25.0000	–49.0000	25.0000	0	0
0	0	25.0000	–49.0000	25.0000	0
0	0	0	25.0000	–49.0000	25.0000
0	0	0	0	0	1.0000

15.0000
–0.2000
–0.4000
–0.6000
–0.8000
25.0000

TFDM =

15.0000
18.7325
21.7077
23.7986
24.9136
25.0000

errFDM =

1.0e-03 *

0 0.3435 0.4652 0.4375 0.2833 0

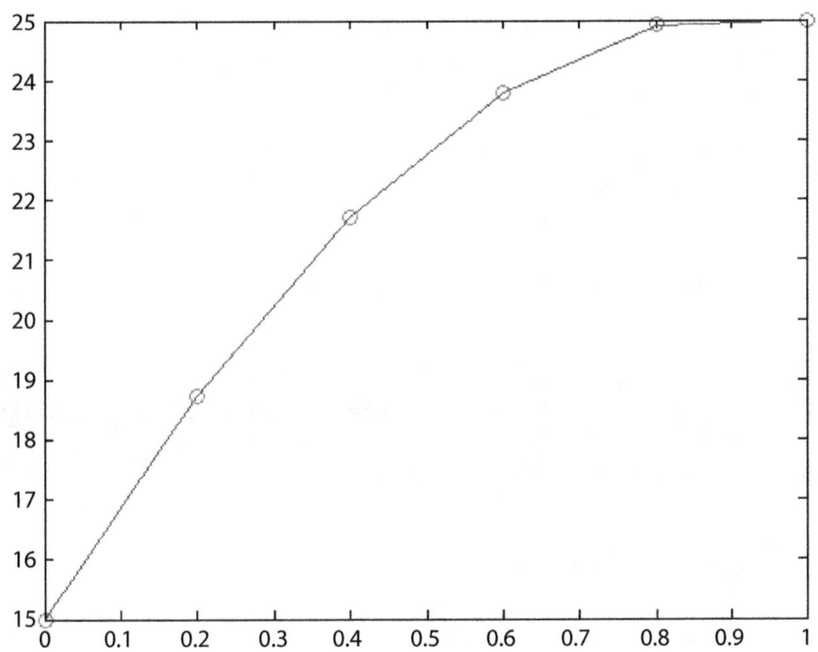

2. Finite Volume Method (FVM)

```
%FINITE VOLUME METHOD
%number of grid points
il=6;
%mesh spacing
```

```
deltax=1/(il-1);

T0=15;
TL=25;

%x location of the ith grid point
for i=1:il
    x(i)=(i-1)*deltax;
end

%FVM coefficients
a=(1/8)+(1/(deltax^2));
b=(3/4)-(2/(deltax^2));
c=(1/8)+(1/(deltax^2));

%initialize problem
A=zeros(il,il);
d=zeros(il,1);

%load FVM equations

%load left boundary equation
A(1,1)=1;
d(1,1)=T0;

%load interior FDE's
for i=2:il-1
    A(i,i-1)=a;
    A(i,i)=b;
    A(i,i+1)=c;
    d(i,1)=-(i-1)*deltax;
end

%load right hand side temperature boundary conditions FDE
A(6,6)=1;
d(6)=TL;

%solve
TFVM=inv(A)*d

%compute relative error
T(1)=15;
T(2)=18.72608;
T(3)=21.69763;
T(4)=23.78822;
T(5)=24.90653;
T(6)=25;

for i=1:il
```

```
    errFVM(i)=abs(TFVM(i)-T(i))/abs(T(i));
end
errFVM

plot(x,TFVM,'or');
hold on;
plot(x,T);

save('results.mat','TFVM','T')
```

TFVM =

15.0000
18.7229
21.6926
23.7831
24.9030
25.0000

errFVM =

1.0e-03 *

0 0.1701 0.2300 0.2162 0.1401 0

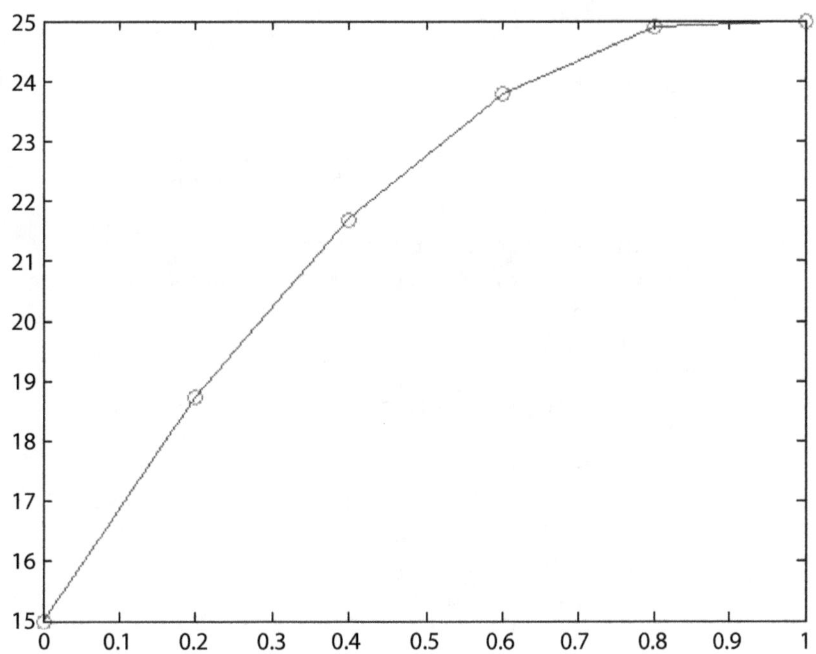

3. Finite Element Method (FEM)

```
%FINITE ELEMENT METHOD

%number of grid points
il=6;
%mesh spacing
dx=1/(il-1);

T0=15;
TL=25;

%x location of the ith grid point
for i=1:il+1
 xx(i)=(i-1)*dx;
end

%define linear finite element shape functions
%see phi function
dphi1x=-1/dx;
dphi2x=1/dx;

%assemble FEM equations
%initialize problem

A=zeros(il,il);
d=zeros(il,1);

%Generate individual element equations and load matrices
%see Kf function

%Assemble stiffness matrix and RHS force(load) vector

%load left boundary equation
[K11,K12,K21,K22,f1,f2] = Kf(1,xx,dx);
A(1,1)=K11;
A(1,2)=K12;
d(1,1)=f1;
%load interior FEM equations

for i=2:il-1
    [K11,K12,K21,K22,f1,f2] = Kf(i,xx,dx);
    [K11_1,K12_1,K21_1,K22_1,f1_1,f2_1] = Kf(i-1,xx,dx);
    A(i,i-1)=K21_1;
    A(i,i)=K22_1+K11;
    A(i,i+1)=K12;
    d(i,1)=f2_1+f1;
end
```

```
%load right boundary FEM equation
[K11,K12,K21,K22,f1,f2] = Kf(6,xx,dx);
[K11_1,K12_1,K21_1,K22_1,f1_1,f2_1] = Kf(5,xx,dx);
A(6,5)=K21_1;
A(6,6)=K22;
d(6,1)=f2;

disp(A);
disp(d);

%Rearrange to impose boundary conditions

%modify force vector by adding negative of first
%column times temperature at left wall and negative of
%last column times temperature to account for temperature
boundary
%conditions

dd=d-(A(:,il)*TL+A(:,1)*T0);
dd(1)=T0;
dd(il)=TL;

%modify first and last rows and columns of stiffness matrix to
account for
%temperature boundary conditions

A(1,:)=[1,0,0,0,0,0];
A(il,:)=[0,0,0,0,0,1];
A(:,1)=[1;0;0;0;0;0];
A(:,il)=[0;0;0;0;0;1];

%Modified equations read

disp(A);
disp(dd);

%solve

TFEM=A\dd

%exact solution at grid points

TE=[15;18.72608;21.69763;23.78822;24.90653;25];

%compute relative error
for i=1:il
    errFEM(i)=abs(TFEM(i)-TE(i))/abs(TE(i));
end
errFEM
```

```
plot(xx(1:6),TFEM,'or');
hold on;
function [phi1,phi2] = phi(i,xx,x)
phi1=(xx(i+1)-x)/(xx(i+1)-xx(i));
phi2=(xx(i)-x)/(xx(i)-xx(i+1));
end
```

```
function [K11,K12,K21,K22,f1,f2] = Kf(i,xx,dx)
syms x
[phi1,phi2]=phi(i,xx,x);
dphi1x=-1/dx;
dphi2x=1/dx;

K11=int(dphi1x^2-phi1^2,x,(i-1)*dx,i*dx);
K12=int(dphi1x*dphi2x-phi1*phi2,x,(i-1)*dx,i*dx);
K21=K12;
K22=int(dphi2x^2-phi2^2,x,(i-1)*dx,i*dx);

f1=int(phi1*x,x,(i-1)*dx,i*dx);
f2=int(phi2*x,x,(i-1)*dx,i*dx);
end
```

```
plot(xx(1:6),TE);
```

4.9333	–5.0333	0	0	0	0
–5.0333	9.8667	–5.0333	0	0	0
0	–5.0333	9.8667	–5.0333	0	0
0	0	–5.0333	9.8667	–5.0333	0
0	0	0	–5.0333	9.8667	–5.0333
0	0	0	0	–5.0333	4.9333

0.0067
0.0400
0.0800
0.1200
0.1600
0.1133

1.0000	0	0	0	0	0
0	9.8667	–5.0333	0	0	0
0	–5.0333	9.8667	–5.0333	0	0
0	0	–5.0333	9.8667	–5.0333	0
0	0	0	–5.0333	9.8667	0
0	0	0	0	0	1.0000

```
    15.0000
    75.5400
     0.0800
     0.1200
   125.9933
    25.0000
TFEM =
    15.0000
    18.7197
    21.6876
    23.7779
    24.8996
    25.0000
errFEM =
   1.0e-03 *
   0    0.3400    0.4600    0.4325    0.2802    0
```

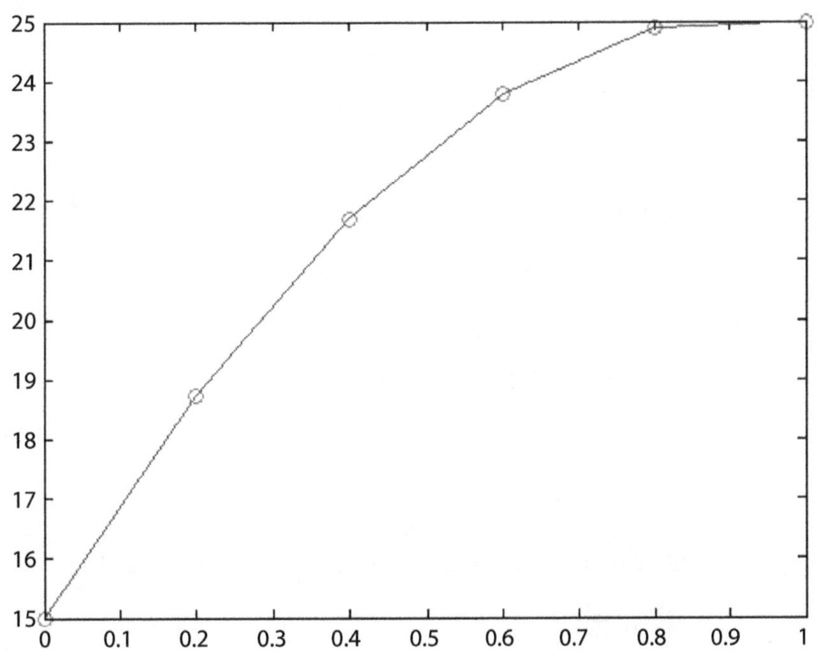

4. Boundary Element Method (BEM)

```
%BOUNDARY ELEMENT METHOD

%Computing the fundamental solution

L=1;
il=6;
dx=1/(il-1);
```

```
for i=1:il
    xx(i)=(i-1)*dx;
end
ws=inline('-0.5*sin(abs(y-x))','y','x');

H=[0.5-dws(0,0) dws(0,1);
    -dws(1,0) 0.5-dws(1,1)]

H=vpa(H);

G=[-ws(0,0) ws(0,1);
    -ws(1,0) ws(1,1)]

%G=vpa(G);

%impose first kind boudary conditions and re-arrange to gen-
erate
%right hand side vector

T0=15;
TL=25;

Tlim=[T0;TL];

fun1=@(x,y) -0.5*sin(abs(y-x));

a1= dblquad(fun1,0,0,0,L);
a2= dblquad(fun1,1,1,0,L);

a=[a1;a2];

b=H*Tlim-a;

%Solve the equations for the normal derivatives at x=0 and x=L

q=inv(G)*b;
q0=q(1,1);
qL=q(2,1);

%Compare with the exact boundary dervivatives
Q0=20.267;
QL=-2.132;

%Compute solution at interior points - use TBEM function

Tbem_20=vpa(TBEM(0.20,q0,qL));
Tbem_40=vpa(TBEM(0.40,q0,qL));
Tbem_60=vpa(TBEM(0.60,q0,qL));
Tbem_80=vpa(TBEM(0.80,q0,qL));
```

```
T_BEM=[T0; double(Tbem_20); double(Tbem_40); double(Tbem_60);
double(Tbem_80); TL]

%Compare with exact solution
TE=[15;18.72608;21.69763;23.78822;24.90653;25];
plot(xx(1:6),T_BEM,'or');
hold on;
plot(xx(1:6),TE);
```

```
function [z] = dws(x,y)
if (y-x)>0
z=-cos(y-x)/2;
end;
if (y-x)<0
 z=cos(x-y)/2;
end;
if y==x
 z=0;
end;
```

```
return
end
```

```
function [z] = TBEM( x,q0,qL )
ws=inline('-0.5*sin(abs(y-x))','x','y');
z=-25*dws(x,1)+15*dws(x,0);
z=z+ws(x,1)*qL-ws(x,0)*q0;
syms y;
z=z+int(ws(x,y)*y,y,0,1);
return
end
```

H =

0.5000	–0.2702
–0.2702	0.5000

G =

0	–0.4207
0.4207	0

T_BEM =

15.0000
18.5790
21.5599
23.6653
24.8034
25.0000

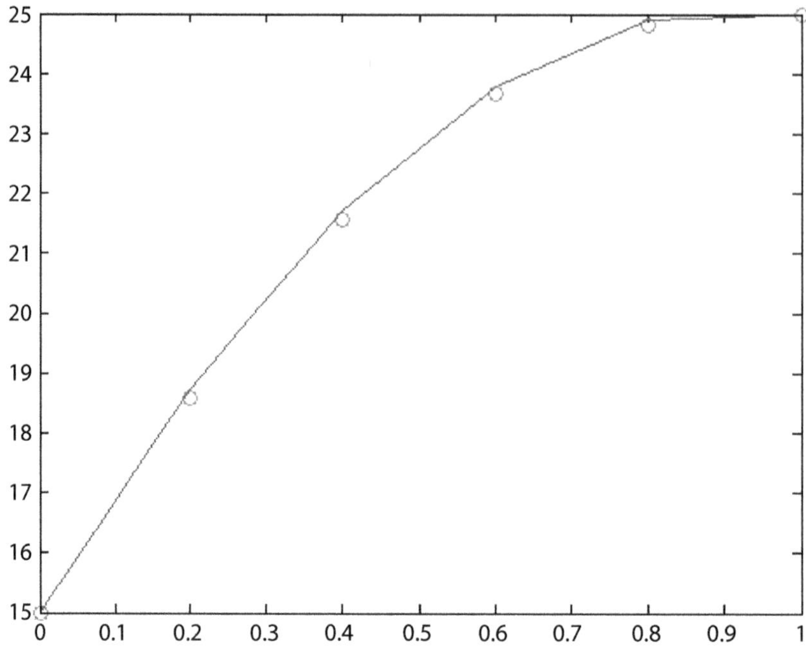

5. Meshless Method (MEM)

```
% MESHLESS METHOD SOLUTION USING MULTIQUADRIC RADIAL BASIS
FUNCTIONS RBF

%number of grid points
il = 6;
%mesh spacing
dx = 1/(il-1);

T0 = 15;
TL = 25;

%x location of the ith grid point
for i = 1:il
 xx(i) = (i-1)*dx;
end

%Define multiquadric interpolant and shape factor S
S = 1000;
n = 1;

chi = @ (y,x) (1+((y-x)^2/(S*dx^2)))^(n-(3/2));

%Second derivative of the RBF
```

```
ddchi= @ (y,x) (3*((x-y)/20)^2/(4*((x-y)^2/40+1)^(5/2))-
(1/(40*((x-y)^2/40+1)^(3/2))));

%Build meshless matrix equations

C = zeros(il,il);
b = zeros(il,1);

for i = 2:il-1
    for k = 1:il
          C(i,k)=ddchi(xx(i),xx(k))+chi(xx(i),xx(k));
          b(i,1)=-xx(i);
          C(il,k)=chi(xx(il),xx(k));
          C(1,k)=chi(xx(1),xx(k));
    end
end

T0=15;
TL=25;

b(1,1)=T0;
b(il)=TL;

%echo matrix
C
fprintf('The Condition Number is %d\n',cond(C))

%Solve for the coefficients

alpha=C\b;

%use meshless expansion to find the solution at interior points

TMEM=zeros(il,1);

for i=1:il
    for j=1:il
         TMEM(i,1)=TMEM(i,1)+alpha(j,1)*chi(xx(i),xx(j));
end
end

TMEM

%Compare with exact solution
TE=[15;18.72608;21.69763;23.78822;24.90653;25];
plot(xx(1:6),TMEM,'or');
hold on;
plot(xx(1:6),TE);
```

6. C =

1.0000	0.9995	0.9980	0.9955	0.9921	0.9877
0.9746	0.9750	0.9746	0.9735	0.9715	0.9688
0.9735	0.9746	0.9750	0.9746	0.9735	0.9715
0.9715	0.9735	0.9746	0.9750	0.9746	0.9735
0.9688	0.9715	0.9735	0.9746	0.9750	0.9746
0.9877	0.9921	0.9955	0.9980	0.9995	1.0000

The Condition Number is 6.145000e+10
TMEM =
15.0000
18.7262
21.6977
23.7883
24.9066
25.0000

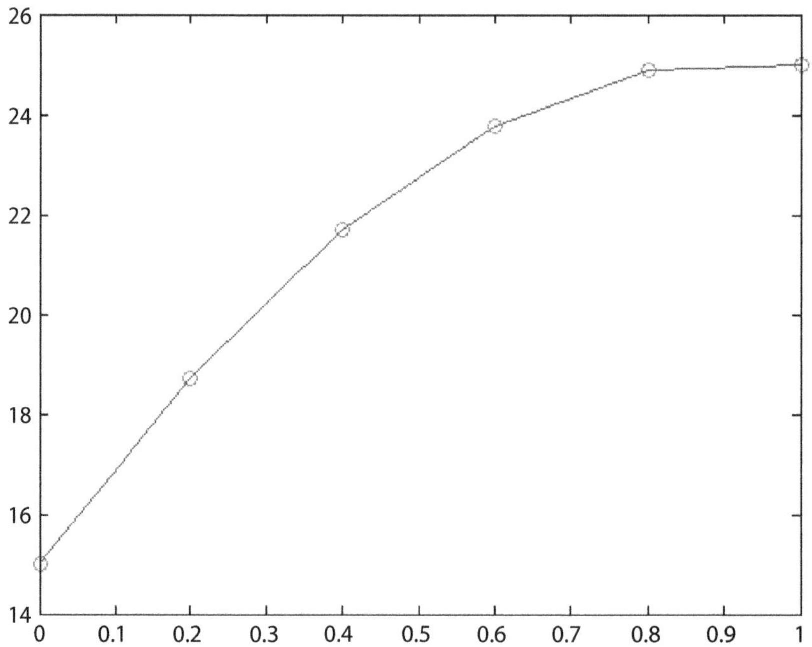

[B1] Chapra, S. C. and Canale, R. P., *Numerical Methods for Engineers*, McGraw-Hill, 7th Ed., NY, NY, 2015.

[B2] Kattan, P., *MATLAB Guide to Finite Elements, An Interactive Approach*, 2nd Ed., Springer, Berlin, 2007.

[B3] Qin, Q-H. and Wang, H., *MATLAB and C Programming for Trefftz Finite Element Methods*, CRC Press, Boca Raton, FL, 2009.

[B4] Coleman, M. P., *An Introduction to Partial Differential Equations with MATLAB*, 2nd Ed., CRC Press, Boca Raton, FL, 2013.

[B5] Attaway, S., *MATLAB, A Practical Introduction to Programming and Problem Solving*, 2nd Ed., Elsevier, Boston, 2012.

APPENDIX C – MAPLE

The following list of 1-D examples closely follow the description of the various techniques used to solve the equation

$$\frac{\partial^2 T(x)}{\partial x^2} + T(x) + x = 0, \quad 0 \le x \le 1$$
$$T(0) = 15$$
$$T(L) = 25$$

The descriptions of the numerical methods using MathCad are very clear and easy to follow in the Introduction. Similar to Appendix B employing MATLAB, we also want to include the use of MAPLE in setting up these schemes, due to its wide use and ease of implementation. MAPLE 18 is used to create the 1-D programs for the FDM, FVM, FEM, BEM, and MEM techniques. The reference by Portela and Charafi [C1] lists various Maple code listings for creating FDM, FEM, and BEM programs using the simple, built-in expressions common to Maple. The code listings included in this appendix incorporate some of these techniques to reduce the complexity and length of the codes.

1. Finite Difference Method (FDM)

```
> restart:
> To:=15:
> TL:=25:
> L:=1:
> x[1],x[2],x[3],x[4],x[5],x[6]:=0,1/5,2/5,3/5,4/5,1:
> fdmcd:=proc(i::integer)
> global T,x;
> local dx2;
```

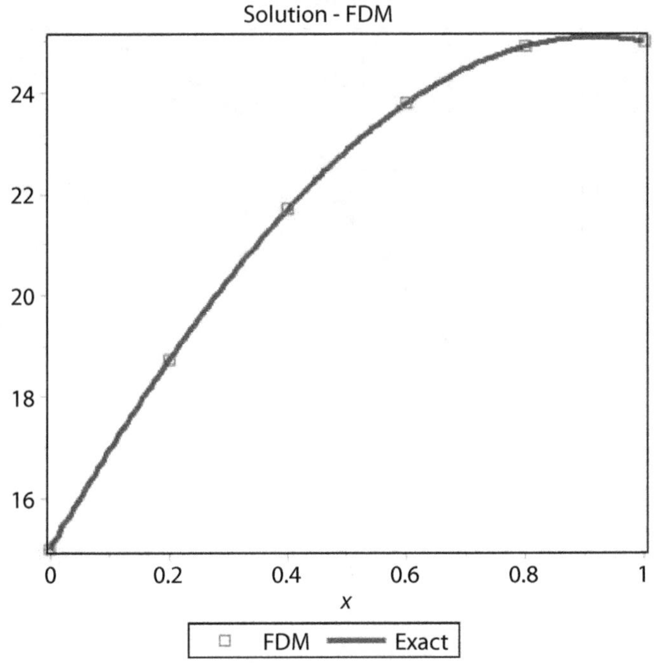

```
> dx2:=x[i+1]-x[i-1];
> 4*(T[i-1]-2*T[i]+T[i+1])/(dx2^2)+T[i]+x[i];
> end proc:
> eqns:=fdmcd(2),fdmcd(3),fdmcd(4),fdmcd(5):
> T[1]:=To:
> T[6]:=TL:
> solve({eqns},{T[2],T[3],T[4],T[5]}):
> assign(%):
> seq(T[i],i=1..6):
> TE:=To*cos(x)+(TL+L-To*cos(L))/sin(L)*sin(x)-x:
> TE:=subs(TE):
> TE:=plot(TE,x=0..1,color=blue,legend="Exact",thickness=3):
> fdm:=[seq([subs(x[i]),subs(T[i])],i=1..6)]:
> fdm:=plots[pointplot](fdm,color=rd,legend="FEM",symbol=box,
> symbolsize=15):
> plots[display](TE,fdm,axes=BOXED,title="Solution - FDM");
```

2. Finite Volume Method (FVM)

```
> restart:
> To:=15:
> TL:=25:
> L:=1:
> x[1],x[2],x[3],x[4],x[5],x[6]:=0,1/5,2/5,3/5,4/5,1:
> fvm:=proc(i::integer)
> global T,x;
> local dx,x1,x2;
> dx:=x[i+1]-x[i];
> x2:=(x[i+1]-x[i])/2;x1:=(x[i]-x[i-1])/2;
```

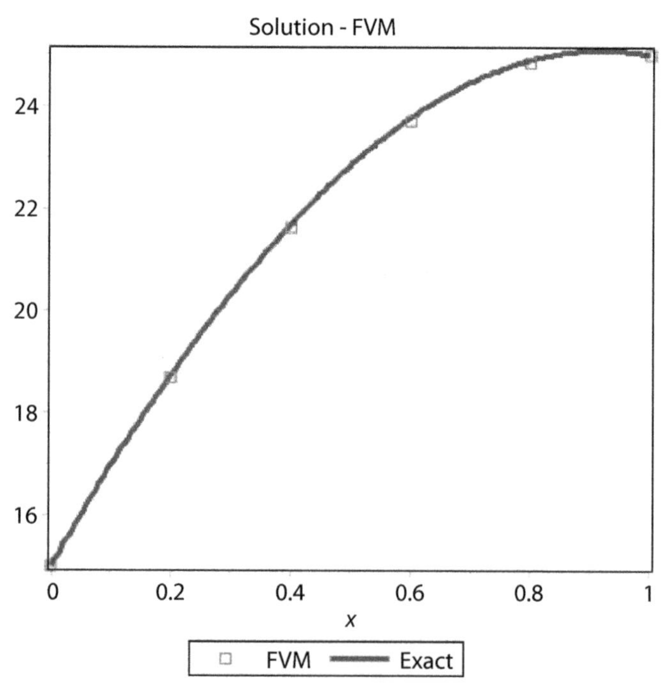

```
> (1/dx^2+1/8)*T[i-1]+(3/4-2/dx^2)*T[i]+(1/dx^2+1/8)*T[i+1]+
(1/2)>*(x2-x1);
> end proc:
> eqns:=fvm(2),fvm(3),fvm(4),fvm(5):
> T[1]:=To:
> T[6]:=TL:
> solve({eqns},{T[2],T[3],T[4],T[5]}):
> assign(%):
> seq(T[i],i=1..6):
> TE:=To*cos(x)+(TL+L-To*cos(L))/sin(L)*sin(x)-x:
> TE:=subs(TE):
> TE:=plot(TE,x=0..1,color=blue,legend="Exact",thickness=3):
> fv:=[seq([subs(x[i]),subs(T[i])],i=1..6)]:
> fv:=plots[pointplot](fv,color=red,legend="FVM",symbol=box,
> symbolsize=15):
> plots[display](TE,fv,axes=BOXED,title="Solution - FVM");
```

3. Finite Element Method (FEM)

```
> restart:
> with(linalg):with(plots):
#FINITE ELEMENT METHOD
> To:=15:TL:=25:il:=6:L:=1:
#define linear finite element shape functions
shape_f:=proc(Xi,Xj)
>    local length;
>    length:=Xj-Xi;
>    (Xj-x)/length,(x-Xi)/length
> end proc:
shape_f(a,b):simplify(%[1]+%[2]);
plot([subs(a=0,b=1,%%[1]),subs(a=0,b=1,%%[2])],x=0.
.1,legend=["Ni","Nj"],axes=BOXED,title="Linear   Shape
Functions",thickness=3);
>
element_k_p:=proc(i::integer,j::integer)
>    global nods,BC;
>    local Xi,Xj,N,k11,k21,k22,e_k,e_p;
>    Xi:=nods[i];Xj:=nods[j];
>    N:=shape_f(Xi,Xj);
>    k11:=diff(N[1],x)^2-N[1]^2;
>    k21:=diff(N[1],x)*diff(N[2],x)-N[1]*N[2];
>    k22:=diff(N[2],x)^2-N[2]^2;
>    e_k:=map(int,array(symmetric,1..2,1..2,[[k11],[k21,k22]
     ]),x=Xi..Xj);
         e_p:=map(int,array([[x*N[1]],[x*N[2]]]),x=Xi..Xj);
         eval(e_k),eval(e_p)
> end proc:

> init_k_p:=proc()
>     global nods,g_k,g_p;
>     local n;
>     n:=nops(nods);
```

```
> g_k:=array(symmetric,1..n,1..n,[seq([seq(0,j=1..i)],i=1
  ..n)]);
> g_p:=array([seq([0],j=1..n)]);
> eval(g_k),eval(g_p)
> end proc:

> global_k_p:=proc()
>    global elems,g_k,g_p;
>    local i,j,e,e_k,e_p;
>    for e in elems do
>        i,j:=e[]:
>        e_k,e_p:=element_k_p(i,j):
>        g_k[i,i]:=g_k[i,i]+e_k[1,1]:
>        g_k[i,j]:=g_k[i,j]+e_k[1,2]:
>        g_k[j,j]:=g_k[j,j]+e_k[2,2]:
>        g_p[i,1]:=g_p[i,1]+e_p[1,1]:
>        g_p[j,1]:=g_p[j,1]+e_p[2,1]:
>        if nargs<>0 then
>           print(`assembling element: `,e);
>           print(g_k,g_p)
>        end if
>    end do;
>    eval(g_k),eval(g_p)
> end proc:

exact_bc:=proc()
>    global nods,BC,g_k,g_p;
>    local m,n,j,i;
>    for m in BC do
>        n:=m[1];
>    for i from 1 to nops(nods) do
>     g_p[i,1]:=g_p[i,1]-g_k[i,n]*m[2]
>    end do;
>    end do;
# reset global matrix and rhs with fixed values
 for m in BC do
>    n:=m[1];
>     for j from 1 to nops(nods) do
>      g_k[n,j]:=0:
>      g_k[n,n]:=1:
>      g_p[n,1]:=m[2]
>    end do;
> end do;
>    eval(g_k),eval(g_p)
> end proc:

> deriv:=proc()
>    global elems,nods,v:
>    local der,e,i,j,Xi,Xj,N,DN,k:
>    der:=[]:
>    for e in elems do
>     i,j:=e[]:
```

```
>      Xi:=nods[i]:Xj:=nods[j]:
>      N:=shape_f(Xi,Xj):
>      DN:=seq(diff(N[k],x),k=1..2):
>      der:=[der[],DN[1]*T[i]+DN[2]*T[j]]
>    end do:
>    der
> end proc:
```

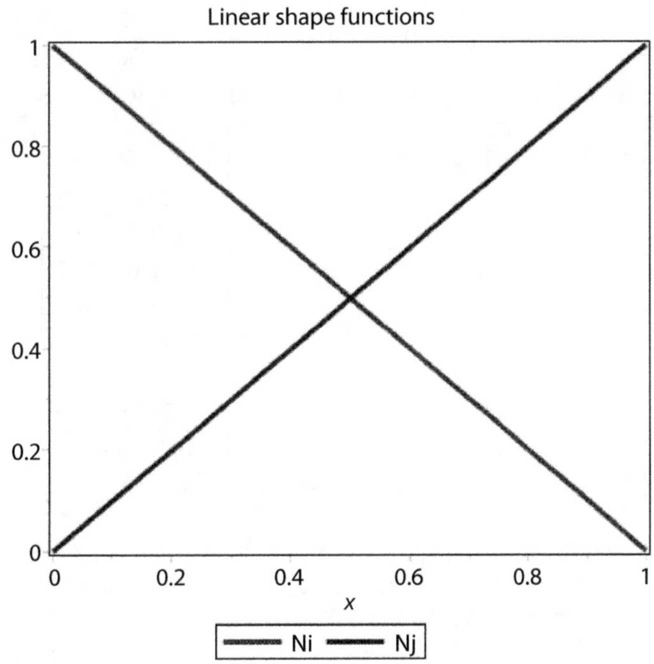


```
> nods:=[0,1/5,2/5,3/5,4/5,1]:
> elems:=[[1,2],[2,3],[3,4],[4,5],[5,6]]:
> BC:=[[elems[1][1],To],[elems[5][2],TL]]:
> init_k_p():
> global_k_p():
> exact_bc();
> linalg[linsolve](g_k,g_p);
> T:=linalg[col](%,1):
```

$$\begin{bmatrix} 1 & 0 & 0 & 0 & 0 & 0 \\ 0 & \frac{148}{15} & -\frac{151}{30} & 0 & 0 & 0 \\ 0 & -\frac{151}{30} & \frac{148}{15} & -\frac{151}{30} & 0 & 0 \\ 0 & 0 & -\frac{151}{30} & \frac{148}{15} & -\frac{151}{30} & 0 \\ 0 & 0 & 0 & -\frac{151}{30} & \frac{148}{15} & 0 \\ 0 & 0 & 0 & 0 & 0 & 1 \end{bmatrix}, \begin{bmatrix} 15 \\ \frac{3777}{50} \\ \frac{2}{25} \\ \frac{3}{25} \\ \frac{18899}{150} \\ 25 \end{bmatrix}$$

$$\begin{bmatrix} 15 \\ \frac{206221207121}{11016259045} \\ \frac{238916761987}{11016259045} \\ \frac{261943842723}{11016259045} \\ \frac{274299919559}{11016259045} \\ 25 \end{bmatrix}$$

```
> TE:=To*cos(x)+((TL+L-To*cos(L))/sin(L))*sin(x)-x:
> TE:=plot(TE,x=0..1,color=blue,legend="Exact",thickness=3):
> dTE:=diff(%%,x):
>  dTE:=plot(dTE,x=0..1,color=green,legend="Exact",thickn
ess=3):
> fem:=[seq([subs(nods[i]),subs(T[i])],i=1..nops(nods))]:
> TFEM:=plots[pointplot](fem,style=line,legend="FEM elements"
,color=red,thickness=3):
> fem:=plots[pointplot](fem,color=red,legend="FEM nodes",sym
bol=box,symbolsize=15):
> plots[display](TE,fem,TFEM,axes=BOXED,title="Function");
deriv():
drv:=subs([seq(seq([nods[elems[i][j]],%[i]],j=1..2),i=1..
nops(elems))]):
> fem_drv:=plot(drv,color=red,legend=["FEM"],thickness=3):
>  plots[display]({dTE,fem_drv},axes=BOXED,title="Derivat
ive");
```

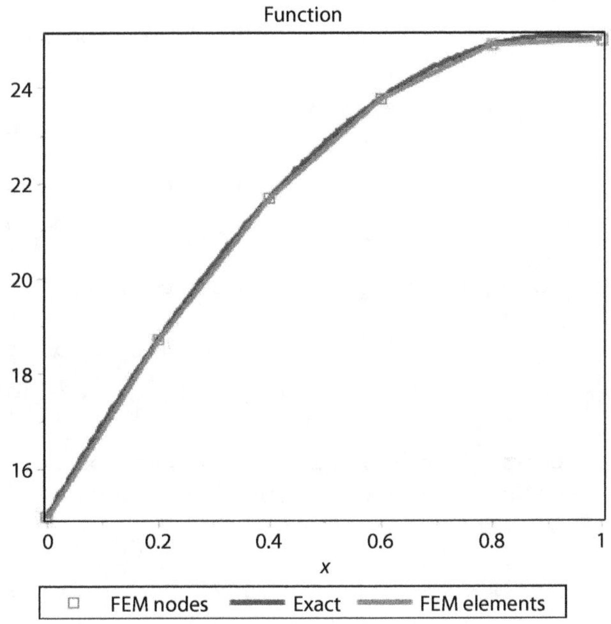

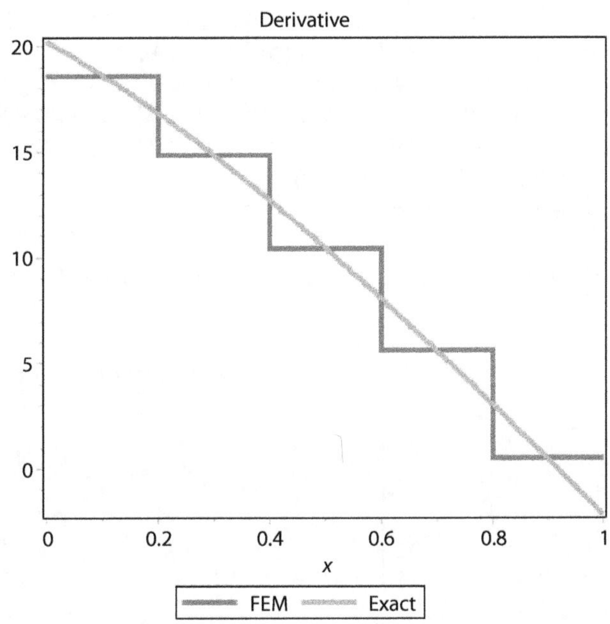

4. Boundary Element Method (BEM)

```
> restart:
eqtn:=diff(W(x,xi),x$2)+W(x,xi)=-Dirac(x-xi);
```

$$eqtn := \frac{\partial^2}{\partial x^2} W(x,\xi) + W(x,\xi) = -\text{Dirac}(x-\xi)$$

```
> W:=(x,xi)->sin(abs(x-xi))/2;
```

$$W := (x,\xi) \to \frac{1}{2}\sin(|x-\xi|)$$

```
> convert(lhs(eqtn),piecewise,x);
```

$$\begin{cases} undefined & x = \xi \\ 0 & otherwise \end{cases}$$

```
> convert(rhs(eqtn),piecewise,x);
```

$$\begin{cases} undefined & x = \xi \\ 0 & otherwise \end{cases}$$

```
> plot([W(x,1),diff(W(x,1),x)],x=-5..6,color=[red,green],thi
ckness=3,legend=["function","derivative"],axes=BOXED,title="
Fundamental Solution");
> nods:=[0,1]:
> r:=(x,xi)->piecewise(xi=nods[1],x,xi=nods[2],1-x);
> W:=r->sin(r)/2;
> D1:=int(x*W(r(x,nods[1])),x=nods[1]..nods[2]);
> D2:=-int(x*W(r(x,nods[2])),x=nods[1]..nods[2]);
> eq:=c*u[i]+q[1]*W(r(x,nods[1]))-q[2]*W(r(x,nods[2]))+D=0;
> eq1:=eval(eq,[c=1,x=nods[1],i=1,D=D1]);
> eq2:=eval(eq,[c=1,x=nods[2],i=2,D=D2]);
> u[1]:=15;u[2]:=25;
> eq1:=eval(eq1);
> eq2:=eval(eq2);
> solve({eq1,eq2},{q[1],q[2]});assign(%):
```

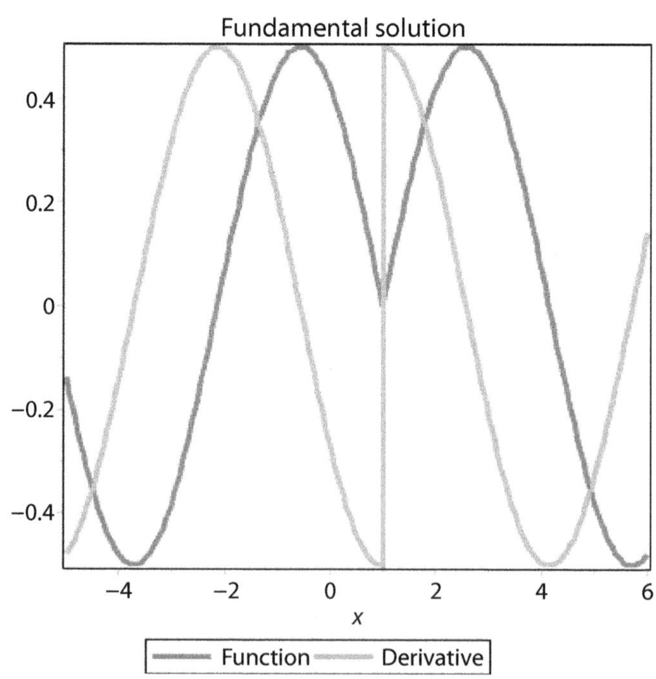

$$r := (x, \xi) \to piecewise\,(\xi = nods_1, x, \xi = nods_2, 1 - x)$$

$$W := r \to \frac{1}{2}\sin(r)$$

$$D1 := \frac{1}{2}\sin(1) - \frac{1}{2}\cos(1)$$

$$D2 := \frac{1}{2}\sin(1) - \frac{1}{2}$$

$$eq := cu_i + \frac{1}{2}q_1\sin(x) + \frac{1}{2}q_2\sin(x-1) + \mathrm{D} = 0$$

$$eq1 := u_1 - \frac{1}{2}q_2\sin(1) + \frac{1}{2}\sin(1) - \frac{1}{2}\cos(1) = 0$$

$$eq2 := u_2 + \frac{1}{2}q_1\sin(1) - \frac{1}{2} + \frac{1}{2}\sin(1) = 0$$

$$u_1 := 15$$

$$u_2 := 25$$

$$eq1 := 15 - \frac{1}{2}q_2\sin(1) + \frac{1}{2}\sin(1) - \frac{1}{2}\cos(1) = 0$$

$$eq2 := \frac{49}{2} + \frac{1}{2}q_1\sin(1) + \frac{1}{2}\sin(1) = 0$$

$$\left\{q_1 = -\frac{\sin(1)+49}{\sin(1)}, q_2 = \frac{\sin(1)-\cos(1)+30}{\sin(1)}\right\}$$

```
>v:=xi->q[1]*W(r(x,nods[1]))-q[2]*W(r(x,nods[2]))-
int(x*W(r(x,nods[2])),x=xi..nods[2])+int(x*W(r(x,nods[1])),
x=nods[1]..xi);
```

$$v := \xi \to q_1W\big(r(x,nods_1)\big) - q_2W\big(r(x,nods_2)\big) - \left(\int_{\xi}^{nods_2} xW\big(r(x,nods_2)\big)\mathrm{d}x\right) +$$

$$\int_{nods_1}^{\xi} xW\big(r(x,nods_1)\big)\mathrm{d}x$$

```
> dv:=xi->q[1]*diff(W(r(x,nods[1])),x)-q[2]*diff(W(r(x,nods
[2])),x)-int(x*diff(W(r(x,nods[2])),x),x=xi..nods[2])-int(x*
diff(W(r(x,nods[1])),x),x=nods[1]..xi);
> To:=15:TL:=25:L:=1:
> TE:=To*cos(x)+((TL+L-To*cos(L))/sin(L))*sin(x)-x:
> TEP:=plot(TE,x=0..1,color=blue,legend="Exact",thickness=3):
> dTE:=diff(%%,x):
> dTEP:=plot(dTE,x=0..1,color=green,legend="Exact",thickn
ess=3):
bemP:=plot(subs(-v(x)),x=0..1,color=red,legend="BEM",thickn
ess=3):
```

```
> plots[display]([TEP,bemP],axes=BOXED,title="Function");
bemP:=plot(subs(-dv(x)),x=0..1,color=red,legend="BEM",thickn
ess=3):
> plots[display]([dTEP,bemP],axes=BOXED,title="Derivative");
#At x=1/2
> BEM=-evalf(subs(x=1/2,v(1/2)));
> EXACT=evalf(subs(x=1/2,TE));
> DBEM=-evalf(subs(x=1/2,dv(1/2)));
> DTE=evalf(subs(x=1/2,dTE));
```

$$dv := \xi \to q_1\left(\frac{\partial}{\partial x}W(r(x,nods_1))\right) - q_2\left(\frac{\partial}{\partial x}W(r(x,nods_2))\right) - \left(\int_{\xi}^{nods_2} x\left(\frac{\partial}{\partial x}W(r(x,nods_2))\right)\mathrm{d}x\right) - \left(\int_{nods_1}^{\xi} x\left(\frac{\partial}{\partial x}W(r(x,nods_1))\right)\mathrm{d}x\right)$$

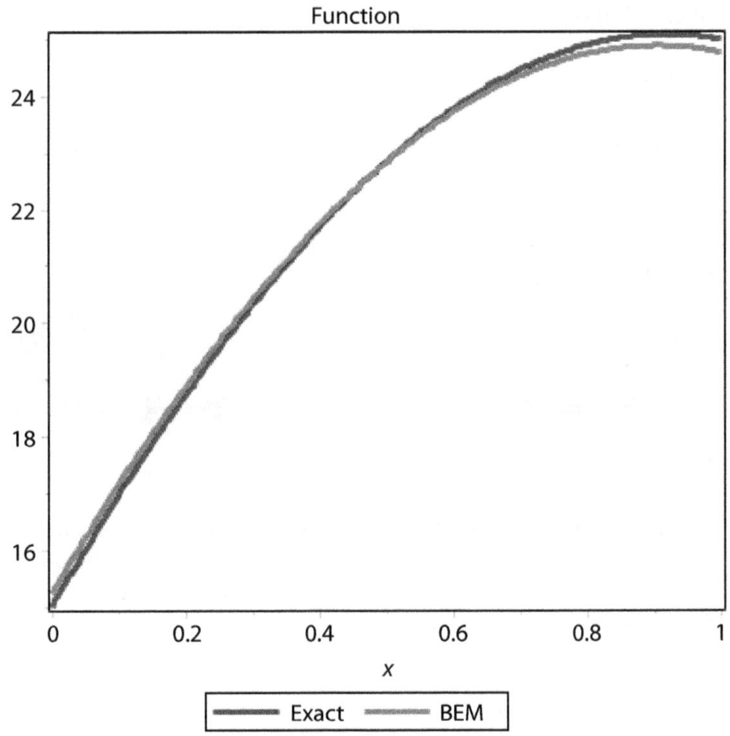
Function
24
22
20
18
16
0
0.2
0.4
0.6
0.8
1
x
Exact
BEM

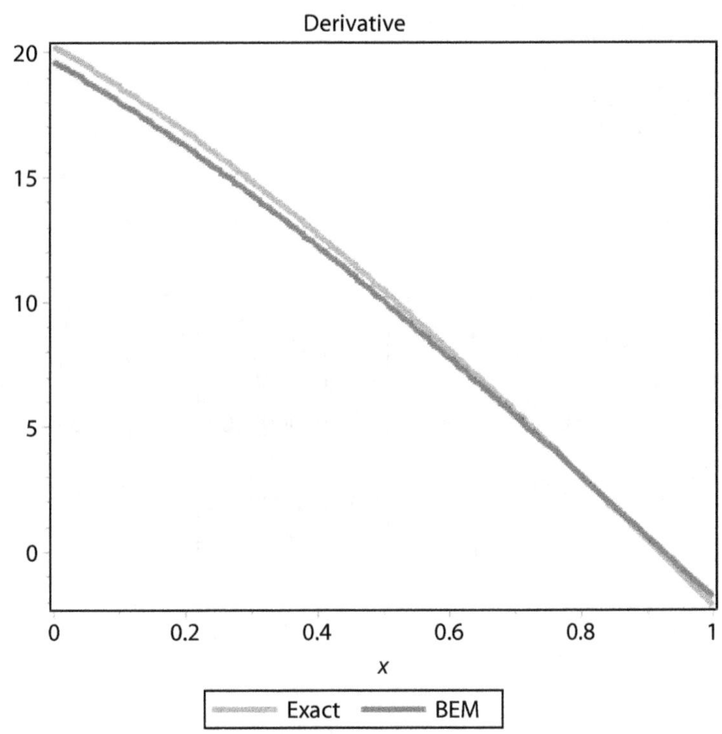
Derivative
20
15
10
5
0
0
0.2
0.4
0.6
0.8
1
x
Exact
BEM

5. Meshless Method (MEM)

```
> restart:
> with(linalg):with(plots):
# MESHLESS METHOD SOLUTION USING MULTIQUADRIC RADIAL BASIS
FUNCTIONS (RBF)
> il:=6:To:=15:TL:=25:L:=1:
> x:=[0,1/5,2/5,3/5,4/5,1]:
> S:=1000:n:=1:dx:=1/(il-1):
>
C:=array(1..il,1..il):phi:=array(1..il,1..il):LM:=array(1..
il,1..il):
> b:=Vector(1..il):TM:=Vector(1..il):alpha:=Vector(1..il):

> for i from 1 to il do
> for j from 1 to il do
> phi[i,j]:=(1+(x[i]-x[j])^2/(S*dx^2))^(n-3/2):
       d2phi:=3*(x[j]/20-x[i]/20)^2/(4*((x[j]-
x[i])^2/40+1)^(5/2))-1/(40*((x[j]-x[i])^2/40+1)^(3/2)):
 LM[i,j]:=d2phi+phi[i,j]:
> end do;
> end do;
>
> for i from 2 to il-1 do
> for j from 1 to il do
> C[i,j]:=LM[i,j];
> C[1,j]:=phi[1,j];
 C[il,j]:=phi[il,j];
> end do:
> b[i]:=-x[i]:
> end do:
> b[1]:=To:b[il]:=TL:
> evalf(b);
```

$$\begin{bmatrix} 15. \\ -0.2000000000 \\ -0.4000000000 \\ -0.6000000000 \\ -0.8000000000 \\ 25. \end{bmatrix}$$

```
> Cond(C):
> alpha:=linalg[linsolve](C,b):
> TM[1]:=To:TM[6]:=TL:
> for i from 2 to il-1 do
>    for j from 1 to il do
>         TM[i]:=TM[i]+alpha[j]*(1+(x[i]-x[j])^2/
          (S*dx^2))^(n-3/2);
>    end do:
> end do:
```

```
> evalf(TM);
```

$$\begin{bmatrix} 15. \\ 18.48 \\ 21.5 \\ 23.6 \\ 24.67 \\ 25. \end{bmatrix}$$

```
> TE:=To*cos(xx)+(TL+L-To*cos(L))/sin(L)*sin(xx)-xx:
> TE:=subs(TE):
>   TE:=plot(TE,xx=0..1,color=blue,legend="Exact",thickn
ess=3):
> MEM:=[seq([subs(x[i]),subs(TM[i])],i=1..6)]:
>
T:=plots[pointplot](MEM,style=line,color=red,legend="MEM",th
ickness=3):
MEM:=plots[pointplot](MEM,color=red,legend="MEM",symbol=box,
symbolsize=15):
> plots[display](TE,MEM,T,axes=BOXED,title="Solution - MEM");
```

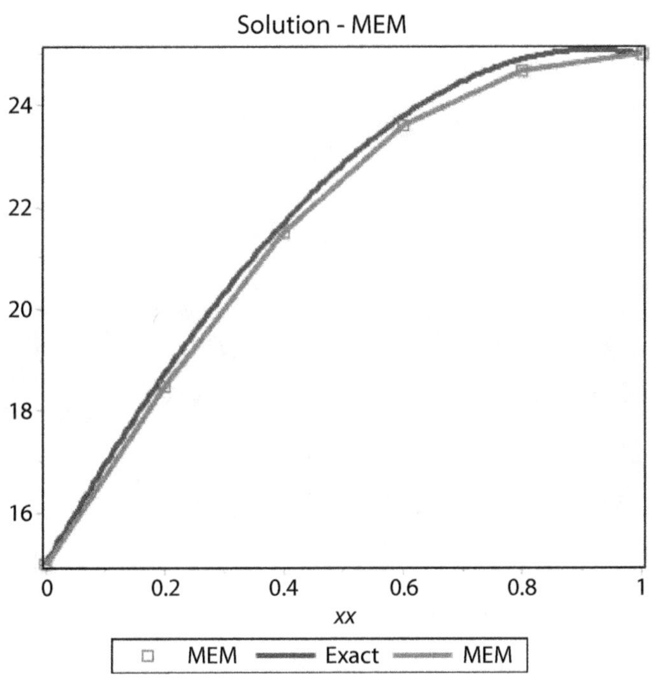

[C1] Portela, A. and Charafi, A., *Finite Elements Using Maple, A Symbolic Programming Approach*, Springer-Verlag, Berlin, Germany, 2002.

PART I
THE FINITE ELEMENT METHOD

Chapter 1

INTRODUCTION

The equations of fluid motion and accompanying boundary conditions are difficult to formulate and solve. For most practical problems, the equations and accompanying boundary conditions must be numerically solved. The types of numerical methods used by the majority of researchers and applications oriented engineers to solve these partial differential equations fall into three categories: 1) finite difference (or finite volume) methods (FDM), 2) finite element methods (FEM), and 3) other approaches (boundary integrals, hybrids, analytical, spectral, etc.). The FDM has been used for a wide variety of problems; nearly all of the early numerical simulations dealing with heat transfer and fluid flow revolved around various solution strategies for the FDM.

As structural analyses using FEM became more routine, researchers began to apply the method to more difficult problem areas, particularly those fields dealing with fluid flow. Some of the earliest work in fluid simulation with finite elements can be traced to the mid-1960s; a comprehensive review is given in Zienkiewicz and Taylor [1]. Based on this early pioneering work, the numerical simulation of fluid flow with finite elements began to proliferate by the 1970s. Today, the FEM is a strong contender for simulating all modes of fluid flow processes, rivaling performance standards associated with FDM.

The two most often used ways to formulate the FEM are the Rayleigh-Ritz variational method and the Galerkin Method of Weighted Residuals (MWR). Both approaches use a combination of appropriate functions to approximate the solution. The unknown coefficients are determined using integral statements in such a way as to approximately satisfy the original differential equations. However, there is a major difference between the Rayleigh-Ritz method and the Galerkin method. The Rayleigh-Ritz method finds the unknown coefficients through an energy minimization process; this process requires a minimum principle. The Galerkin method is based on making the projection of the error in the approximating functions vanish in the finite dimensional space spanned by the functions. This approach allows the Galerkin method to be used in situations when minimum principles do not exist. Such cases occur when convection is the dominant transport mechanism in a fluid system. The Galerkin method is therefore the method of choice in problems involving fluid flow.

In general, the following steps are needed in any finite element approximation to the solution of a differential equation: 1) the equation (or system of equations) and its boundary and initial conditions must be defined to ensure that a well-posed problem is formulated, 2) an element type must be chosen to define the approximation functions to be used in the solution, 3) a mesh must be created that adequately refines regions where large changes in the solution are expected, and that allows the boundary conditions to be properly imposed, 4) the finite element algorithm must be formulated and used to solve the system of algebraic equations, and 5) the error in the approximation must be calculated to determine if the solution is converged or if a more refined solution is needed.

Chapter 2

GOVERNING EQUATIONS

2.1 MASS CONSERVATION

The most general form of the mass conservation or *continuity* equation, using indicial form, is

$$\frac{\partial \rho}{\partial t} + \frac{\partial \rho u_j}{\partial x_j} = 0 \tag{1}$$

where ρ is the fluid density, u_j is velocity (here j = 1, 2, 3, denotes u, v, and w in the x-, y- and z-directions, respectively), and t denotes time. If the fluid is incompressible, Eq. (1) becomes

$$\frac{\partial u_j}{\partial x_j} = 0 \tag{2}$$

which is also called the *dilatation* term and corresponds to the change of volume of a fluid particle.

2.2 NAVIER-STOKES

The Navier-Stokes equations for a Newtonian viscous fluid, also known as the momentum equations, can be written as

$$\rho\left(\frac{\partial u_i}{\partial t} + u_j \frac{\partial u_i}{\partial x_j}\right) = -\frac{\partial p}{\partial x_i} + \frac{\partial}{\partial x_i}\left[\mu\left(\frac{\partial u_i}{\partial x_j} + \frac{\partial u_j}{\partial x_i}\right) + \lambda \frac{\partial u_k}{\partial x_k}\delta_{ij}\right]\rho B_i \tag{3}$$

Replacing λ by – 2/3μ yields the usual form for compressible flow. If the flow is incompressible, Eq. (3) reduces to

$$\rho\left(\frac{\partial u_i}{\partial t} + u_j \frac{\partial u_i}{\partial x_j}\right) = -\frac{\partial p}{\partial x_i} + \frac{\partial}{\partial x_j}\left(\mu \frac{\partial u_i}{\partial x_j}\right)\rho B_i \tag{4}$$

For compressible flows the equation of state is required, given by the relation

$$p = \rho RT \tag{5}$$

where R is the gas constant and T denotes temperature.

2.3 ENERGY CONSERVATION

The energy equation follows from the first law of thermodynamics. In terms of internal energy per unit mass, conservation of energy is written as

$$\rho\left(\frac{\partial e}{\partial t} + u_j \frac{\partial e}{\partial x_j}\right) = -\frac{\partial q}{\partial x_j} + \sigma_{ij}\frac{\partial u_i}{\partial x_j} + f \tag{6}$$

The heat transfer q is given by Fourier's law, $q = -k\nabla T$, where k is the material's conductivity and f is a volumetric heat source or sink. For a calorically perfect fluid, we can relate the internal energy to temperature through the thermodynamic equation

$$de = c_V\, dT \tag{7}$$

where c_V is the specific heat at constant volume. With this assumption, Eq. (7) becomes

$$\rho c_v\left(\frac{\partial T}{\partial t} + u_j\frac{\partial T}{\partial x_j}\right) = \frac{\partial}{\partial x_j}\left(k\frac{\partial T}{\partial x_j}\right) + \sigma_{ij}\frac{\partial u_i}{\partial x_j} + f \tag{8}$$

For an incompressible fluid, Eq. (8) is written in the form

$$\rho c_v\left(\frac{\partial T}{\partial t} + u_j\frac{\partial T}{\partial x_j}\right) = \frac{\partial}{\partial x_j}\left(k\frac{\partial T}{\partial x_j}\right) + \phi + f \tag{9}$$

where ϕ is the viscous heat dissipation function and is defined as

$$\phi = \left[\mu\left(\frac{\partial u_i}{\partial x_j} + \frac{\partial u_j}{\partial x_i}\right) + \lambda\frac{\partial u_k}{\partial x_k}\delta_{ij}\right]\frac{\partial u_i}{\partial x_j} \tag{10}$$

In many applications involving incompressible fluids, the dissipation term is small and is neglected.

2.4 MASS TRANSPORT

The equation for mass transport is written as

$$\frac{\partial c}{\partial t} + u_i\frac{\partial c}{\partial x_i} = \frac{\partial}{\partial x_i}\left(D\frac{\partial c}{\partial x_i}\right) + S \tag{11}$$

where c is the species component, D is the mass diffusion coefficient and S represents sources and/or sinks. In general, one equation of the form of Eq. (11) is needed for each component in the fluid. The coupling between the equations can be quite complicated, particularly in the case of flows with chemical reactions.

2.5 BOUNDARY CONDITIONS

The boundary conditions must be physically realistic and are dependent on the particular geometry, the materials involved, and the values of pertinent parameters. At a solid boundary, the velocity is zero, i.e., a *no-slip* boundary condition. However, the no-slip boundary condition is only valid when the continuum hypothesis is justified; it is not a realistic boundary condition at a solid boundary for a gas with a large mean free path. In this case there is a slip velocity of the gas relative to the solid boundary.

In general, the pressure is not required to satisfy boundary conditions, but one must prescribe a reference value for it to be determined uniquely. In some flow problems the known pressure at a free surface provides the equilibrium condition along that interface, or the prescribed pressure at an open fluid boundary is the driving force in the system. In these cases, the pressure is prescribed along these boundaries.

Chapter 3

THE FINITE ELEMENT METHOD

To illustrate finite element methodology, assume a differential operator (L) exists in two-dimensions such that

$$L \equiv -\nabla \cdot k\nabla = -\frac{\partial}{\partial x}\left(k\frac{\partial}{\partial x}\right) - \frac{\partial}{\partial y}\left(k\frac{\partial}{\partial y}\right) \tag{12}$$

One must now reformulate the basic problem in a way appropriate for the application of the FEM. Let an equation exist of the form

$$Lu(\mathbf{x}) - f(\mathbf{x}) = 0$$

where u and f are functions of (x). Now define the *residual* function as

$$R(u,\mathbf{x}) \equiv Lu(\mathbf{x}) - f(\mathbf{x}) \tag{13}$$

It then follows that if u* is the solution to the differential equation, then $R(u^*,\mathbf{x}) \equiv 0$. However, if u is only an approximation to the solution, the residual provides a measure of the error in the satisfaction of the equation.

If we now multiply Eq. (13) by a weighting function w defined over domain (space) Ω, integrate over Ω, and set the integral equal to zero, we obtain the *weighted residuals* form

$$\int_\Omega w(\mathbf{x})R(u,\mathbf{x})d\Omega = \int_\Omega w(Lu - f)d\Omega = 0 \tag{14}$$

Hence,

$$-\int_\Omega \left[w\left(\frac{\partial}{\partial x}\left(k\frac{\partial T}{\partial x}\right) + \frac{\partial}{\partial y}\left(k\frac{\partial T}{\partial y}\right)\right) + wf \right] d\Omega = 0 \tag{15}$$

We now apply Green's theorem to the second integral terms, i.e.

$$\int_\Omega \left[\frac{\partial}{\partial x}\left(kw\frac{\partial T}{\partial x}\right) + \frac{\partial}{\partial y}\left(kw\frac{\partial T}{\partial y}\right)\right] d\Omega = \int_\Gamma \left(kw\frac{\partial T}{\partial x}dy - kw\frac{\partial T}{\partial y}dx \right) \tag{16}$$

Using the fact that the components of the unit outward normal to Γ are $n_x = dy/d\Gamma$ and $n_y = -dx/d\Gamma$, the line integrals in Eq. (16) become

$$\int_\Gamma \left(kw\frac{\partial T}{\partial x}n_x + kw\frac{\partial T}{\partial y}n_y \right) d\Gamma + \int_\Gamma kw\frac{\partial T}{\partial n}d\Gamma \tag{17}$$

where n denotes the normal to the surface. Hence, Eq. (16) can be rewritten as

$$\int_\Omega \left[k\frac{\partial w}{\partial x}\frac{\partial T}{\partial x} + k\frac{\partial w}{\partial y}\frac{\partial T}{\partial y} - wf \right] d\Omega + \int_\Gamma w\left(-k\frac{\partial T}{\partial n}\right) d\Gamma = 0 \tag{18}$$

Weak weighted residual formulations for any second-order linear differential operator can be obtained in the manner described above. Nonlinear problems must be treated on a case by case basis, but we can always generate a weak form.

3.1 ERROR IN FINITE ELEMENT APPROXIMATION

The mathematical properties of finite element approximation have been thoroughly studied, and there are many excellent books that contain detailed information [2–4]. The amount of work that has been done in the area of stability and error analysis is exhaustive.

To obtain useful error estimates, one can make use of the relation

$$\|u^* - u^h\|_o \le C_1 h^2 \|u^*\|_2 \tag{19}$$

where h, defined as $h = \max\left(|x_i - x_j|, |y_i - y_j|\right)$, is the maximum mesh size in either the x- or y-direction, and the norm $\|\bullet\|_2$, is defined as

$$\|u\|_2 = \left(\int_\Omega \left[u^2 + \left(\frac{\partial u}{\partial x}\right)^2 + \left(\frac{\partial u}{\partial y}\right)^2 + \left(\frac{\partial^2 u}{\partial x^2}\right)^2 + \left(\frac{\partial^2 u}{\partial x \partial y}\right)^2 + \left(\frac{\partial^2 u}{\partial y^2}\right)\right] d\Omega\right)^{\frac{1}{2}} \tag{20}$$

3.2 ONE-DIMENSIONAL ELEMENTS

3.2.1 Linear Element

Consider a piecewise polynomial approximation of the domain $0 < x < 1$. Divide the domain into two equal intervals, and seek a solution that is linear over each of the subintervals. A linear function u between two nodal points x_i and x_{i+1} can be written as

$$u(x) = \left[\frac{x_{i+1} - x}{x_{i+1} - x_i}\right] u_i + \left[\frac{x - x_i}{x_{i+1} - x_i}\right] u_{i+1} \tag{21}$$

Thus,

$$u(x) = \sum N_i(x) u_i = N_1 u_1 + N_2 u_2 + N_3 u_3 \tag{22}$$

where

$$N_1(x) = \begin{cases} 1 - 2x & 0 \le x \le \frac{1}{2} \\ 0 & \text{otherwise} \end{cases} \tag{23a}$$

$$N_2(x) = \begin{cases} 2x & 0 \le x \le \frac{1}{2} \\ 2 - 2x & \frac{1}{2} \le x \le 1 \\ 0 & \text{otherwise} \end{cases} \tag{23b}$$

$$N_3(x) = \begin{cases} 2x - 1 & \frac{1}{2} \le x \le 1 \\ 0 & \text{otherwise} \end{cases} \tag{23c}$$

The functions $N_i(x)$, i = 1, 2, 3, are shown in Fig. 1 and are called *shape* functions, *trial* functions, or *basis* functions.

The functions have *local support*, i.e., they vanish outside a maximum of two elements. The implementation of Dirichlet conditions is trivial, e.g., to impose u(0) = 0, simply set $u_1 = u_1(x_1) = u(0) = 0$. In a Galerkin formulation, set $w_i(x) = N_i$, and the linear shape functions take the generic form over each element shown in Fig. 2.

3.2.2 Quadratic and Higher Order Elements

To obtain higher order elements, we must introduce more nodes in the elements. For example, if we desire to use quadratic polynomials over an element, a function u(x) will be approximated as

$$u(x) \cong a + bx + cx^2 \qquad 0 \le x \le h \tag{24}$$

which contains three unknown parameters. To determine the shape functions we place three nodes within the element, one at each end of the interval and one at the midpoint. Setting the nodes at $x_1 = 0$, $x_2 = h/2$, and $x_3 = h$, the shape functions become

$$N_1(x) = 1 - \frac{3x}{h} + \frac{2x^2}{h^2}, \quad N_2(x) = \frac{4x}{h}\left(1 - \frac{x}{h}\right), \quad N_3(x) = \frac{x}{h}\left(\frac{2x}{h} - 1\right) \tag{25}$$

when $0 \le x \le h$ and zero otherwise.

Fig. 3 shows the local quadratic shape functions. A finite element approximation based on quadratic elements will be more accurate than one based on linear elements. A cubic element consists of two interior nodes located at a distance of h/3 to each end.

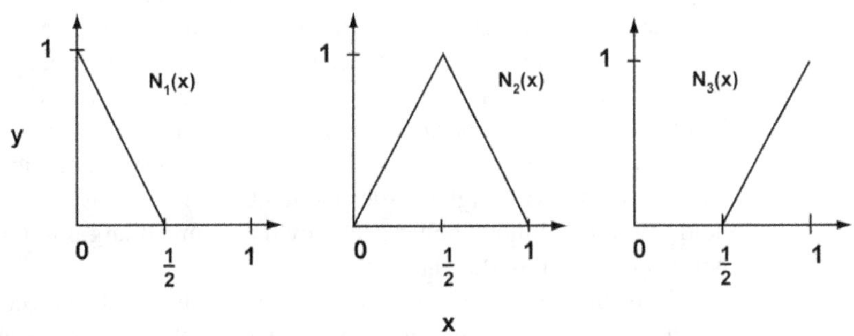

Figure 1. Linear 1-D shape functions for a two element approximation.

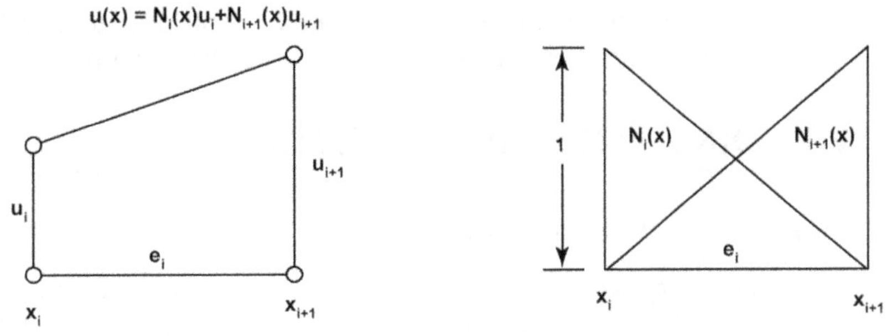

Figure 2. Linear element interpolation and element shape functions.

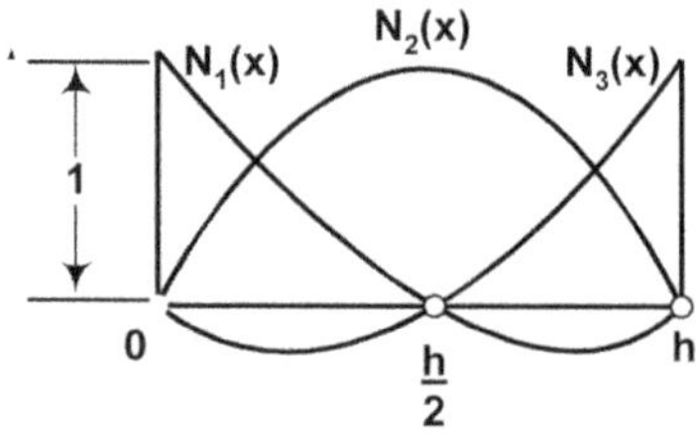

Figure 3. One-dimensional quadratic shape function.

Choosing to approximate the function u with a polynomial of degree n, so that the element has n + 1 nodes located at $x_i = (i-1)h/n$, i = 1, 2, ..., n + 1, the shape functions can be readily obtained using Lagrange's formula

$$N_i(x) = \begin{cases} \dfrac{(x-x_1)(x-x_2)\cdots(x-x_{i-1})(x-x_{i+1})\cdots(x-x_{n+1})}{(x_i-x_1)(x_i-x_2)\cdots(x_i-x_{i-1})(x_i-x_{i+1})\cdots(x_i-x_{n+1})} & \text{for } 0 \le x \le h \\ 0 \quad \text{otherwise} & \text{for } i = 1,\cdots,n+1 \end{cases} \tag{26}$$

It is easy to recover the expressions for linear shape functions when n = 1 and for quadratic elements when n = 2.

There are two ways in which a finite element approximation to any problem can be improved. The first consists in increasing the number of elements used in the mesh, therefore decreasing the size of h and, consequently, the error. This is called the h-method and relies on decreasing the size of the mesh to achieve better accuracy, utilizing always the same element. The second possibility is to keep the number of elements fixed and to increase the degree of the interpolation polynomials in the elements. In this way the number of nodes is increased and so is the order of the element. This is called the p-method. Of course, a combination of both can also be used, and this is referred to as the h-p method.

There is a very significant improvement going from linear to quadratic elements. However, the gains going from quadratic to cubic elements are marginal. On the other hand, the calculation cost increases considerably as we increase the order of the elements. To obtain the element stiffness matrices requires significantly more operations for the higher order elements. However, more important is the fact that the bandwidth of the coefficient matrix becomes larger with higher order elements. In many cases a few quadratic elements will yield solutions of much better accuracy than a much larger number of linear elements, and their use is therefore desirable.

There are other important families of elements based on piecewise polynomial interpolation that are not of the Lagrangian type. An example is Hermite polynomials, which are based on interpolating derivatives as well as the function at the nodes. The cubic Hermite two-noded element is an element for which both the function and its first derivative are continuous over the entire domain Ω. This element is widely used in finite element simulations of beam bending in solid mechanics [5]. It has also been successfully used to simulate transport in aquifers [6]. Smooth finite element approximations can also be obtained using splines as shape functions.

3.3 TWO-DIMENSIONAL ELEMENTS

3.3.1 Triangular Elements

The simplest two-dimensional figure that defines an area is the triangle. The simplest triangular element is obtained defining a linear interpolation field of the form

$$u(x,y) \cong a + bx + cy \tag{27}$$

and placing the nodes at the corners of the triangle. The shape functions can be written in terms of the nodal coordinates as

$$N_1(x,y)=\frac{1}{2A}\left[x_2y_3-x_3y_2+(y_2-y_3)x+(x_3-x_2)y\right]$$

$$N_2(x,y)=\frac{1}{2A}\left[x_3y_1-x_1y_3+(y_3-y_1)x+(x_1-x_3)y\right] \tag{28}$$

$$N_3(x,y)=\frac{1}{2A}\left[x_1y_2-x_2y_1+(y_1-y_2)x+(x_2-x_1)y\right]$$

where the area A is given by

$$2A=(x_2y_3-x_3y_2)+(x_3y_1-x_1y_3)+(x_1y_2-x_2y_1) \tag{29}$$

and the nodes are numbered counterclockwise as in Fig. 4. These elements are discussed by Pepper and Heinrich [7].

The shape functions can be more easily obtained if we use area coordinates. Joining any point P in the triangle to the vertices of the triangle, the three areas, A_1, A_2, and A_3, can be defined as shown in Fig. 4.

A coordinate system that uniquely represents every point in the triangle is given by

$$L_i=\frac{A_i}{A} \qquad i=1,2,3 \tag{30}$$

If the nodes are uniformly distributed along the element sides, the shape functions are easily constructed in this coordinate system, also called the *natural* coordinate system for the triangle.

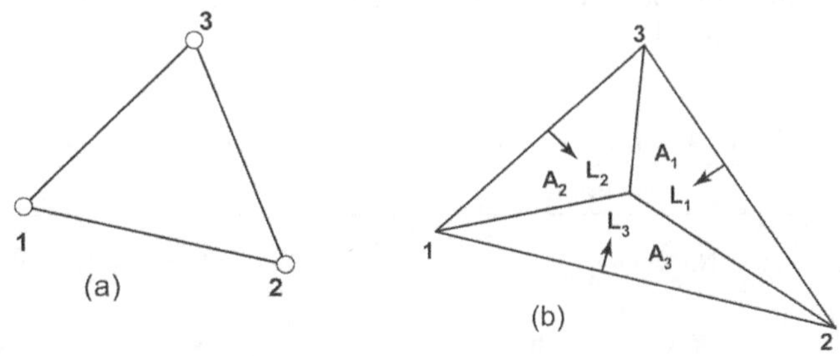

Figure 4. Linear triangular element (a) and area natural coordinate system (b).

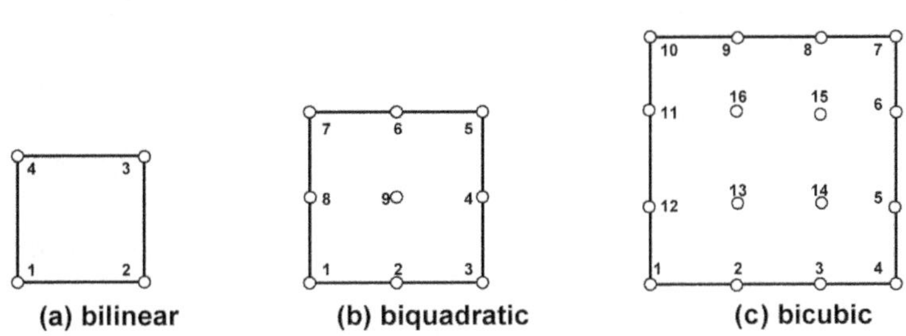

Figure 5. 2-D quadrilateral elements, (a) bilinear, (b) biquadratic, and (c) bicubic.

The shape functions in natural coordinates are independent of the shape of the triangle. This is particularly appealing when we are dealing with highly irregular geometries that may require a large variety of very differently shaped triangles. The ability of the triangle to discretize any kind of geometric figure with relative ease is the main reason for the wide use of triangular elements. From this point of view, triangular elements are always better than quadrilateral elements. Very powerful mesh generators have been developed based on the triangular geometry that can automatically discretize extremely complex regions. Although in the last years much progress has been made in this area using quadrilateral elements, these mesh generators still lack the versatility and degree of automation of those based on triangles.

3.3.2 Quadrilateral Elements

A quadrilateral element is defined by four corner points and therefore is no longer linear, which makes it more complex than a triangular element. However, there is a greater variety of quadrilateral elements, and in general, they can offer many advantages over the use of triangles.

The simplest way to obtain rectangular elements consists in taking the product (also referred to as the tensor product) of one-dimensional elements. In this fashion we generate the family of Lagrangian elements that are bilinear, biquadratic, etc., and contain 2^2, 3^2, 4^2, ... nodes as shown in Fig. 5. To obtain the shape functions we only need to know the form of the shape functions, in one dimension, and the two-dimensional function at a node is obtained as the product of the one-dimensional functions that would correspond to that node in the x and y directions, respectively.

For the 4-node bilinear and nine-node biquadratic elements, the shape functions are defined in Table 1.

Another important family of rectangular elements is known as the *serendipity* elements. These elements differ from the Lagrangian family in that they do not contain any interior nodes. Examples of the eight-noded quadratic and the 12-noded cubic elements are shown in Fig. 6.

TABLE 1. Bilinear and biquadratic element shapes for rectangular element

Bilinear	Biquadratic
$N_1(x,y)=\frac{1}{4ab}(b-x)(a-y)$	$N_1(x,y)=\bar{N}_1(x)\bar{N}_1(y)=\frac{1}{4a^2b^2}xy(x-a)(y-b)$
$N_2(x,y)=\frac{1}{4ab}(b+x)(a-y)$	$N_2(x,y)=\bar{N}_2(x)\bar{N}_1(y)=\frac{1}{2a^2b^2}(a^2-x^2)y(y-b)$
$N_3(x,y)=\frac{1}{4ab}(b-x)(a+y)$	$N_3(x,y)=\bar{N}_3(x)\bar{N}_1(y)=\frac{1}{4a^2b^2}xy(x+a)(y-b)$
$N_4(x,y)=\frac{1}{4ab}(b-x)(a+y)$	$N_4(x,y)=\bar{N}_3(x)\bar{N}_2(y)=\frac{1}{2a^2b^2}x(x+a)(b^2-y^2)$
	$N_5(x,y)=\bar{N}_3(x)\bar{N}_3(y)=\frac{1}{4a^2b^2}xy(x+a)(y+b)$
	$N_6(x,y)=\bar{N}_2(x)\bar{N}_3(y)=\frac{1}{2a^2b^2}(a^2-x^2)(y+b)$
	$N_7(x,y)=\bar{N}_1(x)\bar{N}_3(y)=\frac{1}{4a^2b^2}(x-a)(y+b)$
	$N_8(x,y)=\bar{N}_1(x)\bar{N}_2(y)=\frac{1}{2a^2b^2}x(x-a)(b^2-y^2)$
	$N_9(x,y)=\bar{N}_2(x)\bar{N}_2(y)=\frac{1}{a^2b^2}(a^2-x^2)(b^2-y^2)$

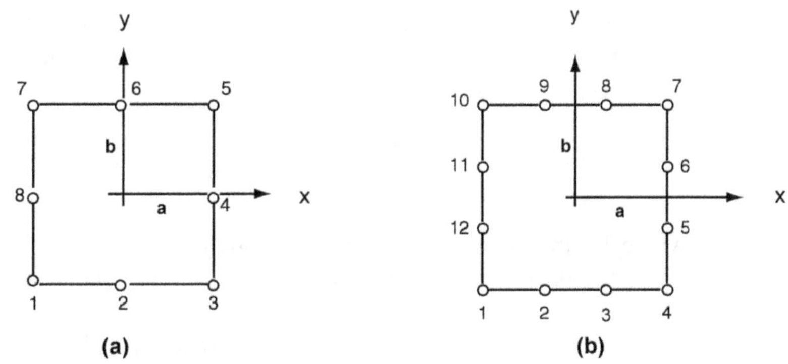

Figure 6. Serendipity elements for (a) quadratic and (b) cubic.

The shape functions for these elements were initially found by a trial and error process, thus the name serendipity. However, with the development of blending function interpolation techniques, a systematic way to construct the shape functions is now available (see [8]). For the eight-noded quadratic element of Fig. 6(a) the shape functions are given by

$$
\begin{aligned}
N_1(x,y) &= \frac{-1}{4a^2b^2}(a-x)(b-y)(bx+ay+ab) \\
N_2(x,y) &= \frac{1}{2ab^2}\left(a^2-x^2\right)(b-y) \\
N_3(x,y) &= \frac{1}{4a^2b^2}(a+x)(b-y)(bx-ay-ab) \\
N_4(x,y) &= \frac{1}{2a^2b}(a+x)\left(b^2-y^2\right) \\
N_5(x,y) &= \frac{1}{4a^2b^2}(a+x)(b+y)(bx+ay-ab) \\
N_6(x,y) &= \frac{1}{2ab^2}\left(a^2-x^2\right)(b+y) \\
N_7(x,y) &= \frac{-1}{4a^2b^2}(a-x)(b+y)(bx-ay+ab) \\
N_8(x,y) &= \frac{1}{2a^2b}(a-x)\left(b^2-y^2\right)
\end{aligned}
\tag{32}
$$

The shape functions for the 12-noded cubic element can be found in many finite element textbooks, e.g., Reddy [5]. The linear triangle and bilinear quadrilateral exhibit exactly the same rate of convergence. In practice, better *accuracy* is obtained using quadrilateral elements than triangular elements.

3.3.3 Isoparametric Elements

Rectangular elements offer advantages over triangular elements. However, rectangular geometry is very restrictive and general quadrilateral elements must be used to deal with more general geometry. To resolve this difficulty, the concept of *isoparametric* transformations was introduced by Irons [9] to general rectangular elements. The idea is based on performing a local (element by element) transformation between a general quadrilateral element in the global coordinate system and a "parent" rectangular element defined in a $\xi - \eta$ coordinate system in the square $-1 \le \xi,\eta \le 1$ as depicted in Fig. 7 for a four-noded element.

Transformation of integrals must be done, i.e.,

$$\iint_e F\left(u(x,y),\frac{\partial u}{\partial x},\frac{\partial u}{\partial y}\right)dxdy = \int_{-1}^{1}\int_{-1}^{1} F\left(u(x(\xi,\eta),y(\xi,\eta)),\frac{\partial u}{\partial x},\frac{\partial u}{\partial y}\right)|\det J|\,d\xi d\eta \qquad (33)$$

where **J** is the Jacobian of the transformation.

The isoparametric transformation itself is easily obtained using the relation

$$\begin{bmatrix} x \\ y \end{bmatrix} = \sum_{i=1}^{N_e} N_i(\xi,\eta)\begin{bmatrix} x_i \\ x_i \end{bmatrix} \qquad (34)$$

where N_e is the number of nodes in the element and $N_i(\xi,\eta)$ are the shape functions for the corresponding parent element, in the square $-1 \le \xi,\eta \le 1$. Actually, this transformation is the inverse of what is really needed, since it maps the parent element, not the actual element. It is clearly defined once the coordinates of the nodes (x_i, y_i) in the global system and the shape functions $N_i(\xi,\eta)$ in the square parent system of coordinates are known.

Given any function, $\phi(x,y)$, the derivatives of ϕ with respect to x and y in the parent coordinate system can be found immediately. Applying the chain rule (using matrix formulation),

$$\begin{bmatrix} \frac{\partial \phi}{\partial \xi} \\ \frac{\partial \phi}{\partial \eta} \end{bmatrix} = \begin{bmatrix} \frac{\partial x}{\partial \xi} & \frac{\partial y}{\partial \xi} \\ \frac{\partial x}{\partial \eta} & \frac{\partial y}{\partial \eta} \end{bmatrix}\begin{bmatrix} \frac{\partial \phi}{\partial x} \\ \frac{\partial \phi}{\partial y} \end{bmatrix} = \mathbf{J}\begin{bmatrix} \frac{\partial \phi}{\partial x} \\ \frac{\partial \phi}{\partial y} \end{bmatrix} \qquad (35)$$

or

$$\begin{bmatrix} \frac{\partial \phi}{\partial x} \\ \frac{\partial \phi}{\partial y} \end{bmatrix} = \mathbf{J}^{-1}\begin{bmatrix} \frac{\partial \phi}{\partial \xi} \\ \frac{\partial \phi}{\partial \eta} \end{bmatrix} = \frac{1}{\det \mathbf{J}}\begin{bmatrix} \frac{\partial y}{\partial \eta} & -\frac{\partial y}{\partial \xi} \\ -\frac{\partial x}{\partial \eta} & \frac{\partial x}{\partial \xi} \end{bmatrix}\begin{bmatrix} \frac{\partial \phi}{\partial \xi} \\ \frac{\partial \phi}{\partial \eta} \end{bmatrix} \qquad (36)$$

Table 2 gives the shape functions for the eight- and nine-noded biquadratic elements.

Similarly, cubic isoparametric elements can be defined; however, in fluid flow applications, quadratic isoparametric elements are used much less than bilinear elements and cubics are hardly ever considered.

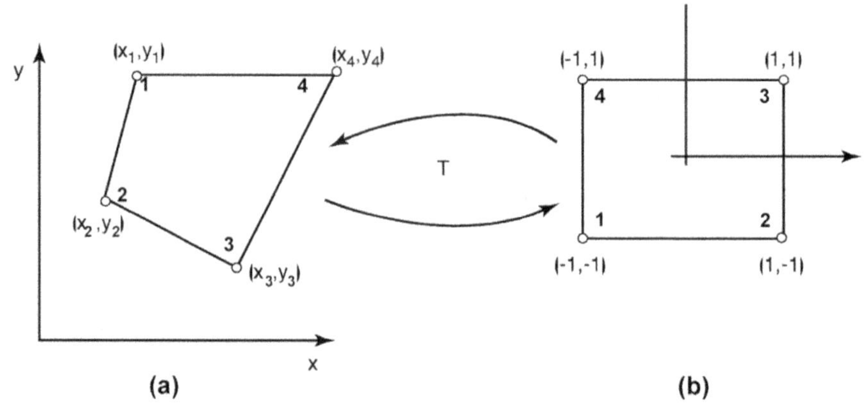

Figure 7. Bilinear isoparametric elements, (a) actual element and (b) parent element.

For triangular elements the concepts described above are applied using the right triangle shown in Fig. 8 as the parent element; in this case, $0 \le \xi \le 1$ and $0 \le \eta \le 1$. The shape functions for the linear and quadratic parent elements are not difficult to find. For the linear element these are

$$
\begin{aligned}
N_1(\xi,\eta) &= 1-\xi-\eta \\
N_2(\xi,\eta) &= \xi \\
N_3(\xi,\eta) &= \eta
\end{aligned}
\tag{37}
$$

TABLE 2. Shape functions for bilinear nine- and eight-noded quadratic isoparametric elements

$N_1(\xi,\eta)=1/4\xi\eta(\xi-1)(\eta-1)$	$N_1(\xi,\eta)=-1/4(1-\xi)(1-\eta)(1+\xi+\eta)$
$N_2(\xi,\eta)=1/2\eta(1-\xi^2)(\eta-1)$	$N_2(\xi,\eta)=1/2(1-\xi^2)(1-\eta)$
$N_3(\xi,\eta)=1/4\xi\eta(\xi+1)(\eta-1)$	$N_3(\xi,\eta)=1/4(1+\xi)(1-\eta)(\xi-\eta-1)$
$N_4(\xi,\eta)=1/2\xi(\xi+1)(1-\eta^2)$	$N_4(\xi,\eta)=1/2(1+\xi)(1-\eta^2)$
$N_5(\xi,\eta)=1/4\xi\eta(\xi+1)(\eta+1)$	$N_5(\xi,\eta)=1/4(1+\xi)(1+\eta)(\xi+\eta-1)$
$N_6(\xi,\eta)=1/2\eta(1-\xi^2)(\eta+1)$	$N_6(\xi,\eta)=1/2(1-\xi^2)(1+\eta)$
$N_7(\xi,\eta)=1/4\xi\eta(\xi-1)(\eta+1)$	$N_7(\xi,\eta)=1/4(1-\xi)(1+\eta)(1+\xi-\eta)$
$N_8(\xi,\eta)=1/2\xi(\xi-1)(1-\eta^2)$	$N_8(\xi,\eta)=1/2(1-\xi)(1-\eta^2)$
$N_9(\xi,\eta)=(1-\xi^2)(1-\eta^2)$	

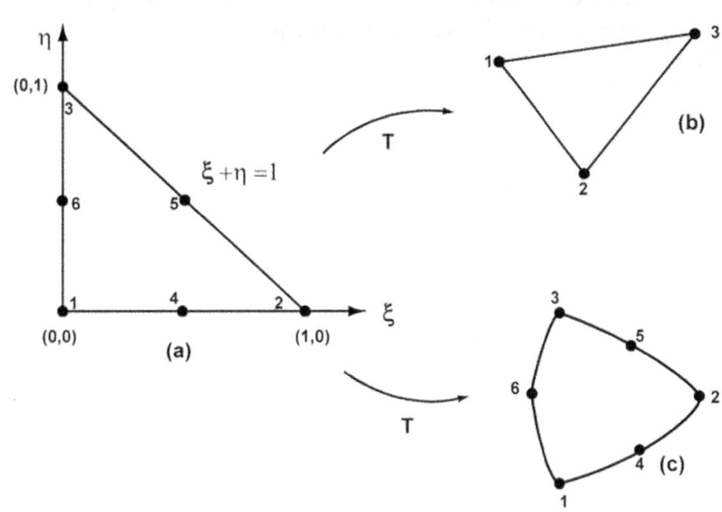

Figure 8. Triangular isoparametric elements for (a) parent element, (b) linear element, and (c) curved quadratic triangle.

and for the quadratic triangle,

$$\begin{aligned}
N_1(\xi,\eta) &= \left[2(1-\xi-\eta)-1\right](1-\xi-\eta) \\
N_2(\xi,\eta) &= \xi(2\xi-1) \\
N_3(\xi,\eta) &= \eta(2\eta-1) \\
N_4(\xi,\eta) &= 4\xi\eta \\
N_5(\xi,\eta) &= 4\eta(1-\xi-\eta) \\
N_6(\xi,\eta) &= 4\xi(1-\xi-\eta)
\end{aligned} \tag{38}$$

The accuracy decreases if the elements are distorted too much. Mathematically, det $|\mathbf{J}| > 0$ at all points in the parent element. This condition leads to different restrictions for every set of shape functions and must be examined for each individual element. For the bilinear transformation det $\mathbf{J} \neq 0$ if and only if the quadrilateral is convex, that is, all its interior angles must be less than π. Fig. 9 shows sample permissible and nonpermissible elements.

For higher order elements these conditions are more difficult to determine and are very sensitive to the location of midside and interior nodes. Serendipity elements do not contain interior nodes, and for this reason the quadratic serendipity element, in particular, has been a very popular element to deal with curved boundaries.

A *subparametric* transformation uses lower order interpolation functions for the geometric transformations than for interpolation. For example, a biquadratic element can be associated with a bilinear transformation if the sides of the elements are always straight lines. In the same note, *superparametric* elements use geometric transformations that are higher order than the interpolation.

When curved boundaries are fitted, curved isoparametric elements can produce enormous improvements in some classes of problems [10]. However, for problems such as convective flows and heat and mass transfer, such techniques must be applied carefully. Remember that the use of isoparametric elements requires *numerical integration* because the determinant of the Jacobian transforms the integral into a rational function, even though the shape functions are polynomials.

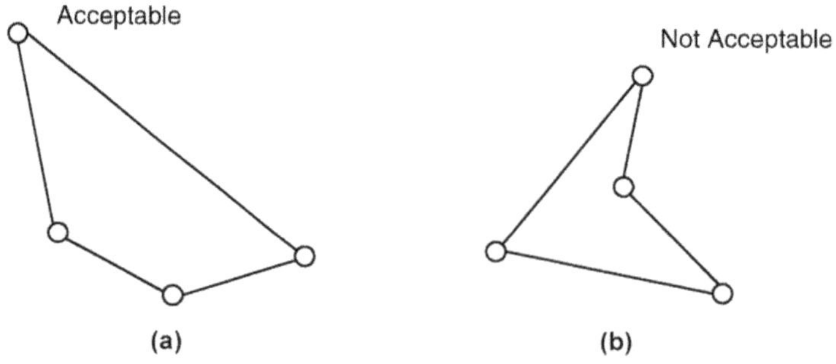

Figure 9. Examples of (a) permissible quadrilateral elements and (b) illegal quadrilateral elements.

3.4 THREE-DIMENSIONAL ELEMENTS

The transition to three-dimensional problems does not involve any new concepts. Area and line integrals are replaced with volume and surface integrals. The natural extension of the one-dimensional linear and two-dimensional bilinear elements is the eight-noded trilinear or brick element shown in Fig. 10. There is also a Lagrangian element (with a node at the centroid) and the shape functions are easily obtained as products of one-dimensional linear functions, as was the case for the bilinear element. Define

$$\bar{N}_1(\alpha) = \left(1 - \frac{\alpha}{\Delta\alpha}\right) \quad \text{and} \quad \bar{N}_2(\alpha) = \frac{\alpha}{\Delta\alpha}$$

where α is x, y, or z. The trilinear shape functions become

$$\begin{aligned}
N_1(x,y,z) &= \bar{N}_1(x)\bar{N}_1(y)\bar{N}_1(z) \\
N_2(x,y,z) &= \bar{N}_2(x)\bar{N}_1(y)\bar{N}_1(z) \\
N_3(x,y,z) &= \bar{N}_2(x)\bar{N}_2(y)\bar{N}_1(z) \\
N_4(x,y,z) &= \bar{N}_1(x)\bar{N}_2(y)\bar{N}_1(z) \\
N_5(x,y,z) &= \bar{N}_1(x)\bar{N}_1(y)\bar{N}_2(z) \\
N_6(x,y,z) &= \bar{N}_2(x)\bar{N}_1(y)\bar{N}_2(z) \\
N_7(x,y,z) &= \bar{N}_2(x)\bar{N}_2(y)\bar{N}_2(z) \\
N_8(x,y,z) &= \bar{N}_1(x)\bar{N}_2(y)\bar{N}_2(z)
\end{aligned} \tag{39}$$

Three-dimensional tetrahedral elements are the natural extension of two-dimensional triangles. The simplest of these elements are linear and quadratic tetrahedrons depicted in Fig. 11. A natural coordinate system is defined for tetrahedral elements by means of the four internal volumes determined when any interior point is connected with the four vertices of the tetrahedron – in a manner similar to that used in two dimensions to define the area coordinates. In this case, they are referred to as volume coordinates. These are discussed by Pepper and Heinrich [7].

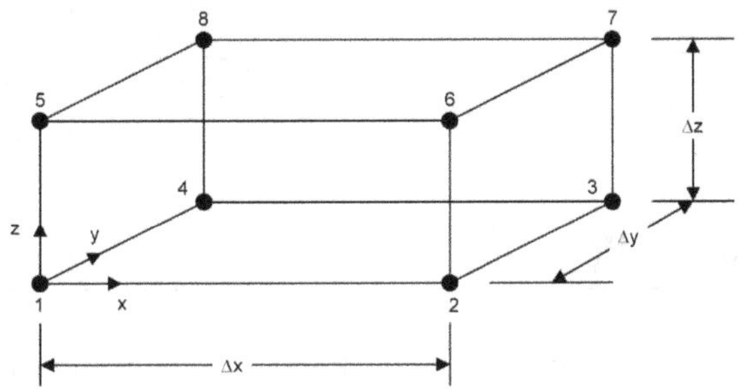

Figure 10. Trilinear "brick" hexahedral element.

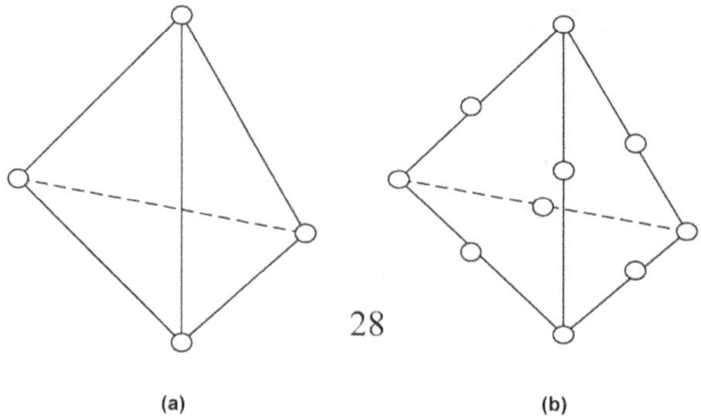

Figure 11. Linear and quadratic tetrahedral elements.

Isoparametric transformations are readily defined for the three-dimensional elements as

$$\begin{bmatrix} x \\ y \\ z \end{bmatrix} = \sum_{i=1}^{N_e} N_i(\xi,\eta,\zeta) \begin{bmatrix} x_i \\ y_i \\ z_i \end{bmatrix} \tag{40}$$

The Jacobian matrix is

$$J = \begin{bmatrix} \frac{\partial x}{\partial \xi} & \frac{\partial y}{\partial \xi} & \frac{\partial z}{\partial \xi} \\ \frac{\partial x}{\partial \eta} & \frac{\partial y}{\partial \eta} & \frac{\partial z}{\partial \eta} \\ \frac{\partial x}{\partial \zeta} & \frac{\partial y}{\partial \zeta} & \frac{\partial z}{\partial \zeta} \end{bmatrix} \tag{41}$$

and the derivatives of the shape functions are obtained from

$$\begin{bmatrix} \frac{\partial N_i}{\partial x} \\ \frac{\partial N_i}{\partial y} \\ \frac{\partial N_i}{\partial z} \end{bmatrix} = J^{-1} \begin{bmatrix} \frac{\partial N_i}{\partial \xi} \\ \frac{\partial N_i}{\partial \eta} \\ \frac{\partial N_i}{\partial \zeta} \end{bmatrix} \tag{42}$$

3.5 QUADRATURE

In finite element approximations using isoparametric elements, the integrals are always performed over an element in the parent coordinate system $-1 \leq \xi \leq 1$; hence

$$I(g)\int_a^b g(x)dx \rightarrow I(f)\int_{-1}^1 f(\xi)d\xi \tag{43}$$

Approximations of the form

$$I_n(g) \cong \sum_{i=1}^{n} w_i g(x_i) \tag{44}$$

are called *numerical quadrature* or *numerical integration formulae*. The points x_i are called *quadrature points*, and the coefficients w_i are called *quadrature weights*.

A quadrature formula with degree of precision m integrates polynomials of degree ≤ m exactly. The maximum degree of precision that can be achieved with a quadrature rule that uses n integration points is m = 2n – 1, and the quadrature formulae that achieve this accuracy are known as Gaussian quadratures. The integration formulae are referred to as Gauss-Legendre quadratures. A list of the integration points and weights for Gauss quadratures up to order n = 4 is shown in Table 3.

In two and three dimensions, one must evaluate double and triple integrals over areas and volumes, respectively. The easiest way is to evaluate the variables one by one in succession, e.g., fix the independent variable η and define

$$F(\eta) \equiv \sum_{i=1}^{n} w_i f(\xi_i, \eta) \cong \int_{-1}^{1} f(\xi, \eta) d\xi \tag{45}$$

Then

$$\int_{-1}^{1} F(\eta) d\eta \cong \sum_{j=1}^{m} w_j F(\eta_j) = \sum_{j=1}^{m} w_j \left(\sum_{i=1}^{n} w_i f(\xi_i, \eta_j) \right) \tag{46}$$

Hence,

$$I_{nm} = \sum_{i=1}^{n} \sum_{j=1}^{m} w_i w_j f(\xi_i, \eta_j) \tag{47}$$

where m and n are not necessarily equal. A quadrature formula with degree of precision 3 is obtained if one uses a Gauss quadrature with n = 2 in each direction, that is,

$$\int_{-1}^{1} \int_{-1}^{1} f(\xi, \eta) d\xi d\eta \cong I_{22} = \sum_{i=1}^{2} \sum_{j=1}^{2} w_i w_j f(\xi_i, \eta_j) \tag{48}$$

where the sampling points ξ_i, η_j in each direction and the weights w_i are given in Table 2 for n = 2. Fig. 12 shows the approximate location of the Gauss points in two and three dimensions for n = m = 2 and n = m = 3.

TABLE 3. Sampling points and weight coefficients for Gaussian quadrature formulae, $\int_{-1}^{1} f(\xi) d\xi \cong \sum_{i=1}^{n} w_i f(\xi_i)$

n	ξ_i	w_i
1	0.0	2.0
2	–0.57735026918963 0.57735026918963	1.0 1.0
3	–0.77459666924148 0.0 0.77459666924148	0.55555555555556 0.88888888888889 0.55555555555556
4	–0.86113631159405 –0.33998104358486 0.33998104358486 0.86113631159405	0.34785484513745 0.65214515486255 0.65214515486255 0.34785484513745

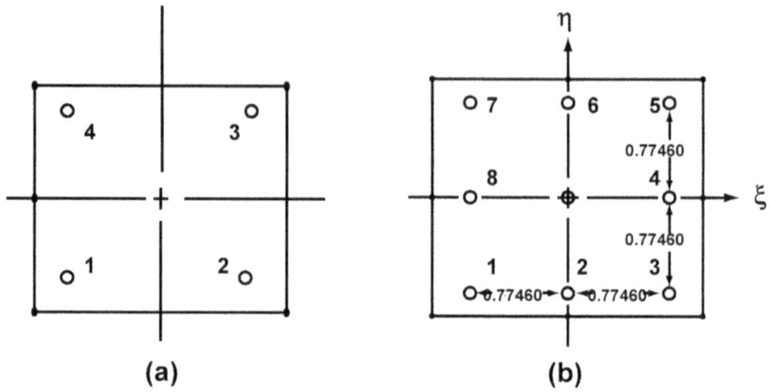

Figure 12. Location of Gauss points for two-dimensional integration in the square $1 \leq \xi, \eta \leq 1$ for (a) 2×2 quadrature and (b) 3×3 quadrature.

TABLE 4. Quadrature formulae for triangles (integration point numbers are indicated in Fig. 12)

Number of Points	Degree of Precision	Point Number i	Area Coordinates of Quadratures $(L_1)_i$	$(L_2)_i$	$(L_3)_i$	Weight w_i
1	1	1	1/2	1/2	1/2	1
3	2	1	1/2	1/2	0	1/3
		2	0	1/2	1/2	1/3
		3	1/2	0	1/2	1/3
4	3	1	1/3	1/3	1/3	–27/48
		2	11/15	2/15	2/15	25/48
		3	2/15	11/15	2/15	25/48
		4	2/15	2/15	11/15	25/48

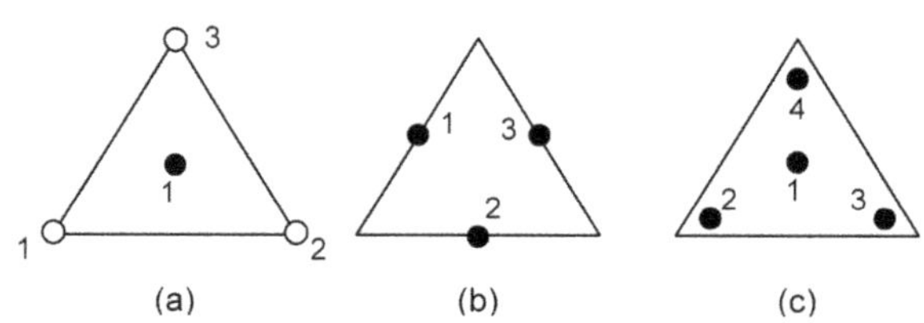

Figure 13. Location of the integration points in triangles for the quadrature given in Table 3 (a) one-point rule, (b) three-point rule, and (c) four-point rule.

For triangular elements, convenient integration formulae were first derived by Hammer et al. [11]. Later, Cowper [12] added integration formulae for triangles with degree of precision up to 7. Table 4 gives the sampling points and weights for integration formulae with degree of precision up to 5 are given. The positions of the quadrature points are shown in Fig. 13.

3.6 REDUCED INTEGRATION

If full integration requires n Gauss points in each coordinate direction, normally, reduced integration will use one point less, i.e., n – 1 Gauss points in each coordinate direction, as shown in Fig. 14.

(a) For bilinear rectangular elements, the 2×2 Gauss quadrature is necessary for full integration. The reduced integration quadrature will use one Gauss point in each direction.

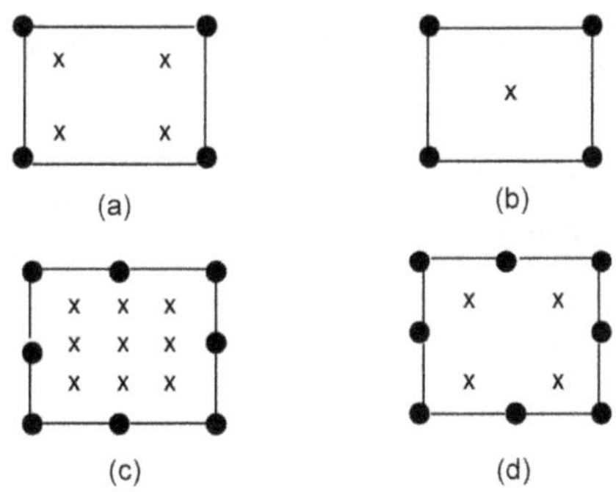

Figure 14. Gauss quadrature points for (a) bilinear element, full integration, (b) bilinear element, reduced integration, (c) biquadratic element, full integration, and (d) biquadratic element, reduced integration.

(b) A biquadratic element requires the 3 × 3 Gauss quadrature for full integration. Fourth-degree polynomials must be integrated exactly, and a degree of precision of at least 4 is required. The reduced integration formula is the 2 × 2 Gauss quadrature, which has degree of precision m = 3.

3.7 TIME DEPENDENCE

3.7.1 The θ Method

The most commonly used time integration algorithm is the θ method, which consists in approximating the time derivative by the backward difference

$$\dot{\mathbf{T}} \cong \frac{1}{\Delta t}\left(\mathbf{T}^{n+1} - \mathbf{T}^{n}\right) \tag{49}$$

where $\mathbf{T}^n \equiv \mathbf{T}(x,t_n)$ denotes the value of a variable, e.g., temperature, at time $t = t_n$, Δt is the time step increment, and $t_{n+1} = t_n + \Delta t$. The temperature **T** is then defined by

$$\mathbf{T} = \theta\mathbf{T}^{n+1} + (1-\theta)\mathbf{T}^{n} \tag{50}$$

where the relaxation parameter θ is normally specified to be a value between 0 and 1 and is used to control the accuracy and stability of the algorithm. This method falls in the general category of *one-step* methods, in which the solution at each step is advanced to time t_{n+1} from known values at time step t_n.

Substituting Eqs. (49) and (50), the one-dimensional transient heat diffusion equation becomes

$$\left(\frac{1}{\Delta t}\mathbf{C} + \theta\mathbf{K}\right)\mathbf{T}^{n+1} = \left(\frac{1}{\Delta t}\mathbf{C} - (1-\theta)\mathbf{K}\right)\mathbf{T}^{n} + \theta\mathbf{Q}^{n+1} + (1-\theta)\mathbf{Q}^{n} \tag{51}$$

where **Q** has been assumed to be a function of time and approximated over the interval $t_n \le t \le t_{n+1}$ using Eq. (50), and the mass matrix, **C**, and stiffness matrix, **K**, are evaluated from the integral relations

$$\mathbf{C} = \int N_i N_j \tag{52}$$

$$\mathbf{K} = k\int \frac{\partial N_i}{\partial x}\frac{\partial N_j}{\partial x}dx$$

If 1-D, linear elements are used in the space discretization and Q is independent of x, the resulting element equations are

$$\left\{\frac{\rho c_v h}{6\Delta t}\begin{bmatrix}2 & 1\\ 1 & 2\end{bmatrix}+\frac{\theta k}{h}\begin{bmatrix}1 & -1\\ -1 & 1\end{bmatrix}\right\}\begin{bmatrix}T_1^{n+1}\\ T_2^{n+1}\end{bmatrix}=\left\{\frac{\rho c_v h}{6\Delta t}\begin{bmatrix}2 & 1\\ 1 & 2\end{bmatrix}-\frac{(1-\theta)k}{h}\begin{bmatrix}1 & -1\\ -1 & 1\end{bmatrix}\right\}\begin{bmatrix}T_1^{n}\\ T_2^{n}\end{bmatrix}+\frac{\theta h Q^{n+1}}{2}\begin{bmatrix}1\\ 1\end{bmatrix}+\frac{(1-\theta)hQ^{n}}{2}\begin{bmatrix}1\\ 1\end{bmatrix} \tag{53}$$

where h denotes the element size. The values $\theta = 0$, 0.5, and 1.0 are most commonly used and, except for the presence of the mass matrix, correspond to the Euler, Crank-Nicolson, and backward implicit methods, respectively. However, the appearance of the mass matrix modifies the algorithms. Therefore they are referred to as Euler-Galerkin when $\theta = 0$, Crank-Nicolson-Galerkin when $\theta = 0.5$, and backward implicit Galerkin when $\theta = 1.0$.

A truncation error analysis shows that the methods converge as first-order methods $O(\Delta t)$ when $\theta = 0.0$ and 1.0; the Crank-Nicolson-Galerkin method is second-order $O(\Delta t^2)$ in time, and for other values of θ between 0 and 1, convergence takes place at intermediate rates between first- and second-order.

A stability analysis shows that:

1. If $1/2 \le \theta < 1$, the method is *unconditionally stable*. Hence the Crank-Nicolson-Galerkin and backward implicit Galerkin methods are stable for any Δt.
2. If $0 \le \theta < 1/2$ there is a time step limitation given by

$$\Delta t < \frac{2}{\lambda(1-2\theta)}$$

where λ is the largest eigenvalue of the generalized eigenvalue problem

$$(\mathbf{K} - \lambda\,\mathbf{C})\,\mathbf{X} = \mathbf{0} \tag{54}$$

3.7.2 Mass Lumping

An important case of the θ method is the Euler-Galerkin algorithm, corresponding to $\theta = 0$. In this case, Eq. (51) takes the form

$$\frac{1}{\Delta t}\mathbf{C}\mathbf{T}^{n+1} = \mathbf{Q}^{n} + \left(\frac{1}{\Delta t}\mathbf{C} - \mathbf{K}\right)\mathbf{T}^{n} \tag{55}$$

Notice that this is not an explicit scheme, since the mass matrix **C** must be inverted to find the solution. In fact, *fully explicit* algorithms do not arise from the standard form of the Galerkin finite element method. To obtain explicit algorithms, one must diagonalize the matrix **C**. This is known as *mass lumping* and is a procedure of great practical importance in finite element modeling. It is also the subject of much controversy because of the effect that mass lumping may have on the accuracy of the algorithm.

The simplest method of mass lumping consists in defining a lumped mass matrix $\bar{\mathbf{C}}$ as

$$\bar{\mathbf{C}} \equiv \left[\bar{c}_{ij}\right] = \begin{cases}\sum_k c_{ik} & \text{if } i = j\\ 0 & \text{if } i \ne j\end{cases}$$

The diagonal elements of $\bar{\mathbf{C}}$ are the row sums of **C**, and the off-diagonal elements are zero.

Replacing the mass matrix **C** by the lumped mass matrix in the $\bar{\mathbf{C}}$ Euler-Galerkin method gives

$$\frac{1}{\Delta t}\bar{\mathbf{C}}\mathbf{T}^{n+1} = \mathbf{Q}^n + \left(\frac{1}{\Delta t}\bar{\mathbf{C}} - \mathbf{K}\right)\mathbf{T}^n \tag{56}$$

which can be solved explicitly as

$$\mathbf{T}^{n+1} = \Delta t \bar{\mathbf{C}}^{-1}\left[\mathbf{Q}^n\left(\frac{1}{\Delta t}\bar{\mathbf{C}} - \mathbf{K}\right)\mathbf{T}^n\right] \tag{57}$$

where $\bar{\mathbf{C}}^{-1}$ is a diagonal matrix with entries $\bar{c}_{ii}^{-1} = 1/\bar{c}_{ii}$.

Algorithms based on the consistent mass matrix are generally more accurate than those using lumped masses. However, the consistent mass matrix may introduce undesired oscillations during the early stages of the calculation, especially if the initial data are not smooth. This phenomenon has been called "*noise*" in the numerical solution and occurs when the time step exceeds a *critical* time step related to the solution of the equation. Unfortunately, the critical time step is usually extremely small, and it is not realistic to attempt to eliminate the problem by making the time step smaller.

Another family of time stepping algorithms is the explicit Runge-Kutta methods. These methods have been used extensively in compressible flow calculations and nonlinear transport [13]. The Euler method is based on the linear Taylor polynomial approximation to the dependent variables. To improve on Euler's method we may consider more terms of the Taylor polynomial, but this has the drawback that higher order derivatives of the function would be required. The Runge-Kutta methods provide ways to improve the approximations without resorting to the higher order derivatives, which are approximated using functional values. The details of how this is done can be found in texts such as those of Isaacson and Keller [14] or Yakowitz and Szidarovszky [15].

The greatest advantage of explicit methods is that the solution of linear systems of equations is not required, and hence large numbers of degrees of freedom can be handled without excessive memory requirements. However, the evaluations of the vector functions may become very computationally intensive. In large calculations, vectorization and/or parallel processing are needed to reduce the computation time, particularly in the fourth-order Runge-Kutta method.

Another time integration scheme used in finite element methods is the Newmark algorithm [16]. This algorithm was originally designed for problems of structural dynamics, and later generalized by Hughes et al. [17] for general time-dependent problems. Newmark methods stem from the equations of dynamic equilibrium and therefore are applicable to both parabolic equations and second-order hyperbolic equations. They are based on the use of generalized physical quantities of displacements, velocities, and accelerations, which makes it easy to interpret the algorithms physically.

3.8 PETROV-GALERKIN METHOD

In order to improve accuracy in time, one can construct weighting functions that are parabolic in time, e.g., the time variation T(t) is

$$T(t) = \frac{4t}{\Delta t}\left(1 - \frac{t}{\Delta t}\right) \tag{58}$$

The weighting functions become $N_i(x)T(t)$.

In the one-dimensional case, the truncation error can be interpreted in a difference equation as an added diffusion, and obtain improved algorithms by introducing a balancing diffusion dependent on a parameter α. In this case, a balancing dispersion term of the form $\beta d(\partial^3\phi/\partial x^2\partial t)$ can be added, where the coefficient d must be proportional to $uh\Delta t$ for dimensional consistency. Looking at the modified advection-diffusion equation, one obtains

$$\frac{\partial\phi}{\partial t}+u\frac{\partial\phi}{\partial x}-\left(D+\frac{\alpha uh}{2}\right)\frac{\partial^2\phi}{\partial x^2}+\beta d\frac{\partial^3\phi}{\partial x^2\partial t}=0 \tag{59}$$

Now apply the Galerkin method and operate on the weak form. The Petrov-Galerkin weights are

$$w_i(x,t)=M_i(x,t)+\frac{\alpha h}{2}\frac{\partial M_i(x,t)}{\partial x}+\frac{\beta h\Delta t}{4}\frac{\partial^2 M_i(x,t)}{\partial x\partial t} \tag{60}$$

The functions $M_i(x,t)$ in $0\le x\le h$, $0\le t\le \Delta t$ are

$$M_1(x,t)=M_4(x,t)=4\left(1-\frac{x}{h}\right)\frac{t}{\Delta t}\left(1-\frac{t}{\Delta t}\right)$$

$$M_2(x,t)=M_3(x,t)=4\frac{x}{h}\frac{t}{\Delta t}\left(1-\frac{t}{\Delta t}\right)$$

The weighting functions become

$$\left.\begin{aligned} w_1(x,t)=w_4(x,t)&=4\left(1-\frac{x}{h}\right)\frac{t}{\Delta t}\left(1-\frac{t}{\Delta t}\right)-2\alpha\frac{t}{\Delta t}\left(1-\frac{t}{\Delta t}\right)-\beta\left(1-\frac{2t}{\Delta t}\right)\\ w_2(x,t)=w_3(x,t)&=4\frac{x}{h}\frac{t}{\Delta t}\left(1-\frac{t}{\Delta t}\right)+2\alpha\frac{t}{\Delta t}\left(1-\frac{t}{\Delta t}\right)+\beta\left(1-\frac{2t}{\Delta t}\right)\end{aligned}\right\} \tag{61}$$

After some algebra (see [8]), α and β are

$$\alpha=\coth\frac{\gamma}{2}-\frac{2}{\gamma}\qquad \beta=\frac{c}{3}-\frac{2\alpha}{\gamma c} \tag{62}$$

If $\beta = 0$, the algorithm reduces to applying the Petrov-Galerkin weights for the steady state equation with a second-order time-stepping scheme, and in this case it is only second-order accurate in space. In the limiting case when $\gamma\to 0$, the expression for β is undefined. Physically u must go to zero. In this case the algorithm reduces to the Crank-Nicolson-Galerkin scheme because $\gamma\to 0$ if $u\to 0$.

There are several Petrov-Galerkin algorithms that have been proposed for the transient problems, e.g., Westerink and Shea [18], Cardle [19], Perrochet [20] and Idelsohn et al. [21]. Most of these formulations follow arguments similar to the ones presented here and have their own distinct advantages.

Extending the Petrov-Galerkin method to more than one space dimension, the weighting functions, anisotropic diffusion and dispersion are introduced in the direction of flow only. For constant velocity **V** the weighting functions can be expressed in the form

$$w_i(\mathbf{x},t)=M_i+\frac{h}{2\|\mathbf{V}\|}\left(\alpha+\frac{\beta\Delta t}{2}\frac{\partial}{\partial t}\right)\mathbf{V}\cdot\nabla M_i \tag{63}$$

where the functions M_i are quadratic in time with the time variation taking the form of Eq. (58).

The parameters α and β are

$$\alpha = \coth\frac{\bar{\gamma}}{2} - \frac{2}{\bar{\gamma}}, \qquad \beta = \frac{\delta}{3} - \frac{2\alpha}{\bar{\gamma}\delta} \tag{64}$$

with

$$\bar{\gamma} = \frac{\|\mathbf{V}\|\bar{h}}{D}, \quad \delta = \frac{\|\mathbf{V}\|\Delta t}{\bar{h}} \tag{65}$$

where $\bar{h}$ is defined as before. When α and β are both different from zero, the stability limit in the two-dimensional case takes the form

$$\Delta t \leq \frac{1}{\frac{|u|}{\Delta x} + \frac{|v|}{\Delta y}} \tag{66}$$

3.9 TAYLOR-GALERKIN METHOD

An important method developed for convective transport is the Taylor-Galerkin method [22]. The idea is to approximate ϕ_t^n by a truncated Taylor series in time as

$$\phi_t^n = \frac{\phi^{n+1} - \phi^n}{\Delta t} - \frac{\Delta t}{2}\phi_{tt}^n - \frac{\Delta t^2}{6}\phi_{ttt}^n + O\left(\Delta t^3\right) \tag{67}$$

where $\phi_t^n = \delta\phi/\delta t(x, t_n)$ with the space variable x continuous. From the differential equation,

$$\phi_t = -u\phi_x \tag{68}$$

and differentiating this expression with respect to time,

$$\phi_{tt} = u^2\phi_{xx} \quad \text{and} \quad \phi_{ttt} = u^2\phi_{txx} \tag{69}$$

From Eqs. (67) and (69), we can write

$$\frac{\phi^{n+1} - \phi^n}{\Delta t} - u^2\frac{\Delta t^2}{6}\phi_{txx}^n + u\phi_x - u^2\frac{\Delta t}{2}\phi_{txx}^n = 0 \tag{70}$$

This equation can now be discretized in time and space to yield a number of different algorithms. For example, replacing the time derivative in Eq. (70) by a forward difference leads to the Euler-Taylor-Galerkin form

$$\frac{\phi^{n+1} - \phi^n}{\Delta t} - u^2\frac{\Delta t}{6}\left(\phi_{xx}^{n+1} - \phi_{xx}^n\right) + u\phi_x - u^2\frac{\Delta t}{2}\phi_{xx}^n = 0 \tag{71}$$

which can be discretized in space using linear or higher order elements. Notice that the first-order equation, Eq. (69), was transformed into a second-order equation in space. Also notice that Eq. (72) will not approach the correct limit as the solution approaches steady state. There are some controversial aspects of these methods. However, the Taylor-Galerkin methods are accurate and simple to implement and have been widely used, particularly in the numerical approximation of solutions to the Euler equation for compressible flows.

Chapter 4

MESH GENERATION

There are basically two types of meshes: structured and unstructured. A structured mesh consists of horizontal and vertical lines that cross orthogonally at intersections called nodes. This constraint is best achieved by discretizing a physical domain that is defined by square or rectangular boundaries – the physical domain becomes the computational domain as well. Much of the early numerical simulations were conducted on problems that were first reduced to rectangular physical systems of interest.

In unstructured meshes, a 2-D physical domain is discretized by a set of seemingly randomly placed nodes that are connected to other nodes via triangular or quadrilaterally shaped subdomains, or elements. The most common types of elements are linear three-noded triangles or linear four-noded quadrilaterals, as discussed earlier. Other popular types of elements include quadratic and cubic triangles and quadrilaterals. In three dimensions these become tetrahedrals and hexahedrals, respectively. The generation of unstructured meshes requires more thought and effort than structured meshes. In general, one puts more nodes (or elements) near surfaces and in regions where activity (or steep gradients) is likely to occur. Many times, the user must guess as to where the most nodes should be placed, ultimately necessitating the generation of a second mesh (and comparing solutions for accuracy). It is up to the user to specify the mesh density (number of nodes and elements), which is best achieved through experience.

In fluid flow problems, one generally tries to use the *lowest* order element to reduce bandwidth and storage. When one elects to use higher order elements to discretize a problem domain, some care should be exercised in selecting the choice of element. For linearly distorted elements, the serendipity family of $O(4)$ or greater will yield quadratic convergence typically $O(h^{p+1})$, where h is the element size and p is the degree of polynomial expansion. The nine-noded Lagrangian element represents better Cartesian polynomials than the eight-noded serendipity element and is the preferred choice in modeling smooth solutions. When there is no distortion, both elements yield similar results. The most accurate 2-D quadratic element for fluid flow simulation is the nine-node Lagrangian; this has been amply demonstrated in numerous examples reported in the literature. In three dimensions the 27-noded element yields the most accurate convergence. However, the amount of computational storage and effort increases dramatically over the use of bilinear quadrilaterals. It is often preferable to use more lower order elements.

4.1 MESH GENERATION GUIDELINES

There is no reason one should end up using severely distorted elements to discretize a domain. In fact, the interior of most domains can be meshed using non-distorted elements; as one approaches the boundaries, several slightly distorted elements can be constructed. Curved sides should only be employed on curved boundaries, and the curvature should be rather mild ($\leq 30°$ arc). When this is not possible, more elements should be utilized.

If the physical domain has all boundary sides straight, with no internal curved surface (e.g., hole), any type of element will match the boundary exactly. Likewise, boundaries defined by higher degree polynomials can also be matched exactly with corresponding higher order elements. However, nonpolynomial curvature cannot be matched exactly by polynomial elements; hence the domain boundary becomes altered to the outer edge of the defining element. Suppose one wishes to use linear elements to prescribe a boundary, as shown in Fig. 15. In order to reduce the error associated with the area omissions, more linear elements are required. This leads to the decision by the user whether to increase the number of lower order elements or use a higher order element. The quadratic yields about 1% geometric error while the linear element produces about 29% error for a 90° arc [23]. In practice, preprocessing typically requires several refinements of the mesh, including boundary matching, before a suitable solution is achieved.

There may be instances when the user wishes to combine elements of different order (although this is discouraged in fluid flow related problems). Interelement continuity and connectivity must be maintained; otherwise a solution mismatch occurs (but convergence may be achieved). Each element side, or face, must match with its adjacent element. There are several ways to achieve this linking of two different order elements. The first approach utilizes *transition* elements, i.e., specialty elements that have an odd number of nodes prescribed on their boundaries. An example of this type of transition is shown in Fig. 16 for a quadratic to linear transition. A second approach is to impose a linear constraint equation that constrains the higher order midside node to match the nodal solutions in the lower order element. This technique is commonly used when employing h-adaptation (to prevent using higher order elements).

A uniform mesh with identically sized elements is easy to generate, much like a finite difference mesh mentioned previously. In complicated regions and/or when the mesh must be refined in a particular region, a finer element density is generated. This can be accomplished in several ways. The first and simplest method is to use mesh gradation in one or more directions. It is best to make the gradation gradual. The second way is to employ transition elements to transition from a finer region to a coarser region. The use of transition with higher order elements keeps the element size constant. This is a form of p-adaptation. Fig. 17 illustrates three ways in which a region in the lower left corner can be refined using the same type elements (Fig. 17a) and mixed elements (Figs. 17b and 17c).

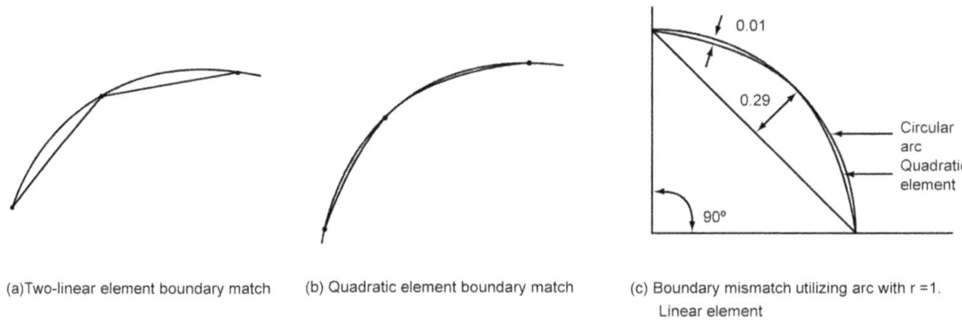

Figure 15. Boundary curvature matching.

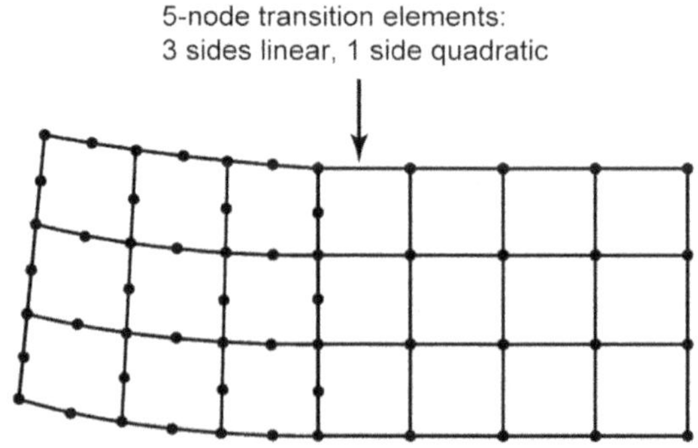

Figure 16. Quadratic to linear transition.

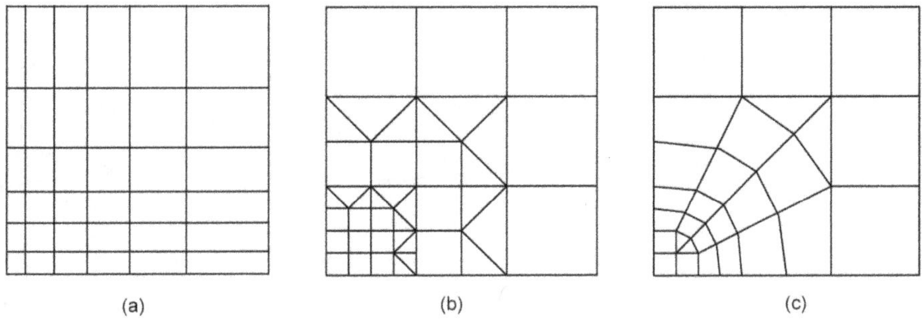

Figure 17. Local mesh refinement approaches.

The use of triangles as interfaces between different sized quadrilaterals is often used in h-adaptation [24].

Elements should not be locally refined to the point that the smallest element is significantly smaller than the largest element, typically 10^2:1 (length) or 10^4:1 (area). Too large a difference can result in an ill-conditioned stiffness matrix and may produce inhibitively small time steps if using explicit marching methods. The more sophisticated (and usually expensive) commercial mesh generation codes, and some preprocessors used with commercial solvers, allow a very refined grid to be generated at the boundaries, utilizing nodal interpolation established from initial nodal values used to create the boundary contours. This procedure generates a uniformly fine mesh within the entire local region of interest and is relatively fast. However, the same error limitations still exist at the boundary.

4.2 BANDWIDTH

In 2-D and 3-D meshes, the node number pattern dictates the bandwidth of the assembled global matrix. Unless the user is employing an explicit marching scheme, the naturally implicit nature of the finite element method creates a banded, sparse matrix that may or may not be symmetric. Hence it behooves the user to minimize the bandwidth of the matrix to reduce storage and computer time. Finding the optimal minimum pattern can be difficult; however, any effort to achieve a near-optimal pattern is worth trying.

There are many algorithms available that automatically renumber a mesh to minimize its bandwidth; for frontal solvers, the wave front is minimized by renumbering the elements; the procedures are similar for nodal or elemental renumbering. The user must create the starting nodes, i.e., an initial mesh, which then gets reordered. These minimization routines are commonly used in many commercial finite element codes.

Renumbering the nodes (or elements) of a mesh allows one to minimize the storage size required by the matrix solver and reduce the number of operations required by the final system, which ultimately reduces the CPU time. There are many methods that perform this renumbering operation, most of them automatic. A detailed discussion on the advantages and disadvantages of the various methods is given by Marro [25] and George [26]. Application of renumbering schemes is described in Carey [27].

Several algorithms exist for constructing Delaunay triangulation (e.g., [28,29]). The most common approach is to utilize the in-circle criterion in a sequential operation. Each point is introduced into an existing Delaunay structure, which is broken and then reconnected to form a new triangulation. An example of Delaunay triangulation is shown in Fig. 18 for two triangles, after an example discussed by Gable et al. [30]. Four points are connected to form a Delaunay triangulation; in Fig. 18a both triangles are the same material

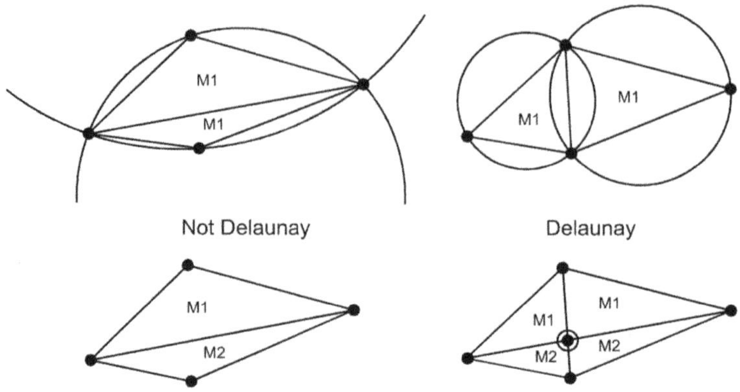

Figure 18. Delaunay triangulation for two triangles (from [30]).

so the triangulation is made Delaunay by flipping the connection. In Fig. 18b, the triangles are different materials; in this case, the triangulation is achieved by adding a point at the interface and making four triangles.

4.3 ADAPTATION

Mesh adaptation is becoming more widely used and has begun to appear in commercial finite element codes (although primarily in structural codes). A few finite element-based fluid flow codes include mesh adaptation – COMSOL (COMSOL, Inc.) is a popular and easy to use commercial multiphysics code; GWADAPT is a code for porous media flow [31]. It has been amply demonstrated that adaptation leads to computations of better solutions, producing optimal meshes in forms of size (number of nodes), nodal positions, and element properties (e.g., shape, orthogonality). Refer to the texts by Babuska et al. [32] and Zienkiewicz and Taylor [33] for detailed discussions on the mathematical aspects of adaptation.

There are basically three types of adaptive techniques in use today: r-refinement, h-refinement, and p-refinement. In r-refinement a fixed mesh is first established; the elements within the mesh are then moved, shrunk, or expanded to accommodate regions where the solution is rapidly changing (or relatively stagnant). This technique has been shown to be effective in some cases; however, the elements can become severely distorted and eventually lead to divergent or less accurate solutions. By far the most popular methods are h- and p-refinement. In h-refinement, elements are subdivided into smaller elements; this technique creates additional nodes and elements, which must be carefully monitored through some form of bookkeeping. In p-refinement the degree of the polynomial is increased to improve the accuracy of the solution, i.e., an element that may have been originally linear is ultimately refined to a cubic, quartic, quintic, or higher order element. A smaller h and a higher p generally yield greater accuracy but slower convergence if too fine a refinement is established. Methods that adapt both h and p together are called h-p refinements. Papers by Shapiro and Murman [34], Ramakrishnam et al. [24], Oden et al. [35], Pelletier and Hetu [36], Zienkiewicz et al. [37], and Pepper and Stephenson [31] are recommended. Other forms of adaptation in the literature include local disenrichment (a form of h-adaptation), which removes one of several points, nested meshes, and multigrid techniques (see [38]).

When using h-adaptation, there are basically two choices to be made, *mesh regeneration* or *element subdivision*. Mesh regeneration, or remeshing, requires completely regenerating the entire mesh, either in regions where there is high error or over the complete domain. The principal advantage of remeshing is that areas can be coarsened where the error is below an allowable amount, thus creating an optimal mesh in which every element has essentially the same level of error. However, the main disadvantage of remeshing is that a high degree of spatial flexibility is necessary when using error estimation procedures. When using element subdivision, every element that exceeds the allowable error threshold is subdivided into smaller elements. This method is particularly effective for four-node quadrilaterals and eight-node hexahedrals. However, the method produces virtual nodes (i.e., constrained midside nodes) that must be handled with care; likewise, only one level of adaptation can be performed at a time.

4.3.1 Mesh Regeneration

The first step one must take in performing mesh regeneration is to define a maximum permissible error, $\bar{\eta}$, which is to be obtained when the analysis is completed, i.e., all the elements in the final mesh must have the same level of error. For example, assume two-dimensional diffusion of variable ϕ. One can begin by calculating the square of the total error,

$$\|q\|^2 \approx \|\hat{q}\|^2 = \sum_{e=1}^{m} \int_{\Omega_e} (\nabla\hat{\phi})^T \kappa \nabla\hat{\phi} d\Omega_e \tag{72}$$

or in two dimensions,

$$\|q\|^2 \approx \|\hat{q}\|^2 = \sum_{e=1}^{m} \int_{\Omega_e} \left[k_x \left(\frac{\partial\hat{\phi}}{\partial x} \right)^2 + k_y \left(\frac{\partial\hat{\phi}}{\partial x} \right)^2 \right] dxdy \tag{73}$$

where m is total number of elements, k_x and k_y are diffusion coefficients, $\hat{q}$ is the calculated square of the error, and $\hat{\phi}$ is the calculated unknown variable. The maximum permissible error for each element is then calculated by distributing $||q||^2$ equally over all the elements,

$$\|\bar{e}\|_e^2 \le \bar{\eta} \left(\frac{\|q\|^2}{m} \right) \tag{74}$$

where $\bar{\eta}$ is the specified maximum value.

The error is now compared for all elements to the maximum permissible error in an element and used to modify the mesh for a second analysis. Let ξ_e be defined as

$$\xi_e = \frac{\|e\|_e}{\|\bar{e}\|_e} \tag{75}$$

where e is the approximate error obtained from smoothed values of ϕ (see [39]),

$$\|e\|_e = \left\{ \int_{\Omega_e} k_x \left[\left(\frac{\partial\tilde{\phi}}{\partial x} - \frac{\partial\hat{\phi}}{\partial x} \right)^2 + k_y \left(\frac{\partial\tilde{\phi}}{\partial x} - \frac{\partial\hat{\phi}}{\partial x} \right)^2 \right] dxdy \right\}^{\frac{1}{2}} \tag{76}$$

and $\tilde{\phi}$ is the smoothed value (a "smoothed value" can be easily obtained by solving the relation $[M]\{\tilde{\phi}\}=\{f\}$). If $\xi_e > 1$, the size of element e must be reduced, thus requiring the mesh to be refined; otherwise, the size of the element must be increased, and the mesh coarsened. The size of the new element is then calculated from the old element as

$$\bar{h}_e = \frac{h_e}{\xi_e^{1/p}}$$

where $\bar{h}_e$ is the predicted size of the element, h_e is the old element, and p is of the order of the shape function.

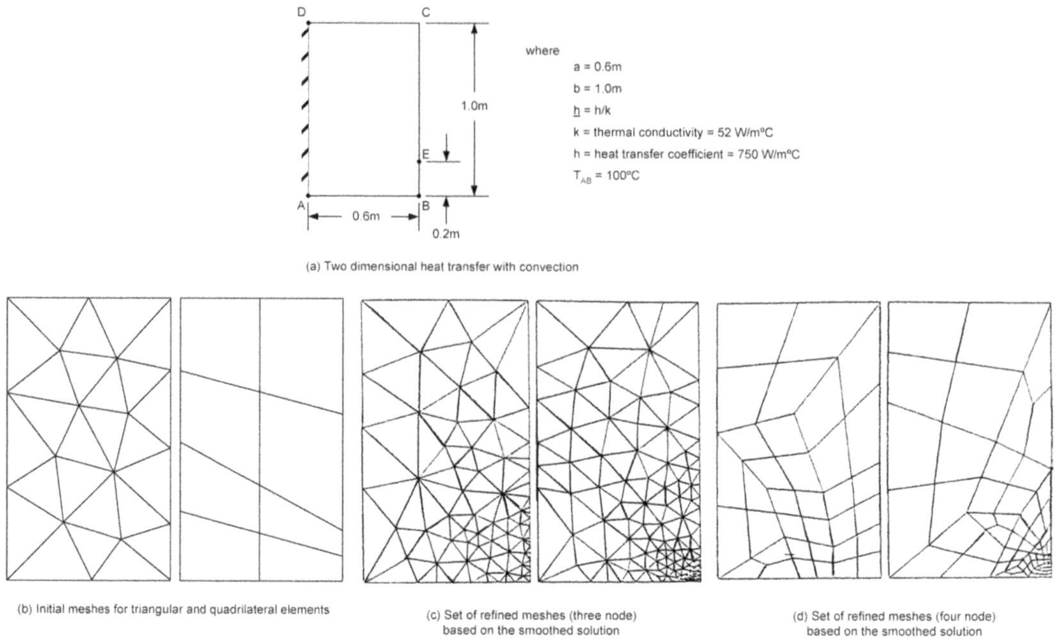

Figure 19. Remeshing example (from [39]).

An example of remeshing for three-node triangles and four-node quadrilaterals is shown in Fig. 19 for a simple heat transfer problem with convective boundary conditions. This benchmark problem is illustrated in more detail in the text by Huang and Usmani [39], which also includes a set of computer programs for generating adaptive meshes based on the remeshing principle. Fig. 19a shows the problem domain with three different boundary conditions. For a uniform mesh of 30 triangular elements (Fig. 19b), a temperature error of −24.6% was obtained. When the adaptive procedure is applied, a first level of adaptation (Fig. 19c) produced a norm error of 14.3% (average temperature error of 2% overall) in the triangular mesh; a second level of adaption produced an error norm of 9.8%. In the quadrilateral case, the initial mesh consisting of eight elements (Fig. 19b) produced a norm error of 21.6%; two mesh refinements, shown in Fig. 19d, yielded 16% and 10%, respectively. If one uses quadratic elements, the error reduces considerably but at a higher computational cost.

4.3.2 Element Subdivision

The starting point for element subdivision is a mesh coarse enough to allow rapid convergence, yet fine enough to allow the flow details to appear. An initial solution is then computed on the crude mesh; it is not necessary to allow this solution to converge completely. The initial solution should not evolve too far before adaptation, or expensive computational time will be used needlessly since the flow features will shift location during the adaptation procedure.

Refinement indicators are computed based on the solution on the initial mesh, and elements that need to be refined are identified. All elements in the mesh that have indicators above a preset refinement threshold value are enriched, whereas those elements that have values below the unrefinement threshold value are coarsened. Refinement proceeds from the coarsest level to the finest level.

After all the mesh changes have been made, the grid geometry is recalculated, the solution is interpolated onto the new grid, and the calculation procedure begun again. For steady-state problems, the entire procedure is repeated until a "converged" mesh is obtained.

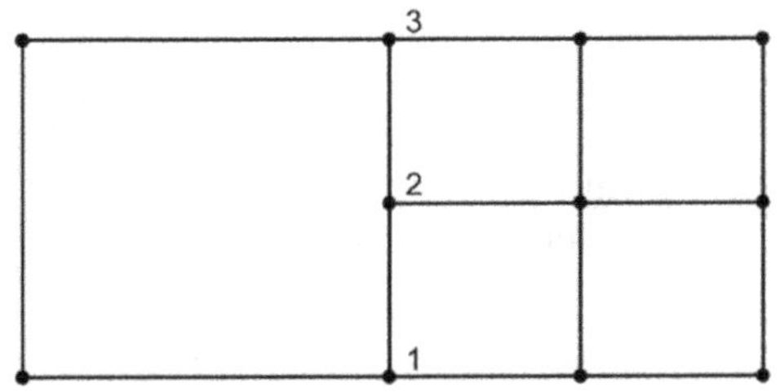

Figure 20. Two-dimensional interface (virtual) node.

A converged mesh is a mesh which no longer changes as the solution progresses. The calculation procedure continues on the converged mesh until each of the dependent variables converge to a criterion of 10^{-4}. In transient problems, the mesh is adapted as needed to properly capture high gradient features as they evolve in time.

In order to decide which elements to refine or unrefine, an adaptation parameter (A_e) must be defined. There is a great deal of literature indicating possible choices for an adaptation parameter. The two most popular refinement criteria are refinement to minimize error and refinement based on gradients. Both criteria are based upon a key variable which is representative of the solution behavior. Refinement criteria based upon the minimization of maximum errors are generally more complex, and are only as accurate as the method of estimating the error.

The adaptive refinement procedure automatically refines all elements that satisfy a criterion $A_e > R$ and unrefines all elements that satisfy $A_e < U$, where R and U are the refine and unrefine threshold values, respectively. The values of R and U are determined experimentally, based on problem geometry and flow features. In the example problems simulated in this project, the values of R and U varied between 0.2–0.4 for R and 0.6–0.8 for U.

The use of quadrilaterals in two dimensions results in midside nodes at the interfaces between the coarse and fine regions of the mesh. In three-dimensions, a face-centered node appears which creates four quadrilateral elements on the face – resulting in eight new hexahedral elements from the original coarse element. These midside nodes are called virtual nodes and require special treatment to obtain a stable, conservative scheme. Fig. 20 shows a typical interface between a locally fine region and a coarser region. The special treatment used is to set the fluxes and the variable value at node 2 equal to the average of the fluxes and value at nodes 1 and 3 after each iteration.

4.3.3 Adaptation Rules

Elements

1. An element may be refined only if its neighbors are at the same refinement level or higher.
2. If a neighbor element of an element to be refined is at a lower level of refinement, it must be refined first.
3. Refinement of an element results in the creation of four subelements, and in the creation of one to five new nodes.
4. To be unrefined, a group of elements must not contain another group of elements, and each element of the group to be unrefined must not be a neighbor to an element with a higher level.

Nodes

1. An embedded virtual node is common to two members of a group only.
2. An embedded node that is created along a domain boundary cannot be a virtual node.

3. If an element and its neighbor, both of which are at the same level, are connected to a third element at a lower level, then the embedded node which exists along the edge common with the third element is a virtual node.
4. If a group of elements is unrefined, then the embedded virtual node along the edge common to an element which is not a member of the group will be eliminated.
5. If a group of elements is unrefined, then the node along the edge common to this group and its neighbor group will become a virtual node.
6. If a group of elements is unrefined, all embedded nodes along a domain boundary will be eliminated.

4.3.4 Mesh Adaptation Example

The rules can be illustrated by considering the uniform quadrilateral mesh of four elements shown in Fig. 21(a). Suppose element A is marked for refinement. By applying element rules 1 and 3, element A is divided into subelements I, II, III, and IV as shown. Application of node rules 1 and 2 shows that the nodes marked with circles are virtual nodes, and those marked with an X are not virtual nodes.

Suppose that element III is marked for further refinement. Element III cannot be refined since one of its neighbors, B, is at a lower level. Refinement of element III before element B

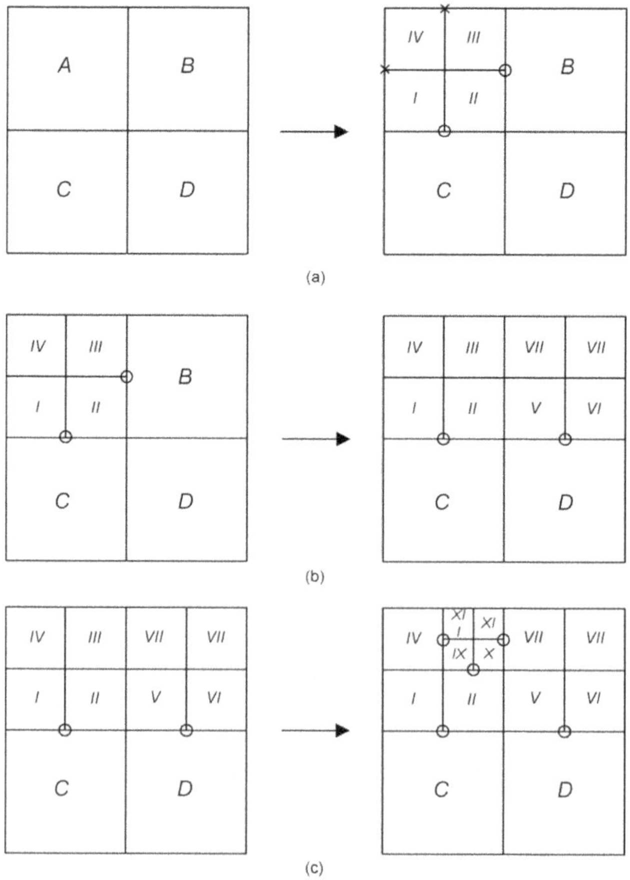

Figure 21. Sequence of mesh refinements.

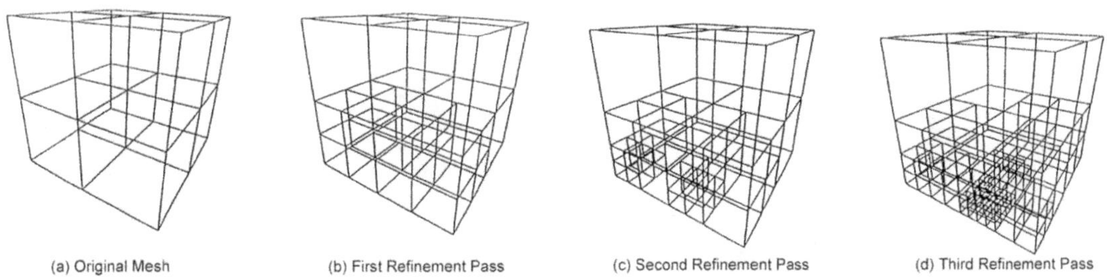

Figure 22. Three-dimensional adaptation of eight elements after three refinements (from [35]).

would violate element rule 1. Therefore, element B is refined as shown in Fig. 21(b). Note that node I is no longer a virtual node, since node rule 1 no longer applies. Node j remains a virtual node. Now that element B has been divided into elements V, VI, VII, and VIII, element rule 1 can be applied. Fig. 21(c) illustrates this division.

Next, suppose that the group of elements V, VI, VII, and VIII shown in Fig. 21(c) is marked for unrefinement. This group is not eligible until the group of elements IX, X, XI, and XII has been unrefined. Element VIII has neighbor X and XI which are at a higher level, and unrefinement would violate element rule 4.

Now let the group of elements IX, X, XI, and XII be marked for unrefinement. Element rule 4 is satisfied and elements IX, X, XI, and XII are replaced by element III. The embedded virtual nodes associated with elements IX, X, XI, and XII are eliminated by node rule 4. The embedded node along the upper domain boundary is eliminated using node rule 6.

An illustration of a 3-D extension of this approach to mesh adaptation is shown in Fig. 22. Here, a single group of eight elements is successively refined until it consists of 13 groups totaling 98 elements. One can see from this figure that a maximum of one level of refinement difference exists between any two adjacent elements. While the 3-D extension of this procedure is conceptually straightforward, the additional bookkeeping is substantial.

Chapter 5

FLUID FLOW APPLICATIONS

The three types of algorithms most commonly used in the solution of the equations for viscous incompressible flow are the mixed formulation, the fractional step method, and the penalty formulation. The latter is the basis for very practical procedures for the solution of the Navier-Stokes equations. Incompressibility is related in a nonexplicit form to the pressure field, which in this case, is a dynamic variable. Mathematically, this is expressed as a consistency condition known as the Ladyzhenkaya, Babuska, and Brezzi (LBB) condition, which establishes the compatibility of the velocity and pressure spaces when the Navier-Stokes equations are discretized directly. A very large number of algorithms have been proposed for the solution of these equations.

A problem commonly used as a benchmark in the numerical solution of the Navier-Stokes equations consists in calculating the flow over a backward facing step, as shown in Fig. 23. A fully developed flow is imposed along the inlet boundary, $x = 0$, $H/2 \le y \le H$; no-slip boundary conditions apply along the solid boundaries. These are all boundary conditions of the Dirichlet type, which are specified on boundary Γ_1.

If $L_2 - L_1$ is large enough, at $x = L_2$ the flow should be fully developed. Therefore the gradients of the velocity components in the direction normal to the boundary must vanish. This is a Neumann boundary condition and constitutes the portion Γ_2 of the boundary.

Imposing the condition $v = 0$ at $x = L_2$ would leave a Neumann condition for the first momentum equation and a Dirichlet condition for the second. This is perfectly acceptable but rarely leads to better approximations and more often will introduce inaccuracies in the calculations by modifying the flow field at the exit – if fully developed conditions have not been reached yet.

In general, no boundary conditions are required for the pressure, except a reference value, which can be provided at any point in the domain. In some cases a known pressure may be the force driving the flow, in which case it can be readily applied at the corresponding boundary.

The weak form of the boundary-value problem can be stated as follows: Find functions u* and v* that satisfy the Dirichlet boundary conditions on Γ_1, and a function p* such that

$$\int_{\Omega}\left\{U\rho\left(\frac{\partial u^*}{\partial t}+u^*\frac{\partial u^*}{\partial x}+v^*\frac{\partial u^*}{\partial y}\right)-\frac{\partial U}{\partial x}p^*+\mu\left(\frac{\partial U}{\partial x}\frac{\partial u^*}{\partial x}+\frac{\partial U}{\partial y}\frac{\partial u^*}{\partial y}\right)+U\rho B_x\right\}d\Omega$$
$$+\int_{\Gamma_2}U\left\{\left(p^*+\mu\frac{\partial u^*}{\partial x}\right)n_x+\mu\frac{\partial u^*}{\partial y}ny\right\}d\Gamma=0 \tag{77}$$

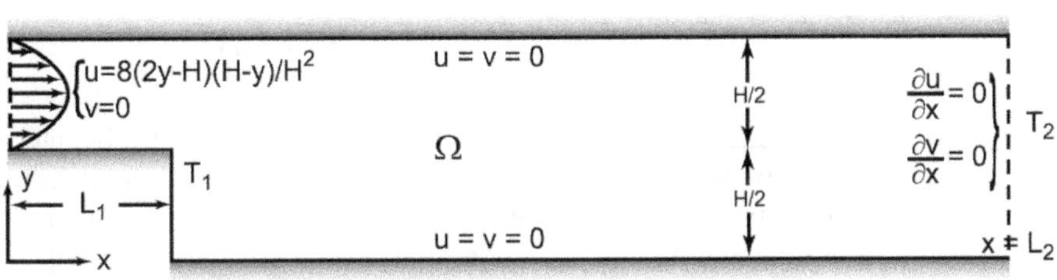

Figure 23. Flow over a backward facing step.

$$\int_\Omega \left\{ V\rho\left(\frac{\partial v^*}{\partial t} + u^*\frac{\partial v^*}{\partial x} + v^*\frac{\partial v^*}{\partial y}\right) - \frac{\partial V}{\partial x}p^* + \mu\left(\frac{\partial V}{\partial x}\frac{\partial v^*}{\partial x} + \frac{\partial V}{\partial y}\frac{\partial v^*}{\partial y}\right) + V\rho B_x \right\} d\Omega$$
$$+ \int_{\Gamma_2} V\left[\mu\frac{\partial v^*}{\partial x}n_x + \left(-p^* + \mu\frac{\partial v^*}{\partial y}\right)n_y\right] d\Gamma = 0 \tag{78}$$

and

$$\int_\Omega Q\left(\frac{\partial u^*}{\partial x} + \frac{\partial v^*}{\partial y}\right) d\Omega = 0 \tag{79}$$

for all weighting pairs of functions U and V and all functions Q. This is called a "mixed variational formulation." The mathematical theory behind this formulation is extremely tedious. For those interested in the mathematical theory of finite element solution to the Navier-Stokes equations, see the works of Teman [40], Girault and Raviart [41], and Carey and Oden [42]. The pressure is sought in a space different from that for the velocity components, therefore the name "mixed" formulation. The relation that the velocity and pressure spaces are "consistent" is the LBB condition. When discretizing a domain using finite elements to approximate velocity and pressure, it is easy to violate the LBB condition.

5.1 CONSTANT-DENSITY FLOWS

5.1.1 Mixed Formulation

Assume constant viscosity and rewrite the Navier-Stokes equations in nondimensional form,

$$\frac{\partial u}{\partial t} + u\frac{\partial u}{\partial x} + v\frac{\partial u}{\partial y} = -\frac{\partial p}{\partial x} + \frac{1}{R_e}\left(\frac{\partial^2 u}{\partial x^2} + \frac{\partial^2 u}{\partial y^2}\right) \tag{80}$$

$$\frac{\partial v}{\partial t} + u\frac{\partial v}{\partial x} + v\frac{\partial v}{\partial y} = -\frac{\partial p}{\partial x} + \frac{1}{R_e}\left(\frac{\partial^2 v}{\partial x^2} + \frac{\partial^2 v}{\partial y^2}\right) \tag{81}$$

where Re is the Reynolds number.

The dependent variables u, v, and p are approximated using shape functions as

$$u(x,y,t) = \sum_j N_j(x,y)u_j(t),\ v(x,y,t)$$
$$= \sum_j N_j(x,y)v_j(t),\ p(x,y,t)$$
$$= \sum_j M_k(x,y)p_k(t)$$

In order to satisfy the LBB condition, the pressure must be interpolated with a lower order polynomial than the velocity components. Fig. 24 shows the simplest and most often used combination of such mixed interpolation. The combinations 1, 3 and 5 involve discontinuous pressure fields. The evidence from numerical experiments indicates that these approximations to the pressure field result in improved accuracy over those combinations using continuous pressures.

Substituting into the weak formulation given by Eqs. (77)–(79), and setting the weighting functions U, V, and P equal to the shape functions, N_i, N_i, and M_i, respectively, yields

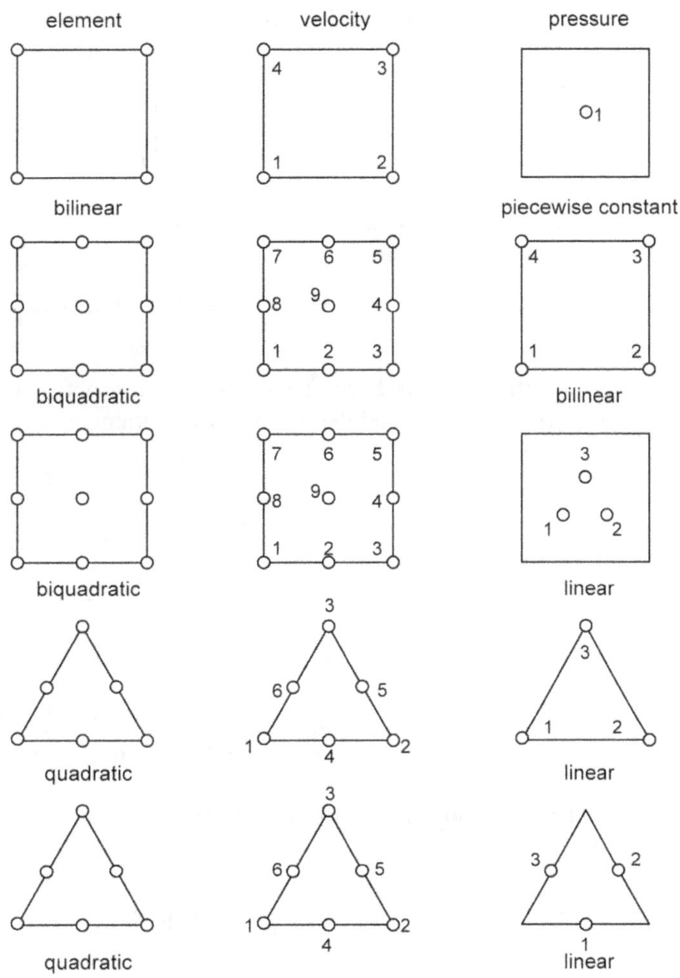

Figure 24. Elements for mixed interpolation of velocity and pressure.

the discrete Galerkin equations. For velocity interpolated to n-noded elements and the pressure using m points,

$$\mathbf{M\dot{d}} + \mathbf{Kd} + \mathbf{F} = \mathbf{0} \tag{82}$$

The dimension of the system is (2n + m) × (2n + m). To simplify the notation, the degrees of freedom are ordered so that

$$\mathbf{d} = \left[u_1 u_2 \ldots u_n v_1 v_2 \ldots v_n p_1 \ldots p_m \right]^T = \begin{bmatrix} \mathbf{u} \\ \mathbf{v} \\ \mathbf{p} \end{bmatrix}$$

For the bilinear velocity-constant pressure and biquadratic velocity-linear pressure elements we have

$$\mathbf{d} = [u_1 u_2 u_3 u_4 v_1 v_2 v_3 v_4 p_1]^T$$

and

$$\mathbf{d} = [u_1 u_2 \ldots u_9 v_1 v_2 \ldots v_9 p_1 p_2 p_3]^T$$

respectively.

The mass matrix **M** is defined by

$$\mathbf{M} = \begin{bmatrix} \mathbf{A} & \mathbf{0} & \mathbf{0} \\ \mathbf{0} & \mathbf{A} & \mathbf{0} \\ \mathbf{0} & \mathbf{0} & \mathbf{0} \end{bmatrix} \tag{83}$$

where **A** is an n × n mass matrix given by

$$\left[a_{ij}\right] = \left[\int_\Omega N_i N_j d\Omega\right] \tag{84}$$

The stiffness matrix **K** contains only the linear part of the operator, i.e., the viscous terms, the pressure terms, and the continuity constraint, and is given by

$$\mathbf{K} = \begin{bmatrix} \mathbf{B} & \mathbf{0} & \mathbf{C}_x \\ \mathbf{0} & \mathbf{B} & \mathbf{C}_y \\ -\mathbf{C}_x^T & -\mathbf{C}_y^T & \mathbf{0} \end{bmatrix} \tag{85}$$

Here **B** is an n × n matrix defined by

$$\left[b_{ij}\right] = \left[\int_\Omega \frac{1}{Re}\left(\frac{\partial N_i}{\partial x}\frac{\partial N_j}{\partial x} + \frac{\partial N_i}{\partial y}\frac{\partial N_j}{\partial y}\right) d\Omega\right] \tag{86}$$

and corresponds to the Laplacian operator. The matrices $\mathbf{C}_x$ and $\mathbf{C}_y$ are both n × m and have the form

$$\left[(c_x)_{ik}\right] = \left[-\int_\Omega \frac{\partial N_i}{\partial x} M_k d\Omega\right],\ \left[(c_y)_{ik}\right] = \left[-\int_\Omega \frac{\partial N_i}{\partial y} M_k d\Omega\right] \tag{87}$$

The function **F** contains the convective terms and is given by

$$\mathbf{F} = \begin{bmatrix} \mathbf{F}_x \\ \mathbf{F}_y \\ \mathbf{0} \end{bmatrix} \tag{88}$$

with the n × 1 vectors F_x and F_y defined by

$$\begin{aligned} \left[(f_x)_i\right] &= \left[\int_\Omega N_i \left\{\left(\sum_j N_j u_j\right)\left(\sum_k \frac{\partial N_k}{\partial x} u_k\right) + \left(\sum_j N_j v_j\right)\left(\sum_k \frac{\partial N_k}{\partial y} u_k\right)\right\} d\Omega\right] \\ \left[(f_y)_i\right] &= \left[\int_\Omega N_i \left\{\left(\sum_j N_j u_j\right)\left(\sum_k \frac{\partial N_k}{\partial x} v_k\right) + \left(\sum_j N_j v_j\right)\left(\sum_k \frac{\partial N_k}{\partial y} v_k\right)\right\} d\Omega\right] \end{aligned} \tag{89}$$

To calculate steady state flows, a direct nonlinear or a time-dependent solution can be used until the solution no longer changes in time. The second approach is preferred when physical instabilities may occur that can change the mode of circulation. A direct iterative solution will normally not be able to branch off to a new mode that is physically more stable. When a time-dependent algorithm is chosen, there are two basic ways to deal with the nonlinear terms.

(i) One method is treating the convective terms explicitly, that is, evaluating the non-linear terms at each time step using the latest computed values and assembling them into the forcing term in the right-hand-side. This introduces a stability condition where the Courant number must be less than 1 for every element in the mesh. If the evolution of the flow field in time is important, this does not constitute a restriction.

(ii) A second method is using an unconditionally stable time integrating algorithm combined with a Newton-Raphson iteration within each time step. This allows for the use of larger time steps in order to achieve steady state more rapidly. However, it can become expensive if the time evolution must be calculated accurately.

When time-dependent algorithms, such as discussed in points (i) and (ii) above, and bilinear elements are used, it suffices to weight the convective terms in the right-hand-side with the weighting functions

$$w_i = N_i + \frac{\alpha \bar{h}}{2|\mathbf{V}|}\left(u\frac{\partial N_i}{\partial x} + v\frac{\partial N_i}{\partial y} \right)$$

where $\alpha = \coth \gamma/2 - 2/\gamma$ as before and γ is now the cell Reynolds number given by

$$\gamma = \mathrm{Re}|\mathbf{V}|\bar{h}$$

The steady state solution to flow over the backward facing step, shown in Fig. 25, is shown for Re = 900, based on the entry length H/2. A mesh of 3000 rectangular bilinear elements is used with piecewise constant pressure, in a mixed formulation. The mesh is irregularly spaced for better resolution close to the solid boundaries and at the entry. Referring to Fig. 23, L_1 = 3H and L_2 = 19H. The solution was obtained using a backward-implicit time-stepping scheme combined with a Newton-Raphson iteration within each time step and Petrov-Galerkin weighting of the convective terms. The steady state streamfunction is shown in Fig. 25a. A second recirculation cell is observed at the top wall that was not present at the low Reflow. The pressure field is shown in Fig. 25b.

Flow over a backward facing step has become one of the benchmark problems to test numerical models [43,44]. Another very popular problem to test methods is the cavity-driven flow introduced by Burggraf [45]. There exists a great amount of information and solutions to this problem in the literature.

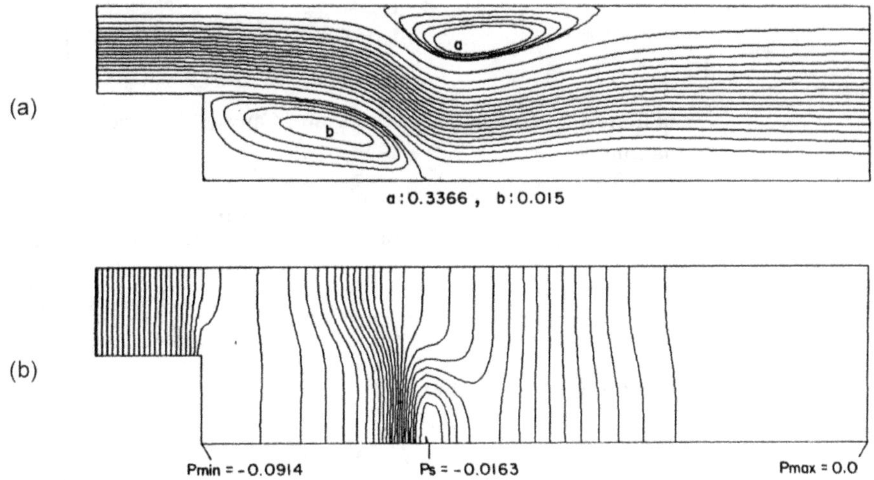

Figure 25. Flow over a backward facing step at Re = 900: (a) streamfunction contours and (b) pressure contours.

In some cases, particularly if the data are not smooth, the calculated pressure field oscillates from one element to the next in what is usually called a "checkerboard" mode. The oscillations occur because the discretization space of the pressure admits functions that are not constant, but for which the discrete divergence operator vanishes.

5.1.2 Fractional Step Method

The method of fractional steps has been used extensively, especially in finite difference discretizations. Some selected references are Chorin [46], Yanenko [47], Schneider et al. [48], Donea et al. [49], and Quartapelle [50]. The method is designed for the solution of the time-dependent equations and consists in performing the solution in two steps as follows.

First, split the velocity components into two parts as

$$u = u^* + u', \quad v = v^* + v'$$

where u^* and v^* satisfy the momentum equations without the pressure terms, i.e., using the vector notation for conciseness,

$$\frac{\partial u^*}{\partial t} = -\mathbf{V}\cdot\nabla u + \frac{1}{Re}\nabla^2 u, \qquad \frac{\partial v^*}{\partial t} = -\mathbf{V}\cdot\nabla v + \frac{1}{Re}\nabla^2 v \tag{90}$$

Using an explicit forward Euler scheme to solve for u and v in Eq. (90),

$$\frac{u^*_{n+1} - u^n}{\Delta t} = -\mathbf{V}^n\cdot\nabla u^n + \frac{1}{Re}\nabla^2 u^n, \qquad \frac{v^*_{n+1} - v^n}{\Delta t} = -\mathbf{V}^n\cdot\nabla v^n + \frac{1}{Re}\nabla^2 v^n \tag{91}$$

The velocity components calculated from Eq. (91) will not satisfy the continuity equation. Therefore corrections u' and v' need to be computed with the help of the pressure field. To do this, one must first find the pressure field. Differentiating the u component in Eq. (90) with respect to x and the v component with respect to y and adding,

$$\nabla^2 p = \frac{\partial}{\partial x}\left(-\mathbf{V}\cdot\nabla u + \frac{1}{Re}\nabla^2 u\right) + \frac{\partial}{\partial y}\left(-\mathbf{V}\cdot\nabla v + \frac{1}{Re}\nabla^2 v\right) \tag{92}$$

Substituting the expressions in Eq. (91) into Eq. (92),

$$\nabla^2 p^{n+1} = \frac{1}{\Delta t}\left(\frac{\partial u^*_{n+1}}{\partial x} + \frac{\partial v^*_{n+1}}{\partial y}\right) \tag{93}$$

which must be solved for the pressure. Once p^{n+1} is known, from the complete discretized equations,

$$\frac{u^{n+1} - u^n}{\Delta t} = -\mathbf{V}^n\cdot\nabla u^n + \frac{1}{Re}\nabla^2 u^n - \frac{\partial p^{n+1}}{\partial x} \tag{94}$$

$$\frac{v^{n+1} - v^n}{\Delta t} = -\mathbf{V}^n\cdot\nabla v^n + \frac{1}{Re}\nabla^2 v^n - \frac{\partial p^{n+1}}{\partial y} \tag{95}$$

Substituting Eqs. (90) and (91)

$$u'_{n+1} = -\Delta t\frac{\partial p^{n+1}}{\partial x}, \quad v'_{n+1} = -\Delta t\frac{\partial p^{n+1}}{\partial y}$$

The new velocity components $u^{n+1} = u^*_{n+1} + u'_{n+1}$ and $v^{n+1} = v^*_{n+1} + v'_{n+1}$ satisfy Eqs. (94) and (95) as well as the incompressibility condition.

It is not hard to imagine that many variations of the present methodology can be obtained once this basic form is understood. The pressure may be interpolated using the same shape functions as for the velocity components. This has led some authors to call them "equal-order interpolation" schemes.

The advantage of this method is that only the solution of the Poisson equation for the pressure must be solved implicitly. The disadvantage is that the time step can become extremely small to satisfy stability, and calculations become prohibitively lengthy.

5.1.3 Penalty Function Formulation

The basic idea of the penalty method consists in expressing the pressure through the pseudoconstitutive relation

$$p = p_s - \lambda \nabla \cdot \mathbf{V} \tag{96}$$

in which λ is a large number. Equation (96) is then substituted into the momentum equations,

$$\frac{\partial u}{\partial t} + \mathbf{V} \cdot \nabla u = -\frac{\partial p_s}{\partial x} + \lambda \frac{\partial}{\partial x}(\nabla \cdot \mathbf{V}) + \frac{1}{Re}\nabla^2 u \tag{97}$$

$$\frac{\partial v}{\partial t} + \mathbf{V} \cdot \nabla v = -\frac{\partial p_s}{\partial y} + \lambda \frac{\partial}{\partial y}(\nabla \cdot \mathbf{V}) + \frac{1}{Re}\nabla^2 u \tag{98}$$

and the continuity equation is no longer necessary. The discretizations of Eqs. (97) and (98) result in the solution of a discrete system of equations that involve only the velocity degrees of freedom; which is about 15% smaller for bilinear elements. Furthermore, there is no occurrence of zero diagonal elements of the final linear matrices, although the system will remain ill conditioned due to the large value of the penalty parameter. Hence, direct equation solvers are required.

For constant density flows, the static component p_s in Eq. (96) is eliminated through a redefinition of the pressure, so that the hydrostatic pressure due to gravity is canceled out. For this reason, the body force terms are not included even though one of these directions might represent the vertical direction aligned with gravity. In this case, the penalty formulation may be written as

$$p = -\lambda \nabla \cdot \mathbf{V} \tag{99}$$

which is the standard form found in most references. However, this form of the penalty method is incorrect when applied to stratified flows and flows with free surfaces.

The two-dimensional, steady state Stokes equations are written as,

$$\mu\left(\frac{\partial^2 u}{\partial x^2} + \frac{\partial^2 u}{\partial y^2}\right) = \frac{\partial p}{\partial x}, \quad \mu\left(\frac{\partial^2 v}{\partial x^2} + \frac{\partial^2 v}{\partial y^2}\right) = \frac{\partial p}{\partial y} \tag{100}$$

The Galerkin penalty function formulations of Eq. (100) are

$$\begin{gathered}
\left[\int_\Omega \mu\left(\frac{\partial N_i}{\partial x}\frac{\partial N_j}{\partial x} + \frac{\partial N_i}{\partial y}\frac{\partial N_j}{\partial y}\right) + u_j d\Omega\right] + \left[\int_\Omega \lambda \frac{\partial N_i}{\partial x}\left(\frac{\partial N_j}{\partial x}u_j + \frac{\partial N_j}{\partial y}v_j\right) d\Omega\right] = 0 \\
\left[\int_\Omega \mu\left(\frac{\partial N_i}{\partial x}\frac{\partial N_j}{\partial x} + \frac{\partial N_i}{\partial y}\frac{\partial N_j}{\partial y}\right) + u_j d\Omega\right] + \left[\int_\Omega \lambda \frac{\partial N_i}{\partial y}\left(\frac{\partial N_j}{\partial x}u_j + \frac{\partial N_j}{\partial y}v_j\right) d\Omega\right] = 0
\end{gathered} \tag{101}$$

where all the line integrals are disgarded since they are irrelevant to this discussion. The final system of linear equations can be written as

$$[\mu \mathbf{K}_1 + \lambda \mathbf{K}_2]\, \mathbf{d} = \mathbf{F} \tag{102}$$

where the vector **F** is generated from the boundary conditions and, in this case, the vector **d** contains only velocity degrees of freedom. Assume now that the matrix $\mathbf{K}_2$ is nonsingular and that λ is increased more and more in an effort to satisfy incompressibility better. Because μ and $\mathbf{K}_1$ remain constant as λ increases, they become negligible, and the solution to Eq. (102) can be written as

$$\mathbf{d} = \frac{1}{\lambda}\mathbf{K}_2^{-1}\mathbf{F} \tag{103}$$

Because $\mathbf{K}_2$ and **F** are also constant, $\mathbf{d} \rightarrow 0$ as $\lambda \rightarrow \infty$. This phenomenon is called "locking" and is directly related to satisfaction (or the lack thereof) of the LBB condition.

The penalty matrix $\mathbf{K}_2$ must be singular. This is achieved using selective reduced integration of the penalty term and a quadrature rule with degree of precision lower than required to guarantee an optimal convergence rate in the matrix $\mathbf{K}_1$. Mathematical techniques to analyze elements to be used with the reduced integration penalty term are discussed in Oden et al. [51], Carey and Krishnan [52], and Idelsohn et al. [53].

Computational experience shows that penalty calculations must be performed using double-precision 64 bit words (if running on 32-bit PCs). Under these circumstances, a penalty parameter λ between 10^7 and 10^9 will be adequate in most practical situations.

5.1.4 Calculation of Pressure

In penalty calculations the pressure P may be recovered regardless of whether P represents the modified pressure or the total pressure. For bilinear interpolation of the velocities,

$$P^e = -\frac{\lambda}{A^e} \int_e \left(\frac{\partial u}{\partial x} + \frac{\partial v}{\partial y} \right) de \tag{104}$$

where the superscript e denotes the restriction to the element under consideration and A^e is the element area. In this case, a one-point reduced integration quadrature formula is employed to evaluate Eq. (104). The pressures are piecewise constant pressures – the nodal pressures are obtained with the least squares procedure.

5.1.5 Open Boundaries

The issue of boundary conditions at outflow boundaries has been treated by many authors (see, e.g., [54]), although in a rather ad hoc way. An analysis performed by Heinrich et al. [55] points out that pressure boundary conditions should not be applied – only a reference pressure must be prescribed at some point in the domain. Therefore the difficulties at the open boundaries can also be eliminated by removing the pressure from the boundary condition, so that only

$$\frac{\partial u}{\partial x} = \frac{\partial v}{\partial x} = 0$$

are used. To achieve this, the line integrals

$$\int_\Gamma U p n_x \, d\Gamma \qquad \int_\Gamma V p n_y \, d\Gamma$$

must be discretized and retained in the weighted residuals formulation. If a mixed formulation is being used, these integrals will contribute to the matrices $\mathbf{C}_x$ and $\mathbf{C}_y$ in Eq. (87), but no contribution will be added to $\mathbf{C}_x^T$ and $\mathbf{C}_y^T$ in the last row. Thus $\mathbf{K}$ will become an asymmetric matrix even in the case of Stokes flow.

If a penalty formulation is used, retrieving the integrals is only possible in a time-dependent formulation where the integrals are evaluated explicitly on the right-hand side. To try to include them implicitly by means of the penalty term will result in an ill-conditioned stiffness matrix.

For a similar benchmark problem involving heat fluxes along channel boundaries that has been solved with a variety of existing numerical models, the interested reader should look at the work by Blackwell and Pepper [44].

5.2 FREE SURFACE FLOWS

Consider the flow of a fluid with a free surface that is subjected to small-amplitude gravity waves. Assume that the flow is isothermal. The fluid layer is two-dimensional with a reference depth h_0, as shown in Fig. 26. The deviation of the depth from the reference depth is denoted by h, so that the position of the free surface S(x, y, t) is

$$S = y - h_0 - h = 0 \tag{105}$$

Rewrite the governing equations of motion as

$$\nabla \cdot \mathbf{V} = 0 \tag{106}$$

$$\frac{D\mathbf{u}}{Dt} = -\frac{1}{\rho_0}\nabla p + v\nabla^2 \mathbf{u} + \mathbf{B} \tag{107}$$

The boundary conditions at the free surface express continuity of stresses and are given by

$$\sigma_{ij} n_j = \left(-p_a + 2\sigma G\right) n_i, \quad \sigma_{ij} t_j = 0 \quad \text{on} \quad S \tag{108}$$

where p_a is the pressure at the interface, e.g., atmospheric pressure at a water-air interface, and is assumed to be constant; σ is the surface tension; G the mean surface curvature; $\mathbf{n}$ the unit outward normal vector and $\mathbf{t}$ the unit tangent vector, as shown in Fig. 26.

A kinematic condition to describe the surface motion is also needed, which is

$$\frac{DS}{Dt} = \frac{\partial S}{\partial t} + \mathbf{V} \cdot \nabla S = 0 \tag{109}$$

Using Eq. (105), this expression becomes

$$\frac{\partial h}{\partial t} + u\frac{\partial h}{\partial x} - v = 0 \tag{110}$$

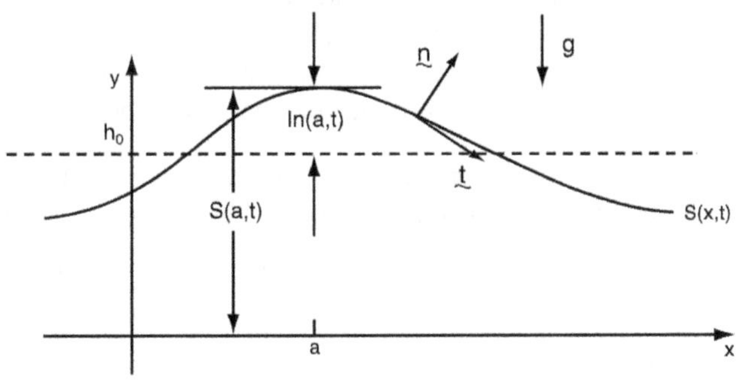

Figure 26. Free surface flow in two dimensions.

Let $H/y_c = 2/3$, where H is the total head and y_c is called the critical depth. Using $\bar{U} = \sqrt{gy_c}$, the velocity of infinitesimal gravity waves as the characteristic velocity, and y_c as the characteristic length, the relevant nondimensional numbers are the Froude number, Fr, and Reynolds number, Re, defined as

$$Fr = \frac{U^2}{gy_c} \qquad Re = \sqrt{\frac{8g}{27}} H^{3/2}/\nu$$

The static pressure p_s becomes

$$p_s = \rho g\,(h + h_0 - y) \tag{111}$$

Hence, the pressure gradients are

$$\frac{\partial p_s}{\partial x} = \rho g \frac{\partial h}{\partial x}, \quad \frac{\partial p_s}{\partial y} = -\rho g \tag{112}$$

The solution of free surface problems becomes complicated by the fact that the location of the free surface must be calculated as part of the problem. At every time step the location of the free surface is recalculated so as to satisfy dynamic equilibrium at the end of each time step, following the basic idea of Rushak [56] to modify the mesh.

There is much literature on free surface flows in which surface tension is important. Finite element algorithms for such flows have been proposed by, among others, Nickell et al. [57], Kistler and Scriven [58] and Chippada et al. [59].

5.3 FLOWS IN ROTATING SYSTEMS

There is an interesting analogy between convective flows and Taylor-Couette flows. To this purpose, rewrite the governing equations in the form

$$\frac{\partial u_i}{\partial x_i} = 0 \tag{113}$$

$$\frac{\partial u_i}{\partial t} + u_j \frac{\partial u_i}{\partial x_j} = -\frac{\partial p}{\partial x_i} + \varepsilon_1 \frac{\partial^2 u_i}{\partial x_j \partial x_j} + f_i \tag{114}$$

$$\frac{\partial \theta}{\partial t} + u_j \frac{\partial \theta}{\partial x_j} = \varepsilon_2 \frac{\partial^2 \theta}{\partial x_j \partial x_j} \tag{115}$$

In the case of an isothermal flow without body forces, $f_i = 0$ and Eq. (114) is either irrelevant or it may model the transport of a species that does not affect the fluid motion. In this case, we set $\varepsilon_1 = 1/Re$. If a Boussinesq fluid is being modeled, Eq. (115) becomes the energy equation and θ the temperature. There are two distinct cases:

(i) In the case of natural convection the appropriate parameters are the Prandtl and Rayleigh numbers. Set $\varepsilon_1 = Pr$, $\varepsilon_2 = 1.0$, $f_1 = 0$, and $f_2 = PrRa\theta$.

(ii) For free and forced convection, the flow is characterized by the Reynolds, Péclet, and Grashof or Froude numbers. Therefore, set $\varepsilon_1 = 1/Re$, $\varepsilon_2 = 1/Pe$, $f_1 = 0$, and $f_2 = \theta/Fr$.

Finally, if we are modeling an isothermal flow in a rotating system (of the Taylor-Couette type), the only relevant parameter is the Taylor number, Ta, given by

$$Ta = \frac{\Omega^2 R d^3}{\nu^2} \tag{116}$$

where Ω is the angular velocity, R is the outer radius, and d the characteristic thickness of the fluid layer. In this case set $\varepsilon_1 = \varepsilon_2 = 1.0$, $f_1 = 0$, and $f_2 = Ta\theta^2$. Equation (115) becomes the equation for the circumferential component of the velocity.

In general, a Petrov-Galerkin discretization will be necessary to guarantee stability of the algorithm. Using isoparametric bilinear elements and the penalty formulation with penalty parameter λ, the weak form of Eqs. (113)–(115) become

$$\begin{aligned}\int_\Omega N_i\left[\frac{\partial u}{\partial t}+\varepsilon_1\left(\frac{\partial N_i}{\partial x}\frac{\partial u}{\partial x}+\frac{\partial N_i}{\partial y}\frac{\partial u}{\partial y}\right)\right]d\Omega+\int_\Omega \lambda\frac{\partial N_i}{\partial x}\left(\frac{\partial u}{\partial x}+\frac{\partial v}{\partial y}\right)d\Omega \\ =\int_\Omega (N_i+P_{i1})\left(u\frac{\partial u}{\partial x}+v\frac{\partial u}{\partial y}\right)d\Omega-\int_{\Gamma_0} N_i p n_x d\Gamma\end{aligned} \tag{117}$$

$$\begin{aligned}\int_\Omega N_i\left[\frac{\partial v}{\partial t}+\varepsilon_1\left(\frac{\partial N_i}{\partial x}\frac{\partial v}{\partial x}+\frac{\partial N_i}{\partial y}\frac{\partial v}{\partial y}\right)\right]d\Omega+\int_\Omega \lambda\frac{\partial N_i}{\partial y}\left(\frac{\partial u}{\partial x}+\frac{\partial v}{\partial y}\right)d\Omega \\ =\int_\Omega\left[(N_i+p_{i1})\left(u\frac{\partial v}{\partial x}+v\frac{\partial v}{\partial y}\right)+N_i f_2\right]d\Omega-\int_{\Gamma_0} N_i p n_y d\Gamma\end{aligned} \tag{118}$$

$$\int_\Omega\left[N_i\frac{\partial \theta}{\partial t}+\varepsilon_2\left(\frac{\partial N_i}{\partial x}\frac{\partial \theta}{\partial x}+\frac{\partial N_i}{\partial y}\frac{\partial \theta}{\partial y}\right)\right]d\Omega=\int_\Omega (N_i+P_{i2})\left(u\frac{\partial \theta}{\partial x}+v\frac{\partial \theta}{\partial y}\right)d\Omega+\int_{\Gamma_1} N_i q d\Gamma \tag{119}$$

Here N_i denotes the bilinear shape functions and P_{ij} are the Petrov-Galerkin perturbations applied to the convective terms only. The open portions of the boundary are denoted by Γ_0, and those for which a heat flux q is prescribed, by Γ_2.

The spatial discretization leads to the system of ordinary differential equations

$$\mathbf{M}\dot{\mathbf{d}}+\mathbf{K}\mathbf{d}=\mathbf{F}(\theta)-\mathbf{N}(\mathbf{d}) \tag{120}$$

$$\mathbf{C}\dot{\theta}+\mathbf{D}\theta=-\mathbf{G}(\mathbf{d},\theta) \tag{121}$$

where **M** and **C** are the mass matrices for the momentum and energy equations, respectively; **d** is the vector of velocity degrees of freedom; **K** contains the contributions of the viscous and penalty terms; **N** is the vector containing the nonlinear convective terms in the momentum equation; **F** contains the contributions of the body forces and the prescribed boundary conditions; **D** is the heat diffusion matrix; and **G** contains the convective terms and contributions from the boundary conditions in the energy equation.

The nonlinear convective terms and the body force terms are calculated explicitly in the right-hand side using the latest available values of the dependent variables. This has the advantage that the system matrices ($\mathbf{M}+\gamma\Delta t\mathbf{K}$) and ($\mathbf{C}+\gamma\Delta t\mathbf{D}$) on the left-hand side are constant throughout the calculation, provided the time step remains constant. Therefore the matrices can be factored at the beginning of the calculation using an LU decomposition method. To advance the solution in time after the first time step, it suffices to update the right-hand side and perform a forward and backward substitution, thereby reducing the computational time significantly. Furthermore, these matrices are symmetric, enhancing computational efficiency.

5.4 ISOTHERMAL FLOW PAST A CIRCULAR CYLINDER

Flow past a circular cylinder is a well-defined example of a time-dependent flow. When Re exceeds about 60, vortices are shed from the two re-circulating cells behind the cylinder. Brooks and Hughes [60] performed calculations at Re = 100 based on the cylinder diameter. The domain and mesh used in their work are shown in Fig. 27(a–c). The mesh contains 1510 nodes and 1436 isoparametric bilinear elements; the time step was fixed at $\Delta t = 0.03$. The simulation was started from zero velocities, and a unit horizontal inlet velocity was applied at t = 0. A symmetric steady state solution is obtained unless the flow is perturbed [61].

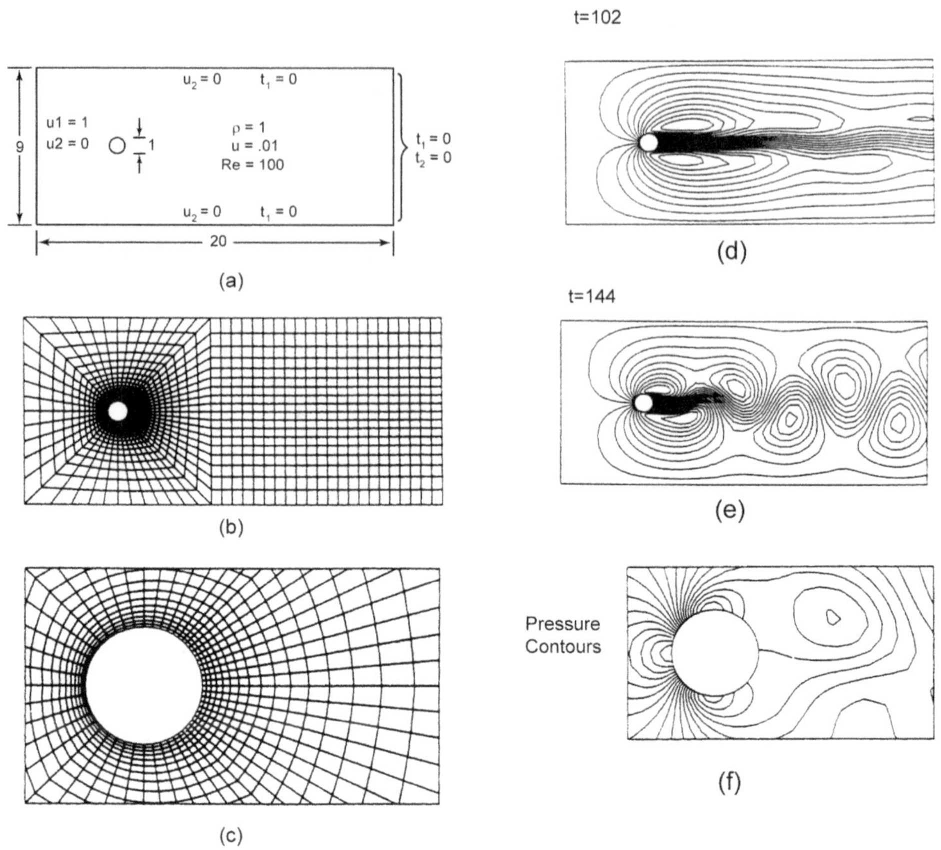

Figure 27. Flow past a circular cylinder at Re = 100: (a) Domain and boundary conditions, (b) finite element mesh, (c) detail of the mesh next to the cylinder, (d) streamlines at t = 100, (e) streamlines at t = 144, and (f) pressure contours close to the cylinder at t = 144.

A perturbation was accomplished by imposing small forces of magnitude 0.0001 to the boundary layer nodes between time t = 54 and t = 58.5. The perturbation effectively destabilized the flow, and vortex shedding began at time t = 96. Stream function plots at times t = 102 and t = 144 are shown in Fig. 27(d–f), together with a pressure plot in the neighborhood of the cylinder at time t = 144.

5.5 TURBULENT FLOW

Most real-life flows are turbulent; however, none of the currently available approaches to turbulence give an answer to the question of how to find instantaneous turbulent velocities, even if they satisfy the momentum equations. This is a formidable problem, which can be somewhat simplified by concentrating on the mean flow and the mean turbulence [62–64].

The velocities and pressure are written as

$$u_i = \bar{u}_i + u_i', \quad p = \bar{p} + p' \tag{122}$$

where u_i' and p' are the velocity and pressure fluctuations about the mean. Substitution into the Navier-Stokes equations yields the equations in terms of the averaged velocities:

$$\frac{\partial \bar{u}_j}{\partial x_j} = 0 \tag{123}$$

$$\frac{\partial \bar{u}_i}{\partial t} + \bar{u}_j \frac{\partial \bar{u}_i}{\partial x_j} = -\frac{1}{\rho}\frac{\partial \bar{p}}{\partial x_i} + v\frac{\partial^2 \bar{u}_i}{\partial x_j \partial x_j} - \frac{\partial}{\partial x_j}(\overline{u_i' u_j'}) \quad (124)$$

The last term in Eq. (124) is called the Reynolds stress term and is defined by

$$\tau_{ij} = -\rho\overline{u_i' u_j'} \quad (125)$$

Equations (124) and (125) are now incomplete because of the new unknown functions τ_{ij}. One must therefore produce a turbulence model for τ_{ij}. This is a very difficult task, and many models have been constructed to deal with different types of flow, ranging from simple algebraic models (Prandtl's mixing length) to multiple-equations model. The most popular are the two-equation models, also known as k-ε and k-ω models. These are generally referred to as Reynolds Averaged Navier-Stokes (RANS) closure schemes. A great variety of turbulence models are presented and discussed by Launder and Spalding [65] and in the review by Rodi [66].

In a k-ε model, k is the turbulence kinetic energy and ε is the kinematic dissipation rate of the turbulence energy, defined as

$$k = \frac{1}{2}\overline{u_i' u_i'}, \qquad \varepsilon = v\left(\frac{\partial u_i'}{\partial x_j}\right)^2 \quad (126)$$

The use of turbulence models is difficult and usually becomes the dominant aspect of the numerical computations. To illustrate the complexity of turbulent flow calculations, the example of turbulent flow over a backward-facing step is used from Sohn [67]. The k-ε turbulence model is taken form Launder and Spalding [68], where the equations for k and ε are given by

$$\frac{\partial k}{\partial t} + \bar{u}_i \frac{\partial k}{\partial x_i} = \frac{1}{\rho}\frac{\partial}{\partial x_i}\left(\frac{\mu_t}{\sigma_k}\frac{\partial k}{\partial x_i}\right) + \frac{\mu_t}{\rho}\left(\frac{\partial \bar{u}_i}{\partial x_j} + \frac{\partial \bar{u}_j}{\partial x_i}\right)\frac{\partial \bar{u}_i}{\partial x_j} - \varepsilon \quad (127)$$

and

$$\frac{\partial \varepsilon}{\partial t} + \bar{u}_i \frac{\partial \varepsilon}{\partial x_i} = \frac{1}{\rho}\frac{\partial}{\partial x_i}\left(\frac{\mu t}{\sigma_\varepsilon}\frac{\partial \varepsilon}{\partial x_i}\right) + \frac{c_1 \mu_t}{\rho}\frac{\varepsilon}{k}\left(\frac{\partial \bar{u}_i}{\partial x_j} + \frac{\partial \bar{u}_j}{\partial x_i}\right)\frac{\partial \bar{u}_i}{\partial x_j} - c_2\frac{\varepsilon^2}{k} \quad (128)$$

The Reynolds stresses are given by

$$\tau_{ij} = \mu_t\left(\frac{\partial \bar{u}_i}{\partial x_j} + \frac{\partial \bar{u}_j}{\partial x_i}\right) \qquad \text{where} \qquad \mu_t = \rho c_\mu \frac{k^2}{\varepsilon} \quad (129)$$

where μ_t is the turbulent eddy viscosity and σ_k, σ_ε, c_1, c_2, and c_μ are empirical constants.

The geometry and boundary conditions are shown in Fig. 28. The Reynolds number based on the inlet velocity and the height of the step were chosen to match the experiments of Kim et al. [69]. The finite element mesh consisted of 22 × 16 nine-noded biquadratic elements refined near the walls.

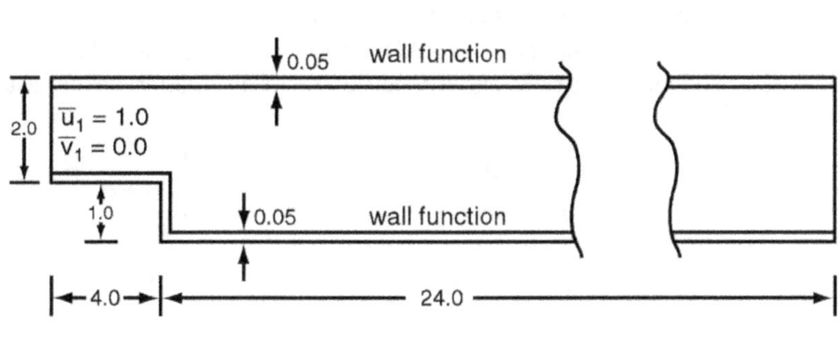

Figure 28. Domain and boundary conditions for the turbulent flow over a backward-facing step.

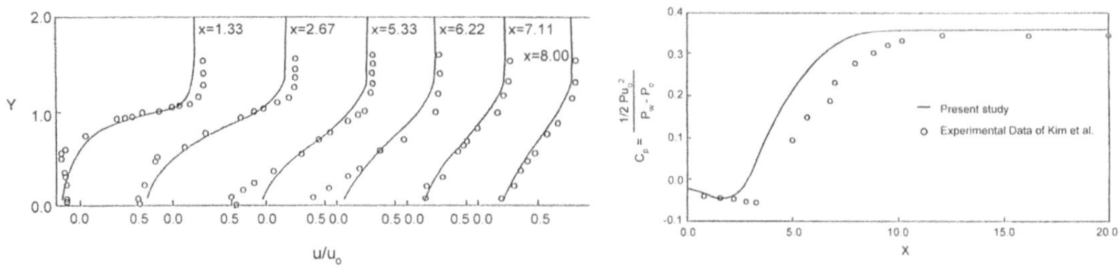

Figure 29. Numerical prediction and experimental measurements of the averaged horizontal velocity and calculated and experimentally measured pressure coefficient for turbulent flow over a backward-facing step (from [67]).

A further complication with the use of turbulence models is that they do not always do a good job over the entire domain. It is known that the k-ε model is not valid near the walls, where laminar sublayers may develop [64]. To correct for this deficiency, the computational domain was truncated at a distance from the wall and a "wall function" was used to describe the velocity in the immediate vicinity of the walls, as shown in Fig. 28. The tangential velocity, u_t, near the walls was calculated as

$$u_t = \begin{cases} wz & 0 < z < 30 \\ \dfrac{w}{\kappa}\ln(Ez) & 30 < z < 100 \end{cases} \tag{130}$$

where z and w are a nondimensional distance from the wall and a frictional velocity, respectively, and are given by

$$z = \frac{wy}{v}, \qquad w = \sqrt{v\frac{\partial u_t}{\partial y}} \tag{131}$$

Here κ is the von Karmàn constant, $\kappa = 0.41$; E is a roughness parameter equal to 9.0. The normal velocity is zero. The rest of the empirical constants in the turbulence model are taken from Launder and Spalding [68] and are $\sigma_k = 1.0$, $\sigma_\varepsilon = 1.3$, $c_1 = 1.44$, $c_2 = 1.92$, and $c_\mu = 0.09$.

Steady state results for the mean horizontal velocity are shown in Fig. 29 along with the calculated pressure coefficient. The experimental results of Kim et al. [69] are presented for comparison. The model does a good job only in certain parts of the domain. The flow is not predicted well in the recirculating zone, and the pressure coefficient is only captured properly near the exit of the region.

The other popular RANS model is the k-ω formulation [70]. In this approach, turbulent kinetic energy, k, and a specific rate of dissipation, ω, are established using

$$\begin{aligned} \frac{\partial(\rho k)}{\partial t} + \frac{\partial(\rho u_j k)}{\partial x_j} &= P - \beta^* \rho \omega k + \frac{\partial}{\partial x_j}\left[\left(\mu + \sigma_k \frac{\rho k}{\omega}\right)\frac{\partial k}{\partial x_j}\right] \\ \frac{\partial(\rho\omega)}{\partial t} + \frac{\partial(\rho u_j \omega)}{\partial x_j} &= \frac{\gamma\omega}{k} - \beta\rho\omega^2 + \frac{\partial}{\partial x_j}\left[\left(\mu + \sigma_\omega \frac{\rho k}{\omega}\right)\frac{\partial \omega}{\partial x_j}\right] + \frac{\rho\sigma_d}{\omega}\frac{\partial k}{\partial x_j}\frac{\partial \omega}{\partial x_j} \end{aligned} \tag{132}$$

where

$$P = \tau_{ij} \frac{\partial u_i}{\partial x_j}$$

$$\tau_{ij} = \mu_t \left(2S_{ij} - \frac{2}{3} \frac{\partial u_k}{\partial x_k} \delta_{ij} \right) - \frac{2}{3} \rho k \delta_{ij} \tag{133}$$

$$S_{ij} = \frac{1}{2} \left(\frac{\partial u_i}{\partial x_j} + \frac{\partial u_j}{\partial x_i} \right)$$

with $\gamma = 13/25$, $\sigma_k = 0.6$, $\sigma_\omega = 0.5$, $\beta^* = 0.09$, $C_{lim} = 7/8$, and

$$\mu_t = \frac{\rho k}{\hat{\omega}}$$

$$\hat{\omega} = \max \left[\omega, C_{lim} \sqrt{\frac{2\bar{S}_{ij}\bar{S}_{ij}}{\beta^*}} \right] \tag{134}$$

$$\bar{S}_{ij} = S_{ij} - \frac{1}{3} \frac{\partial u_k}{\partial x_k} \delta_{ij}$$

where ρ is density and μ is the molecular viscosity. The k-ε model is simple to implement and can produce reasonable predictions for many flows. However, it is not effective when dealing with swirling and rotating flows, flows with strong separation, or flows in non-circular ducts. The numerical behavior of the k-ω model is similar to that of the k-ε models, and suffers from many of the same drawbacks (and assumes that μ_t is isotropic).

5.5.1 Large Eddy Simulation (LES)

Direct numerical simulations (DNS) have been used by some investigators to overcome the shortcomings of RANS in the study of turbulence-driven flows. In this approach, the governing equations are solved exactly, thereby yielding the full spectrum of the length and time scales of a turbulent flow. For example, Versteeg and Neuwsdat [71] and Boudjemadi et al. [72] studied flow in a vertical channel maintained at uniform temperature using DNS. Their results indicated that the turbulent transport process occurring in the fully turbulent regime is able to enhance the heat transfer, thereby making it a desirable means of improving air systems.

Although DNS can resolve the entire spectrum of the length scales of a turbulent flow, the huge computational resources necessitated by the approach implies that it is often (i) limited to cases with low Rayleigh numbers, and (ii) used with the periodic or cyclic boundary condition to simplify the problem. This poses considerable challenge to the simulations of turbulent flows because most flows have moderately high Rayleigh numbers and the 'ideal' periodic boundary condition is seldom encountered in practical engineering applications. The energy cascade is illustrated in Fig. 30.

The development of LES has progressed through a history which dated back to the 1960s with the formulation of the famous Smagorinsky [73] model. Based on SM, the history of LES started to demonstrate extraordinary development in the 1990s when several dynamic variants of the Smagorinsky model were developed, e.g., Germano et al. [74] and Ghosal et al. [75]. As a result, academic and industrial researchers globally became infatuated with these techniques, while realizing the limits of the classical modeling methods based on the RANS approaches.

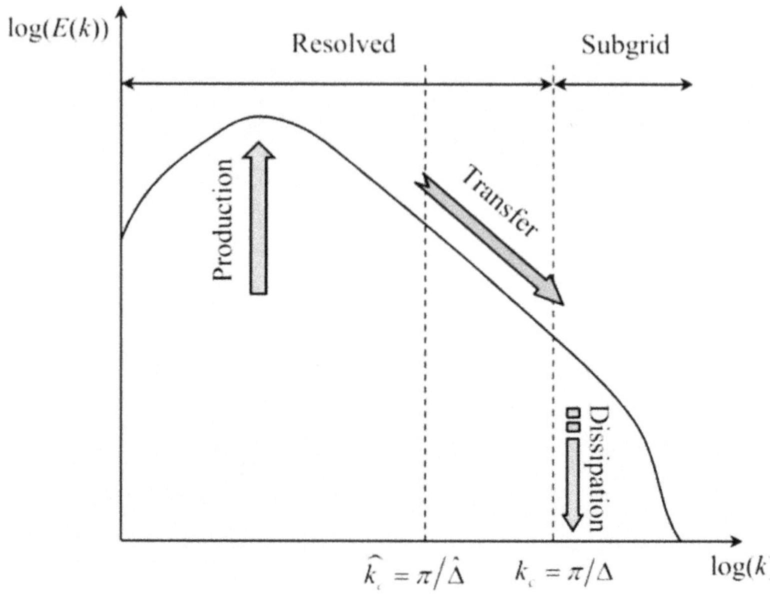

Figure 30. Illustration of the energy cascade of in LES. The larger filter width represents the test filter range.

Even though the concept of LES has experienced profound development over many decades, its implementation in turbulent flows was only actualized at recent times; such dramatic breakthrough was made possible by the rapid development of computational architectures. This has allowed LES to be used as a practical engineering tool for many types of applications, mainly dealing with massively separated flows in complex configurations. For example, the application of LES in studying turbulent buoyant jets [76–80], rotating flows [81,82], vortex breakdown and recirculation flows [83,84], as well as flow over an obstruction [85,86] have shown reasonably accurate prediction of incompressible turbulent flows. This is mainly attributed to the fact that, unlike RANS, the large eddies which are also the most energy-containing eddies have been captured accurately.

In LES, a turbulent flow is separated into large and small scales based on a reference or cut-off length. This cut-off length is commonly assumed to be equal to the size of the computational grid. The scales which have a characteristic size greater than the cut-off length are then called large or resolved scales, and the others are known as the small or subgrid scales.

In order to ensure that the dynamics of the resolved scales remain accurate, the SGS terms have to be considered in the solutions of the governing equations. Since the small scales are more isotropic than the large ones, their effect can be reasonably accounted for by means of an SGS model. Modeling of the SGS terms involves approximating the subgrid tensors based on the information contained in the large or resolved scales.

An LES solution is highly dependent on the choice of SGS model because it accounts for not only energy exchange, but also dissipation between the resolved and subgrid scales. For this reason, the major effort in the development in LES has been concentrated on modeling of the SGS viscosity.

In LES modeling, large eddies are computed and the smallest, subgrid-scale (SGS) eddies are modeled. The underlying premise is that the largest eddies are directly affected by the boundary conditions, carry most of the Reynolds stress, and must be computed. The small-scale turbulence is weaker, contributing less to the Reynolds stresses, and is therefore less critical. Because LES involves modeling the smallest eddies, the smallest cells can be much larger than the Kolmogorov length, and much larger timesteps can be taken than are possible in a DNS. Hence, for a given computing cost, it is possible to achieve much higher Reynolds numbers with LES than with DNS, or conversely to obtain a solution at a given Reynolds number more cheaply.

The primary issue in accuracy remains that of computing derivatives for the smallest scales (highest wavenumbers) resolved. The ultimate test of grid convergence is the requirement that excessive energy must not accumulate in the smallest scales. The primary requirement is to get the dissipation rate right; details of the dissipating eddies are unimportant in LES. DNS nominally requires accurate simulation of the dissipating eddies. A major difficulty in LES is that near a solid surface all eddies are small – to the extent that the stress-bearing and dissipation ranges of eddy size overlap. If one requires LES to resolve most of the stress-bearing range, the grid spacing, and timestep, required by LES gradually fall towards that needed for full DNS as the surface is approached.

A filter provides a formal definition of the averaging process and separates the resolvable scales from the subgrid scales. Filtering is used to derive the resolvable-scale equations. There are many kinds of filters that can be used. One of the simplest types of filter is the volume-average box filter [87]. The filter is a simple integral expression used over a control volume, where a subgrid-scale (SGS) velocity is used along with a filter width based on the control volume dimensions (Δx, Δy, and Δz). Orszag et al. [88] utilized a Fourier cutoff filter based on the Fourier transform of the turbulence equations. Ferziger [89] employed a Gaussian filter. Leonard [90] adapted a filter function to Deardorff's formulation, resulting in a second averaging that yields a different result from the first averaging. This second stress term removes significant energy from the resolvable scales and can be computed directly and need not be modeled. This is sometimes inconvenient, however, depending on the numerical method used. Clark et al. [91] verify that this representation is very accurate, at low Reynolds number, by comparing with DNS results. However, these additional stresses are of the same order as the truncation error when a second-order accurate differencing scheme is used, and they are thus implicitly represented [92]. Many other filters have been proposed and used. In all cases, the filter introduces a scale, Δ, that represents the smallest turbulence scale allowed by the filter.

The governing equations for turbulence flow can be written in a more convenient form, i.e.,

$$
\begin{aligned}
&\frac{\partial \bar{u}_i}{\partial t} + \frac{\partial}{\partial x_i}(\bar{u}_i \bar{u}_j) = -\frac{1}{\rho}\frac{\partial P}{\partial x_i} + \frac{\partial}{\partial x_j}\left[\nu \frac{\partial \bar{u}_i}{\partial x_j} + \tau_{ij}\right] \\
&\tau_{ij} = -(Q_{ij} - \frac{1}{3} Q_{kk}\delta_{ij}) \\
&P = \bar{p} + \frac{1}{3}\rho Q_{kk}\delta_{ij} \\
&Q_{ij} = R_{ij} + C_{ij} \\
&R_{ij} = \left(\overline{u_i' u_j'}\right) \\
&C_{ij} = \overline{\left(\bar{u}_j u_j'\right)} + \overline{\left(\bar{u}_j u_i'\right)}
\end{aligned}
\tag{135}
$$

Note that $\bar{\bar{u}}_i \neq \bar{u}_i$ in filtering. The cross-term stress tensor, C_{ij}, drains significant energy from the resolvable scales. Most current efforts model the sum of C_{ij} and R_{ij}. Clearly, the accuracy of a LES depends critically upon the model used for these terms.

At this point, the fundamental problem of Large Eddy Simulation is evident. Specifically, we must establish a satisfactory model for the SGS stresses. To emphasize the importance of achieving an accurate SGS stress model, consider the following. In simulating the decay of homogeneous isotropic turbulence with $16^3 = 4096$ and $32^3 = 32768$ grid points, Ferziger [89] showed that the SGS turbulence energy was 29% and 20%, respectively, of the total. Thus, the subgrid scales constitute a significant portion of the turbulence spectrum. Models have been postulated that range from a simple gradient-diffusion model [92] to a one-equation model [93], to the analog of a second-order closure model [94]. Nonlinear stress-strain rate relationships have also been postulated [95].

5.5.2 Subgrid-Scale (SGS) Modeling

Smagorinsky [73] was the first to postulate a model for the SGS stresses. The model assumes the SGS stresses follow a gradient-diffusion process, similar to molecular motion. Consequently, τ_{ij} is given by

$$\tau_{ij} = 2\nu_T S_{ij} \quad \text{where} \quad S_{ij} = \frac{1}{2}\left(\frac{\partial \bar{u}_i}{\partial x_j} + \frac{\partial \bar{u}_j}{\partial x_i}\right) \tag{136}$$

where S_{ij} is called the "resolved strain rate" and ν_T is the Smagorinsky eddy viscosity,

$$\nu_T = (C_S\Delta)^2\sqrt{S_{ij}S_{ij}} \tag{137}$$

where C_s is the Smagorinsky coefficient. Note that Equation (7) is akin to a mixing-length formula with mixing length $C_s\Delta$. Obviously the grid scale Δ, of $(\Delta_1\Delta_2\Delta_3)^{1/3}$ if the steps in the three coordinate directions are different, is an overall scale of the SGS motion, but assuming it to be a unique one is crude. If Δ were in the inertial subrange of eddy size, and sufficiently larger than the Kolmogorov viscous length scale, η, that the viscous-dependent part of the SGS motion was a small fraction of the whole, then no other length scale would be relevant and the Smagorinsky constant would be universal. This is rarely the case.

The physical assumption behind the mixing-length formula, that eddies behave like molecules, is simply not true. Nevertheless, just as the mixing-length model can be calibrated for a given class of flows, so can the Smagorinsky coefficient. Its value varies from flow to flow, and from place to place within a flow. In the early days of LES, the basic Smagorinsky subgrid-scale model was widely used, C_s being adjusted to get the best results for each flow [96]. In the critical near-wall region, law-of-the-wall arguments valid in well-behaved flows suggest that C_s should be a function of and an increasingly strong function of $\upsilon_r\gamma/\nu$ as the latter decreases. However there seems to be no record of attempts to calibrate this function – virtually all users of the basic model keep C_s constant throughout the flow. There are two key reasons why the basic Smagorinsky enjoys some degree of success. First, the model yields sufficient diffusion and dissipation to stabilize the numerical computations. Second, low-order statistics of the larger eddies are usually insensitive to the SGS motions.

In an attempt to incorporate some representation of the dynamics of the subgrid scales, Lilly [93] postulated that

$$\nu_T = C_L\Delta q \tag{138}$$

where q^2 is the SGS kinetic energy, and C_L is a closure coefficient. The subgrid scale stress anisotropy now depends on the sign of the resolved strain rate, rather than on its magnitude as in the Smagorinsky formula. An equation for q^2 can be derived from a moment of the Navier-Stokes equation, which involves several terms that must be modeled. This model is very similar to Prandtl's one-equation model, both in spirit and in results obtained. Schumann [97] used the model in his LES research, but found it difficult to conclude that any significant improvement over the Smagorinsky model could be obtained with such a model.

Dynamic models undoubtedly work surprisingly well, even in cases where rigorous justification is not valid. Jimenez [98] points out that the essential feature of an SGS model is to dissipate the kinetic energy cascaded down to it. Jimenez et al. [99] elaborate further on the efficacy of dynamic models. On the one hand, any eddy-viscosity model assumes the subgrid stresses are "perfectly related" to the strain rate, even though they are essentially uncorrelated. On the other hand, dynamic models are "very robust" to this fundamental error in physics, largely because "the formula for their eddy viscosity contains a sensor that responds to the accumulation of energy wave numbers of the spectrum before it contaminates the energy containing range."

A subgrid-scale model with some general similarities to the dynamic model has been suggested by Domaradzki and Saiki [100]. The resolved motion, interpolated on a length scale of half the grid size, and the phases of the resulting subgrid modes, were adjusted to correspond to the phases of the subgrid modes by nonlinear interactions of the resolved mode in the LES.

If LES is to be applied to wall flows at indefinitely high Reynolds numbers, the *viscous sublayer* or *viscous wall region* must be excluded from the main computation. Moreover, the distance of the first LES grid point from the solid surface must be independent of Reynolds number. It must be set at some suitable fraction of the shear-layer thickness, so that the total number of grid points is also independent of Reynolds numbers. The number of grid points required in the normal direction increases proportionally. A factor of 10 increase in Reynolds number means a factor of 30 increase in computer work. The current fashion in RANS modeling is integration to the wall rather than the use of off-the-wall boundary conditions (*wall functions* in RANS modeling). However, if LES is to be a significant improvement over RANS models, it must deal with strongly-nonequilibrium flows, notably separated flows, in which the simple law of the wall is not valid [101,102].

5.6 COMPRESSIBLE FLOW

The governing equations for compressible flow can be written using the nondimensional form given by Anderson [70], which uses the internal energy form. This form is simpler, but the equations will not be in conservation form. The governing equations are

Continuity

$$\frac{\partial \rho}{\partial t} + u_j \frac{\partial \rho}{\partial x_j} + \rho \frac{\partial u_j}{\partial x_j} = 0 \tag{139}$$

Momentum

$$\rho\left(\frac{\partial u_i}{\partial t} + u_j \frac{\partial u_i}{\partial x_j}\right) = -\frac{\partial}{\partial x_i}\left(\frac{1}{\gamma M^2} p - \frac{1}{3}\frac{\mu}{Re}\frac{\partial u_j}{\partial x_j}\right) + \frac{\mu}{Re}\frac{\partial^2 u_i}{\partial x_j \partial x_j} \tag{140}$$

Energy

$$\rho\left(\frac{\partial e}{\partial t} + u_j \frac{\partial e}{\partial x_j}\right) = \frac{\gamma}{Re\,Pr}\frac{\partial}{\partial x_j}\mu\frac{\partial e}{\partial x_j} - \left(\frac{1}{\gamma M^2} p + \frac{2}{3}\frac{\mu}{Re}\frac{\partial u_j}{\partial x_j}\right)\frac{\partial u_j}{\partial x_j} + \frac{\mu}{Re}\left(\frac{\partial u_i}{\partial x_j} + \frac{\partial u_j}{\partial x_i}\right)\frac{\partial u_i}{\partial x_j} \tag{141}$$

State

$$p = \rho e \tag{142}$$

In Eq. (133), the divergence term is kept in the first parentheses on the right-hand side separate from the viscous terms (the second term on the right-hand side), even though they are of the same form and are usually combined into one term. Combining them is misleading, however, because the divergence term may behave very differently, especially when the flow approaches conditions of constant density. Also, they must be treated differently from a numerical point of view. In fact, to avoid pressure oscillations in regions of stagnation or where the flow is locally of constant density, such terms are treated in a manner consistent with an incompressible flow.

The viscosity is considered temperature dependent and can be calculated using Sutherland's law [70]

$$\mu = T^{3/2}\frac{1+S}{T+S} \tag{143}$$

where S is a constant. The quantities ρ_∞, U_∞, e_∞, T_∞, L, μ_∞, and k_∞ are reference values, and the nondimensional parameters are the Mach number, Reynolds number, Prandtl number, and ratio of specific heats:

$$M = \frac{U_\infty}{\sqrt{\gamma\frac{p_\infty}{\rho_\infty}}}, \quad Re = \frac{\rho_\infty U_\infty L}{\mu_\infty}, \quad Pr = \frac{\mu_\infty c_p}{k_\infty}, \quad \gamma = \frac{c_p}{c_v} \tag{144}$$

A stable, consistent Petrov-Galerkin weak formulation using bilinear elements to approximate the density, velocities and temperature, and piecewise-constant pressure can be obtained as follows:
Continuity

$$\begin{aligned}\int_\Omega N_i\frac{\partial\rho}{\partial t}d\Omega = &-\int_\Omega (W_\rho)_i\left(u\frac{\partial\rho}{\partial x}+v\frac{\partial\rho}{\partial y}\right)d\Omega\\ &-\int_\Omega N_i\rho\left(\frac{\partial u}{\partial x}+\frac{\partial v}{\partial y}\right)d\Omega\end{aligned} \tag{145}$$

x-momentum

$$\begin{aligned}\int_\Omega \rho N_i\frac{\partial u}{\partial t}d\Omega = &-\int_\Omega\left[(W_u)_i\left[u\frac{\partial u}{\partial x}+v\frac{\partial u}{\partial y}\right]+\frac{\mu}{Re}\left(\frac{\partial N_i}{\partial x}\frac{\partial u}{\partial x}+\frac{\partial N_i}{\partial y}\frac{\partial u}{\partial y}\right)\right]d\Omega\\ &+\int_\Omega\frac{\partial N_i}{\partial x}\left[\frac{1}{\gamma M^2}p-\frac{1}{3}\frac{\mu}{Re}\left(\frac{\partial u}{\partial x}+\frac{\partial v}{\partial y}\right)\right]d\Omega\\ &+\int_\Gamma N_i\left[-\frac{1}{\gamma M^2}p-\frac{1}{3}\frac{\mu}{Re}\left(\frac{\partial u}{\partial x}+\frac{\partial v}{\partial y}\right)\right]n_x d\Gamma\\ &+\int_\Gamma N_i\frac{\mu}{Re}\left[2\frac{\partial u}{\partial x}n_x+\left(\frac{\partial u}{\partial y}+\frac{\partial v}{\partial x}\right)n_y\right]d\Gamma\end{aligned} \tag{146}$$

y-momentum

$$\begin{aligned}\int_\Omega \rho N_i\frac{\partial v}{\partial t}d\Omega = &-\int_\Omega\left[(W_v)_i\left(u\frac{\partial v}{\partial x}+v\frac{\partial v}{\partial y}\right)+\frac{\mu}{Re}\left(\frac{\partial N_i}{\partial x}\frac{\partial v}{\partial x}+\frac{\partial N_i}{\partial y}\frac{\partial v}{\partial y}\right)\right]d\Omega\\ &+\int_\Omega\frac{\partial N_i}{\partial y}\left[\frac{1}{\gamma M^2}p-\frac{1}{3}\frac{\mu}{Re}\left(\frac{\partial u}{\partial x}+\frac{\partial v}{\partial y}\right)\right]d\Omega\\ &+\int_\Gamma N_i\left[-\frac{1}{\gamma M^2}p+\frac{1}{3}\frac{\mu}{Re}\left(\frac{\partial u}{\partial x}+\frac{\partial v}{\partial y}\right)\right]n_y d\Gamma\\ &+\int_\Gamma N_i\frac{\mu}{Re}\left[\left(\frac{\partial u}{\partial y}+\frac{\partial v}{\partial x}\right)n_x+2\frac{\partial v}{\partial y}n_y\right]d\Gamma\end{aligned} \tag{147}$$

Energy

$$
\begin{aligned}
\int_{\Omega} \rho N_i \frac{\partial e}{\partial t} d\Omega = &-\int_{\Omega} \left[(W_e)_i \left(u\frac{\partial e}{\partial x} + v\frac{\partial e}{\partial y} \right) + \frac{\gamma\mu}{\mathrm{Re}\,\mathrm{Pr}} \left(\frac{\partial N_i}{\partial x}\frac{\partial e}{\partial x} + \frac{\partial N_i}{\partial y}\frac{\partial e}{\partial y} \right) \right] d\Omega \\
&+ \int_{\Omega} N_i \left[\frac{1}{\gamma M^2} p + \frac{2}{3}\frac{\mu}{\mathrm{Re}} \left(\frac{\partial u}{\partial x} + \frac{\partial v}{\partial y} \right) \right] \left(\frac{\partial u}{\partial x} + \frac{\partial v}{\partial y} \right) d\Omega \\
&+ \int_{\Gamma} N_i \frac{\mu}{\mathrm{Re}} \left[2\left(\frac{\partial u}{\partial x} \right)^2 + 2\left(\frac{\partial v^2}{\partial y} \right) + \frac{\partial u}{\partial y} + \frac{\partial v}{\partial x} \right] d\Omega \\
&+ \int_{\Gamma} N_i \frac{\gamma\mu}{\mathrm{Re}\,\mathrm{Pr}} \left(\frac{\partial e}{\partial x} n_x + \frac{\partial e}{\partial y} n_y \right) d\Gamma
\end{aligned}
\tag{148}
$$

State

$$
\int_{\Omega} M_i p \, d\Omega = \int_{\Omega} \rho M_i e \, d\Omega \tag{149}
$$

were N_i are the bilinear shape functions and M_i are piecewise-constant shape functions such that $M_i = 1$ over element e_i and $= 0$ otherwise. The Petrov-Galerkin weighting functions, W_ρ, W_u, W_v, and W_e are given as

$$
(W_a)_i = N_i + \alpha_a \frac{\bar{h}}{2\|\mathbf{V}\|} \mathbf{V} \cdot \nabla N_i \tag{150}
$$

where a = ρ, u, v, and e. The parameter α_a is given by

$$
\alpha_a = \coth\frac{\gamma_a}{2} - \frac{2}{\gamma_a} \tag{151}
$$

and the local numbers, γ_a, are

$$
\gamma_\rho = \infty, \text{ (i.e., } \alpha_\rho = 1), \; \gamma_u = \frac{\rho \mathrm{Re} \|\mathbf{V}\| \bar{h}}{\mu\left(1 + \dfrac{u^2}{3\|V\|^2} \right)}, \; \gamma_v = \frac{\rho \mathrm{Re} \|V\| \bar{h}}{\mu\left(1 + \dfrac{v^2}{3\|V\|^2} \right)}, \; \gamma_e = \frac{\rho \mathrm{Pr}\,\mathrm{Re} \|V\| \bar{h}}{\mu\gamma} \tag{152}
$$

Two examples of the application of the algorithms are examined. In both cases, steady state solutions were sought. These were obtained using mass lumping and a second-order Runge-Kutta method to integrate in time.

5.6.1 Supersonic Flow Impinging on a Cylinder

The computational domain and boundary conditions for this example are shown in Fig. 31. The cylinder has unit radius, the free stream Mach number is $M = 2$, Pr = 0.72, $\gamma = 1.4$, and Re = 1000 based on the radius of the cylinder. In this example the viscosity is assumed to be constant. The finite element mesh shown in Fig. 31 consists of a 40 × 25 node mesh clustered near the cylinder surface [104].

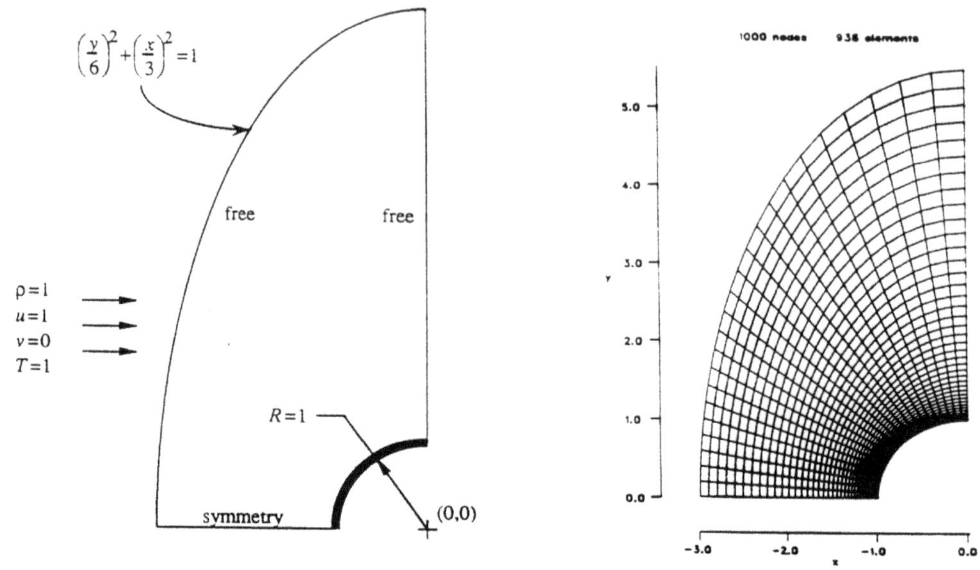

Figure 31. Geometry, boundary conditions and grid for viscous flow impinging on a cylinder of unit radius.

Fig. 32 shows contours of density, pressure, local Mach number, and temperature calculated at steady state. The detached bow shock is evident; three to four elements are needed to accommodate it along the line of symmetry. The formation and development of the boundary layer can be observed in the plots of local Mach number and temperature; a further illustration is given in Fig. 33, where the velocity vectors near the cylinder surface are shown.

5.6.2 Transonic Flow Through a Rectangular Nozzle

This is an example of a three-dimensional compressible flow moving through a nozzle configuration that changes from annular to rectangular studied by Brueckner and Pepper [105]. The study was motivated by an interest in thrust vectoring. The geometry and computational mesh are shown in Fig. 33. Symmetry conditions are applied along the x-y plane. The mesh consists of 71,512 trilinear elements with 77,395 nodes. The reference length was taken to be the half-width in the y-direction. The free stream Mach number is 0.22, the Reynolds number is 500, and the Prandtl number is 0.72. The reference temperature is 865°K, $\gamma = 1.4$, and the pressure ratio p_{inlet}/p_{outlet} is 1.83. The incoming flow is subject to a 20° swirl angle. A plot of the transverse velocity vector at the nozzle exit plane is given in Fig. 34a, where axial vortices that form along the sidewalls of the nozzle as the fluid moves down the transition region can be observed. Fig. 34b also shows temperature contours at various cross sections through the computed plane downstream of the nozzle exit, showing the effect of swirl on the plume.

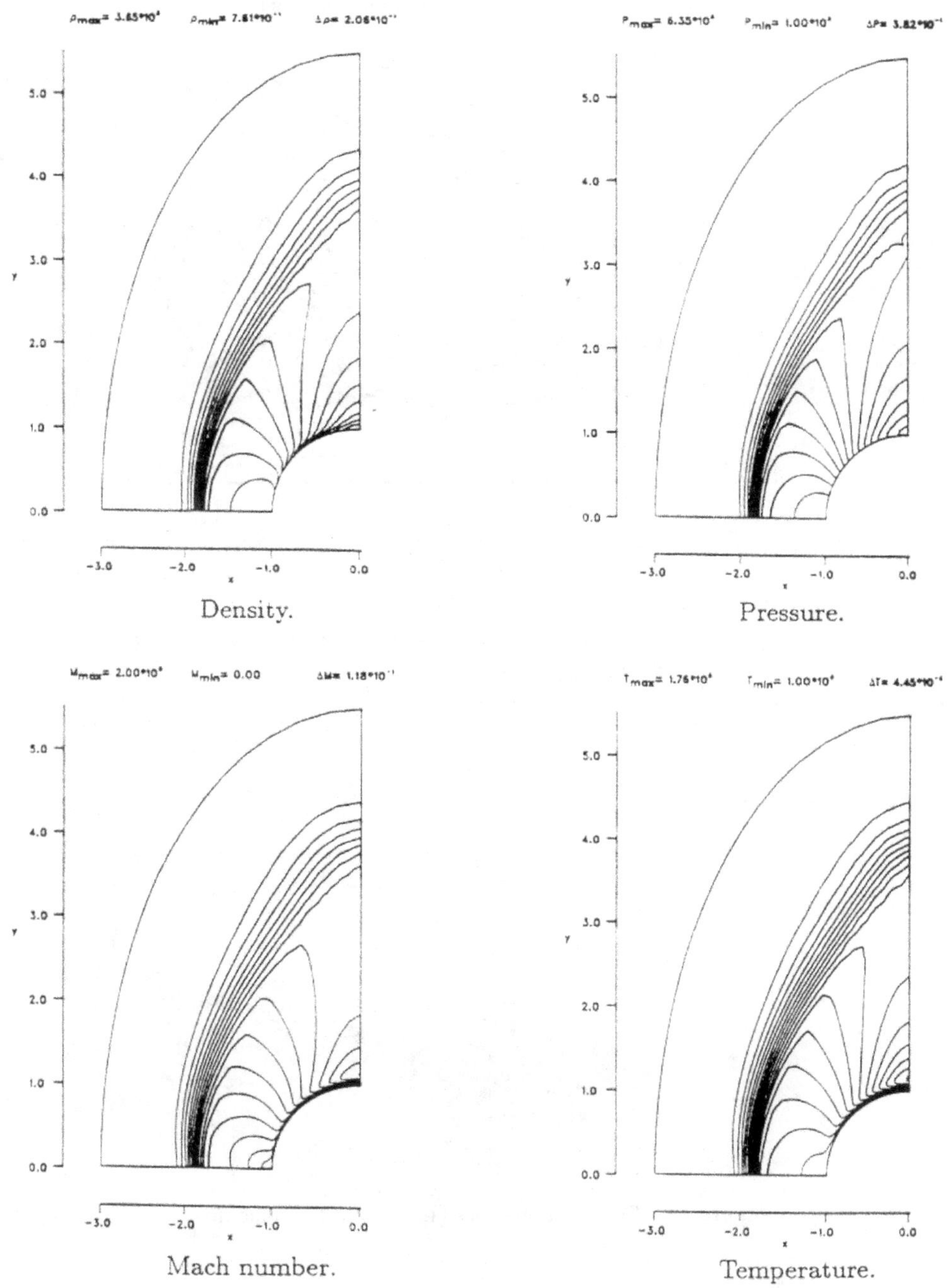

Figure 32. Results for viscous flow impinging on a cylinder at $M = 2$ and $\mathrm{Re} = 1000$.

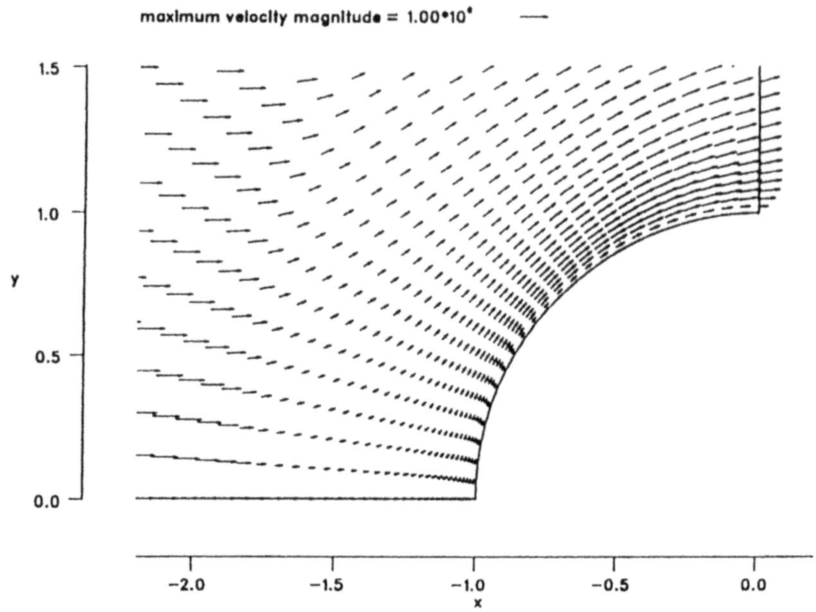

Figure 33. Detail of the velocity field for viscous flow impinging on a cylinder of unit radius at $M = 2$ and Re = 1000.

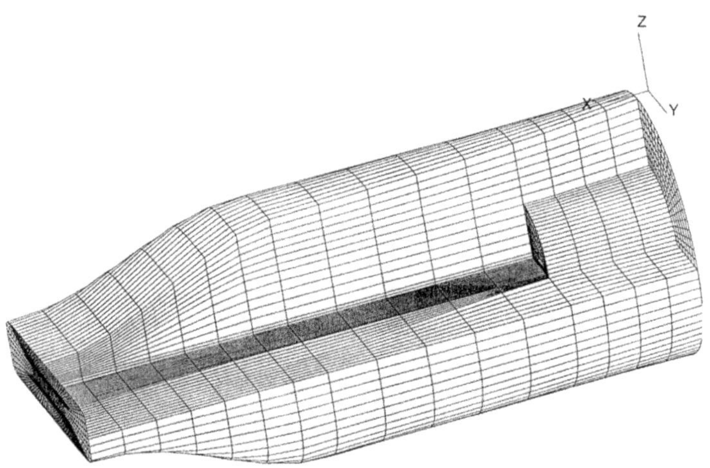

Figure 34. Computational mesh for viscous transonic flow through a nozzle.

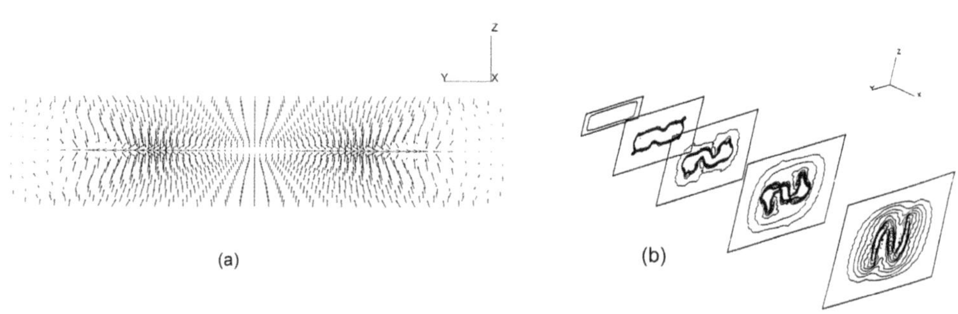

Figure 35. Nozzle plume (a) transverse velocity vector field at the nozzle exit plane and (b) temperature contours at various cross sections.

Chapter 6

LIST OF COMMERCIAL CODES

There are numerous companies selling FEM, as well as CFD-related, software. Many of the smaller companies have either closed or have been absorbed by larger companies. Some of this began with the purchase of Fluid Dynamics International (developers of FIDAP, a penalty approach FEM) by Fluent, Inc., which sold FLUENT, a popular CFD finite volume code. Fluent later obtained NEKTON, a spectral element code developed at MIT. Fluent, Inc. was subsequently purchased by ANSYS, Inc., who had also purchased FLOTRAN (an equal-order FEM code). COMCO, Inc., who had originally developed an hp-adaptive flow code in UT-Austin, and CENTRICS, a high-end finite element CFD code first developed at Stanford, both closed. The choices are numerous, and sometimes confusing for the buyer interested in obtaining a good code, especially one that is FEM based. COMSOL is a multiphysics FEM code that is widely used for its ability to couple various physics problems, e.g., fluid-structural-heat transfer, together into one overall solution strategy. This is a very easy program to use, reasonably priced and runs on PCs, with follow-on support from COMSOL, Inc. Today, even the finite volume based CFD codes use unstructured meshes (generally polygons) once unique to the FEM approach.

One of the biggest issues to face is whether one should develop a FEM code, or purchase one from a vendor. This is best decided by the user as dictated by experience, available resources, time constraints, and purpose. Obviously the big advantage of building one's own FEM code is knowing what's in the code, and having access to the source code in order to modify and upgrade it as needed. The major disadvantages are the time required to develop and test the code, and the ability to create a usable pre- and post-processing capability. Mesh generation is quite complicated, especially for 3-D domains, and the creation of graphical routines for displaying results in 2- and 3-D is equally difficult. Fortunately, there are some fairly good commercial packages now available for handling these I/O problems – leaving the user to modify the FEM solver to fit the data requirements of the pre- and post-processors. Several of the more popular stand-alone graphics codes are TECPLOT, FIELDVIEW, and ENSIGHT. Mesh generators such as PATRAN, CUBIT, FEMAP, FEMGV, and TRUGRID are good stand-alone packages. Nearly all the commercial FEM codes have pre- and post-processors built in to their overall program structure.

When selecting an FEM code, there are several suggested steps that one should follow. These steps are fairly typical of the process one would follow in selecting a commercial code. In addition, there is a small book published by the National Association for Finite Element Methods and Standards (NAFEMS, Glasgow, Scotland) entitled *How to Choose a Finite Element Analysis System*, which may be helpful (although not specifically aimed at CFD). The suggested steps are:

1. **Examine the requirements of the problem – what is to be solved?** Are there other related projects to be solved later, e.g., compressible versus incompressible flows? How wide a range of applications is anticipated – look beyond a one-year horizon.
2. **What resources do you have in-house at the present time?** Many times one group in a company doesn't realize that another group may have obtained a code, and are quite adept at using it. Commercial companies generally don't inform the buyer of users in other parts of the same organization – unless they already obtained an expensive site license. How much technical expertise resides in the

company? Are the intended users novices, or experienced FEM users with multi-physics expertise?

3. **What is the cost of the software, and what platform will it be run on?** Also, how many users will be running the program and on how many machines. Most of the commercial codes require computer id information, which are unique identities per computer (e.g., the ipconfig is commonly used for PCs), thus allowing only one user per machine.
4. **How easy is the code to use, and what is the suggested training time allotted to learn the program?** Support is fairly crucial in the beginning, and the willingness of the company to help the new user without additional charges is important to consider. Some codes may be quite accurate and sophisticated, but exceedingly difficult to learn to use. A not-so-accurate code that is very easy to use may be more satisfactory and useful to a company than a highly detailed, all-encompassing CFD code that takes a long time to solve problems and requires a high-end computer platform.
5. **Have the company send a demo copy of the code to evaluate – some companies send the actual code with a time lock.** A few companies will actually set up a user's specific problem and attempt to solve it for one specific boundary constraint. Very few companies ever release their source code (FLOW3D, a CFD code developed by A. Hirt in Los Alamos, NM, was one of the few exceptions, but it is a finite volume code). Likewise, there are various free FEM packages that can now be downloaded from the web, but it is wise to understand their purposes and limitations.
6. **Check with other users of CFD packages.** If possible, also try to check usage at several universities, where the students and faculty have generally examined various flow and FEM codes and have determined which are the easiest and safest to use – their costs are quite low to obtain software, which allows them to maintain many different sets of codes.
7. **Take a short course if you must come up to speed quickly and/or choose a code.** ASME and AIAA offer both 2-3 day short courses as well as home study courses on FEM methods and their uses for fluid flow. In addition, there are numerous short courses offered by individuals and universities. The most expensive courses are not necessarily the best courses for novices. Some software companies offer their own training courses with the purchase of their software – this is good as far as learning to run the code, but may not be sufficient for learning the fundamentals about FEM.

The following list of free and commercially available FEM codes, including pre- and post-processor software, is not complete, but does include some of the more popular and widely known codes. The company name is listed, along with the code name or programming language, whether it is free, and email address. These companies will be more than happy to send you literature, and put you on their mailing lists.

Company	Code or Language	Free	Web/Email
Adina R&D	ADINA (900 nodes free)		www.adina.com
Advanced Design	GRAITEC		www.graitec.com
Agros2D	C++, Phython	X	www.agros2D.org
ALGOR	See AUTODESK		www.autodesk.com
ALTAIR	HyperWorks		www.altair.com
ANSYS	Various		www.ansys.com
ANSYS	FLUENT (FVM)		www.ansys.com
AUTODESK	AUTOCAD		www.autodesk.com
Calculix	ABAQUS	X	www.calculix.de
CEI	ENSIGHT (graphics)		www.ceintl.com
CSI	Various		www.csiberkeley.com
Code_Aster	Code_Aster	X	www.code-aster.org

Company	Code or Language	Free	Web/Email
COMSOL	COMSOL		www.comsol.com
Deal.II	C++	X	www.dealii.org
Dessault Systems	SIMULIA/ABAQUS		www.3ds.com
Diffpack	Various		www.diffpack.com
DUNE	C++	X	www.dune-project.org
Elmer	FORTRAN, C, C++	X	www.cs.fi/elmer
EMRC	NISA		www.nisasoftware.com
ESI Group	CFD-ACE, -FASTRAN		www.esi-group.com
FEBio	BEBio (biomechanics)	X	www.febio.org
FeniCS Project	C++, Python	X	www.fenicsproject.org
FreeFem++	C++	X	www.freefem.org
GetFem++	C++, MATLAB, Python	X	download.gna.org/getfem/html/homepage/index.html
Hermes Project	C, C++, Python	X	hpfem.org/hermes
jFEM	C++	X	www.thecomputationalphysicist.com
IMTEK Math.	Mathematica	X	https://simulation.uni-freiburg.de/downloads/ims
Intelligent Light	FIELDVIEW (graphics)		www.ilight.com
LAPCAD Eng.	LAPFEA		info@lapcad.com
LMS Int.	PolyFEM		www.lmsintl.com
LSTC	LS-DYNA		www.lstc.com
MoFEM Joseph	C++	X	https://bitbucket.org/likask/mofem-joseph/wiki/Home
MSC	PATRAN/MARC		www.mscsoftware.com
NEI Software	FEMAP/NASTRAN		info@neisoftware.com
OOFEM	C++	X	www.oofem.org
OpenFoam	OpenFOAM	X	www.openfoam.org
OpenSees	C++	X	opensees.berkeley.edu
Predictive Engr.	FEMAP		www.predictiveengineering.com
Sandia Nat. Lab.	CUBIT	X	www.snl.gov
SRAC	COSMOS/M		www.stresscalc.ru
TECPLOT	TECPLOT (graphics)		www.tecplot.com
UNLV/NCACM	GWADAPT		www.ncacm.unlv.edu
XYZ Sci. Appl.	TRUEGRID		www.trugrid.com
Z88 (Aurora)	z88 aurora	X	www.z88.de

Chapter 7

CONCLUSION

In this first section of the book, we have introduced the concept of numerical approximations using the finite element method. We started with the simplest one-dimensional, linear, steady-state conduction problem, expanded the method to two- and three-dimensional elements, and ended with the time-dependent, nonlinear incompressible and compressible Navier Stokes and energy equations. Once the basic concepts are understood, the method can be readily applied to fluid dynamics, heat transfer, wave propagation, structural analysis, species transport, as well as a multitude of other transport related processes. Hopefully you now have an appreciation of the power and versatility of the finite element technique. You are now ready to move to the next section of this book that deals with the Boundary Element Method (BEM).

Understanding the fundamentals of the finite element method significantly assists in grasping the basic formulation, and the power, of the boundary element technique. The BEM is really an extension of the FEM, i.e., the MWR employing integration by parts – twice. For example, let's look at the 1-D advection-diffusion equation,

$$U\frac{d\phi}{dx} = D\frac{d^2\phi}{dx^2} \tag{153}$$

or

$$\frac{d^2\phi}{dx^2} - A\frac{d\phi}{dx} = 0, \quad A = \frac{U}{D}$$

where U is the advective velocity and D is the diffusion. Applying the MWR and evaluating over the interval, a<x<b,

$$\langle W, L\phi \rangle = \int_a^b W\left(\frac{d^2\phi}{dx^2} - A\frac{d\phi}{dx}\right)dx \tag{154}$$

where $L \equiv d^2/dx^2 - Ad/dx$ (differential operator), and < > denotes the *inner product* (a mathematical term for creating a weak statement). W(x) is a weight, or arbitrary function, similar to the weighting in the Galerkin method. Thus, integrating the equation by parts (Green's formula),

$$I = \left[W\frac{d\phi}{dx}\right]_a^b - \left[\left(\frac{dW}{dx} + AW\right)\phi\right]_a^b + \int_a^b \phi\left[\frac{d^2W}{dx^2} + A\frac{dW}{dx}\right]dx \tag{155}$$

We now introduce the *adjoint operator*, $L^* = d^2/dx^2 + Ad/dx$, and can rewrite the integral equation as [106]

$$\langle W, Lu \rangle = B + \langle \phi, L^*W \rangle \tag{156}$$

where

$$B = \left[W\frac{d\phi}{dx}\right]_a^b - \left[\left(\frac{dW}{dx} + AW\right)\phi\right]_a^b$$

To establish a homogenous solution, we set L*W = 0. To obtain a fundamental solution, we can set $L^*W = -\delta(x-\xi)$, where the Dirac delta function is applied at the point $x = \xi$ [106]. Extension to 2-D is straightforward, where the problem reduces from 2-D to a 1-D line integral.

REFERENCES

[1] Zienkiewicz, O.C. and Taylor, R.L., *The Finite Element Method*, Vol. 2, 4th ed., McGraw-Hill, London, 1991.

[2] Hughes, T.J.R., *The Finite Element Method*, Prentice-Hall, Englewood Cliffs, N.J., 1987.

[3] Carey, G.F. and Oden, J.T., *Finite Elements: A Second Course*, Prentice-Hall, Englewood Cliffs, N.J., 1983.

[4] Strang, G. and Fix, G.J., *An Analysis of the Finite Element Method*, Prentice-Hall, Englewood Cliffs, N.J., 1973.

[5] Reddy, R.N., *An Introduction to the Finite Element Method*, 2nd ed., McGraw-Hill, New York, 1993.

[6] Pinder, G.F. and Gray, G., *Finite Element Simulations in Surface and Subsurface Hydrology*, Academic Press, New York, 1977.

[7] Pepper, D.W. and Heinrich, J.C., *The Finite Element Method: Basic Concepts and Applications*, Hemisphere (Taylor and Francis), Washington, D.C., 1992.

[8] Heinrich, J.C. and Pepper, D.W., *Intermediate Finite Element Method: Fluid Flow and Heat Transfer Applications*, Taylor and Francis, Philadelphia, PA, 1999.

[9] Irons, B.M., "Engineering Application of Numerical Integration in Stiffness Method," *AIAA J.*, Vol. 14, pp. 2035–2037, 1966.

[10] Zlamal, M., "Curved Elements in the Finite Element Method," *SIAM J. Numer. Anal., Part I*, Vol. 10, No. 1, pp. 229–240, 1973.

[11] Hammer, P.C., Marlowe, O.P., and Stroud, A.H. "Numerical Integration over Simplexes and Lower," *Math Table Aids Comput.*, Vol. 10, pp. 130–137, 1956.

[12] Cowper, G.R., "Gaussian Quadrature Formulas for Triangles," *Int. J. Numer. Methods Eng.*, Vol. 7, pp. 405–408, 1973.

[13] Fletcher, C.A.J. *Computational Galerkin Methods*, Springer-Verlag, New York, 1984.

[14] Isaacson, E. and Keller, H.B., *Analysis of Numerical Methods*, John Wiley and Sons, New York, 1966.

[15] Yakowitz, S. and Szidarovszky, F., *An Introduction to Numerical Computations*, 2nd Ed., Macmillan, New York, 1989.

[16] Newmark, N.M., "A Method of Computation for Structural Dynamics," *Trans. ASCE*, EM 3, Vol. 85, pp. 67–94, 1959.

[17] Hughes, T.J.R., Liu, W.K., and Brooks, A. "Finite Element Analysis of Incompressible Viscous Flows by the Penalty Function Formulation," *J. Comput. Phys.*, Vol. 30, pp. 1–60, 1979.

[18] Westerink, J.J. and Shea, D., "Consistent Higher Degree Petrov-Galerkin Methods for the Solution of the Transient Convection-Diffusion Equation," *Int. J. Numer. Methods Eng.*, Vol. 28, pp. 1077–1102, 1989.

[19] Cardle, J.A., "A Modification of the Petrov-Galerkin Method for the Transient Convection-Diffusion Equation," *Int. J. Numer. Methods. Eng.*, Vol. 38, pp. 171–181, 1995.

[20] Perrochet, P., "Finite Hyper-Elements: A 4-D Geometrical Framework Using Covariant Bases and Metric Tensors," *Comm. Num. Methods Eng.*, Vol. 11, pp. 525–534, 1995.

[21] Idelsohn, S.R., Nigro, N., Sorti, M.A., and Buscaglia, G., "A Petrov-Galerkin Formulation for Advection-Reaction Diffusion Problems," *Comput. Meth. Appl. Mech. Engr.*, Vol. 136, pp. 27–46, 1996.

[22] Donea, J., "A Taylor-Galerkin Method for Convective Transport," *Int. J. Numer. Methods Eng.*, Vol. 20, pp. 101–119, 1984.

[23] Burnett, D.S., *Finite Element Analysis*, Addison-Wesley, Reading, Mass, 1987.

[24] Ramakrishnam, R., Bey, K.S., and Thornton, E.A., "Adaptive Quadrilateral and Triangular Finite Element Scheme for Compressible Flows," *AIAA J.*, Vol. 28, No. 1, pp. 51–59, 1990.

[25] Marro, L., "Méthodes de Réduction de Largeur de Bande et de Profil Efficace des Matrices Creuses," Thèse, Universite de Nice, France, 1980.

[26] George, P.L., *Automatic Mesh Generation*, John Wiley and Sons, Chichester, UK, 1991.

[27] Carey, G.F., *Computational Grids: Generation, Adaptation, and Solution Strategies*, Taylor and Francis, Washington, D.C., 1997.

[28] Gable, C.W., Trease, H.E., and Cherry, T.A., "Automated Grid Generation From Models of Complex Geologic Structure and Stratigraphy," *3rd Int. Conf./Workshop on Integrating GIS and Environmental Modeling, WWW and CD*, National Center for Geographic Information and Analysis, Santa Barbara, 1996.

[29] Watson, D.F., "Computing the n-Dimensional Delaunay Tessellation with Application to Voronoi Polytopes," *Comput. J.*, Vol. 24, No. 2, pp. 167–172, 1981.

[30] Weatherhill, N.P., "A Method for Generating Irregular Computational Grids in Multiply Connected Planar Domains," *Int. J. Numer. Methods Fluids*, Vol. 8, pp. 181–197, 1988.

[31] Pepper, D.W. and Stephenson, D.E., "An Adaptive Finite-Element Model for Calculating Subsurface Transport of Contaminant," *Ground Water*, Vol. 33, No. 3, pp. 486–496, 1995.

[32] Babuska, I., Zienkiewicz, O.C., Gago, J., and de Oliveira, E.R., *Accuracy Estimates and Adaptive Refinements in Finite Element Computations*, John Wiley and Sons, New York, 1986.

[33] Zienkiewicz, O.C. and Taylor, R.L., *The Finite Element Method*, 4th ed. Vol. 1, *Basic Formulation and Linear Problems*, McGraw-Hill, London, UK, 1989.

[34] Shapiro, R.A. and Murman, E., "Adaptive Finite Element Methods for the Euler Equations," *AIAA Paper 88-0034*, 1988.

[35] Oden, J.T., Strouboulis, T., and Devloo, P., "Adaptive Finite Element Methods for the Analysis of Inviscid Compressible Flow: Part I. Fast Refinement/Unrefinement and Moving Mesh Methods for Unstructured Meshes," *Comput. Methods Appl. Mech. Eng.*, Vol. 59, pp. 327–362, 1986.

[36] Pelletier, D.H. and Hetu, J., "An Adaptive Finite Element Methodology for Incompressible Viscous Flow," in *Advances in Finite Element Analysis in Fluid Dynamics*. ASME Publications, New York, pp. 1–11, 1992.

[37] Zienkiewicz, O.C., Gago, J., and Kelly, D.W., "The Hierarchical Concept in Finite Element Analysis," *Comput. Struct.*, Vol. 16, pp. 53–65, 1981.

[38] Hackbush, W. and Trottenberg, U., "Multigrid Methods," *Lect. Notes Math.*, Vol. 960, 1982.

[39] Huang, A.B. and Usmani, A.F., *The Finite Element Method for Heat Transfer*, Springer-Verlag, London, UK, 1994.

[40] Teman, R., *Navier-Stokes Equations: Theory and Numerical Analysis*, North-Holland, Amsterdam, 1979.

[41] Girault, V. and Raviart, P.A., *Finite Element Approximations of the Navier-Stokes Equations*, Lecture Series in Mathematics No. 749. Springer-Verlag, Berlin, 1979.

[42] Carey, G.F. and Oden, J.T., *Finite Elements, Vol. 6, Fluid Mechanics*, Prentice-Hall, Englewood Cliff, NJ, 1986.

[43] Gartling, D.K., "A Test Problem for Outflow Boundary Conditions – Flow over a Backward Facing Step," *Int. J. Numer. Methods Fluids*, Vol. 11, pp. 957–967, 1990.

[44] Blackwell, B. and Pepper, D.W., (editors), *Benchmark Problems for Heat Transfer Codes.* ASME Book No. G00709, HTD Vol. 222. ASME Publication, New York, 1992.

[45] Burggraf, O.R., "Analytical and Numerical Studies of the Structure of Steady Separated Flows," *J. Fluid Mech.*, Vol. 24, pp. 113–151, 1966.

[46] Chorin, A.J., "A Numerical Method for Solving Viscous Incompressible Flow Problems," *Comput. Phys.*, Vol. 2, pp. 12–26, 1967.

[47] Yanenko, N.N., *The Method of Fractional Steps*, Springer-Verlag, Berlin, 1971.

[48] Schneider, G.E., Raithby, G.D., and Yovanovich, M.M., "Finite Element Solution Procedures for Solving the Incompressible Navier-Stokes Equations Using Equal Order Interpolation," *Numer. Heat Transfer*, Vol. 1, pp. 435–451, 1978.

[49] Donea, J., Giuliani, S., Laval, H. and Quartapelle, L., "Time-Accurate Solutions of Advection-Diffusion Problems by Finite-Elements," *Comp. Math. Appl. Mech. Engr.*, Vol. 45, pp. 123–145, 1984.

[50] Quartapelle, L., *Numerical Solution of the Navier-Stokes Equations*, Basel-Birkhauser Verlag, Boston, 1993.

[51] Oden, J.T., Kikuchi, N., and Song, Y.J., "Penalty-Finite Element Methods for the Analysis of Stokesian Flows," *Comput. Methods Appl. Mech Eng.*, Vol. 31, pp. 183–224, 1982.

[52] Carey, G.F. and Krishnan, A., "Penalty Finite Element Method for Navier-Stokes Equations," *Comput. Methods Appl. Mech. Eng.*, Vol. 42, pp. 183–224, 1984.

[53] Idelsohn, S., Storti, M., and Nigro, N., "Stability Analysis of Mixed Finite Element Formulations with Special Mention of Equal-Order Interpolate," *Int. J. Num. Meth. Fluids*, Vol. 20, pp. 1003–1022, 1995.

[54] Sani, R.L. and Gresho, P.M., "Reviews and Remarks on the Open Boundary Condition Minisymposium," *Int. J. Numer. Methods Fluids*, Vol. 18, pp. 983–1008, 1994.

[55] Heinrich, J.C., Idelsohn, S.R., Oñate, E., and Vionnet, C.A., "Boundary Conditions for Finite Element Simulations of Convective Flows with Artificial Boundaries," *Int. J. Num. Mech. Engr.*, Vol. 39, pp. 1053–1071, 1996.

[56] Rushak, K.J., "A Method for Incorporating Free Boundaries with Surface Tension in Finite Element Fluid Flow Simulators," *Int. J. Numer. Methods. Eng.*, Vol. 15, pp. 639–648, 1980.

[57] Nickell, R.E., Tanner, R.I., and Caswell, B., "The Solution of Viscous Incompressible Jets and Free-Surface Flows Using Finite Element Methods," *J. Fluid Mech.*, Vol. 65, pp. 189–206, 1974.

[58] Kistler, S.F. and Scriven, L.E., "Coating Flow Theory by Finite Element and Asymptotic Analysis of the Navier-Stokes System," *Int. J. Numer. Methods Fluids*, Vol. 4, pp. 207–229, 1984.

[59] Chippada, S., Jue, T.C., and Ramaswamy, B., "Finite Element Simulation of Combined Buoyancy and Thermocapillary Driven Convection in Open Cavities," *Int. J. Numer. Methods Eng.*, Vol. 38, pp. 335–351, 1995.

[60] Brooks, A. and Hughes, T.J.R., "Streamline Upwind/Petrov-Galerkin Formulations for Convection Dominated Flows with Particular Emphasis on the Incompressible Navier-Stokes Equations," *Comput. Methods Appl. Mech. Eng.*, Vol. 32, pp. 199–259, 1982.

[61] Anderson, C.R., Greengard, C., Greengard, L., and Rokhlin, V., "On the Accurate Calculation of Vortex Shedding," *Phys. Fluids A*, Vol. 2, pp. 883–885, 1990.

[62] Arpaci, V.S. and Larsen, P.S., *Convection Heat Transfer*, Prentice-Hall, Englewood Cliffs, N.J., 1984.

[63] Munson, B.R., Young, D.F., and Okiishi, T.H., *Fundamentals of Fluid Mechanics*, John Wiley and Sons, New York, 1994.

[64] Warsi, Z.U.A., *Fluid Dynamics: Theoretical and Computational Aspects*, CRC Press, Boca Raton, Fla., 1993.

[65] Launder, B.E. and Spalding, D.B., *Lectures in Mathematical Models of Turbulence*, Academic Press, London, 1972.

[66] Rodi, W., *Turbulence Models and Their Application in Hydraulics*, International Association for Hydraulic Research Secretariat, Delft, The Netherlands, 1980.

[67] Sohn, J.L., "Evaluation of FIDAP on Some Classical Laminar and Turbulent Benchmarks," *Int. J. Numer. Methods Fluids*, Vol. 8, pp. 1469–1490, 1988.

[68] Launder, B.E. and Spalding, D.B., "The Numerical Computation of Turbulent Flows," *Comput. Methods Appl. Mech. Eng.*, Vol. 3, pp. 269–289, 1974.

[69] Kim, J., Kline, S.J., and Johnson, J.P., "Investigation of a Reattaching Turbulent Shear Layer: Flow over a Backward Facing Step," *J. Fluid Eng., ASME Trans.*, Vol. 102, pp. 302–308, 1980.

[70] Wilcox, D. C., Turbulence Modeling for CFD, 3rd Ed., DCW Industries, Inc., La Canada, CA, 2006.

[71] Versteegh, T. A. M. and Nieuwstadt, F. T. M., A direct numerical simulation of natural convection between two infinite vertical differentially heated walls scaling laws and wall functions, *International Journal of Heat and Mass Transfer*, Vol. 42, pp. 367–393, 1999.

[72] Boudjemadi, B., Maupu, V., Laurence, D. and Le Quéré, P., Budgets of turbulent stresses and fluxes in a vertical slot natural convection flow at Rayleigh Ra = 10^5 and 5.4×10^5, *International Journal of Heat and Fluid Flow*, Vol. 18, pp. 70–79, 1997.

[73] Smagorinsky, J., General Circulation Experiments with the Primitive Equations. I. The Basic Experiment, *Mon. Weather Rev.*, Vol. 91, pp. 99–164, 1963.

[74] Germano, M., Piomelli, U., Moin, P. and Cabot, W. H., A dynamic subgrid-scale eddy-viscosity model, *Physics of Fluids A*, Vol. 3, pp. 1991.

[75] Ghosal, S., Lund, T. S., Moin, P. and Akselvoll, K., A dynamic localization model for large-eddy simulation of turbulent flows, *Journal of Fluid Mechanics*, Vol. 286, pp. 22955, 1995.

[76] DesJardin, P. E. and Frankel, S. H., Large eddy simulation of a non-premixed reacting jet: Application and assessment of subgrid-scale combustion models, *Physics of Fluids*, Vol. 10, pp. 2298–2314, 1998.

[77] Addad, Y., Benhamadouche, S. and Laurence, D., The negatively buoyant wall-jet: LES results, *International Journal of Heat and Fluid Flow*, Vol. 25, pp. 795–808, 2004.

[78] Muller, S. B. and Kleiser, L., Large-eddy simulation of vortex breakdown in compressible swirling jet flow, *Computers & Fluids*, Vol. 37, pp. 844–856, 2008.

[79] Labbe, O., Maglaras, E. and Garnier, F., Large-eddy simulation of a turbulent jet and wake vortex interaction, *Computers & Fluids*, Vol. 36, pp. 772–785, 2007.

[80] Dinesh, K. K. J. R. and Kirkpatrick, M. P., Study of jet precession, recirculation and vortex breakdown in turbulent swirling jets using LES, *Computers & Fluids*, Vol. 38, pp. 1232–1242, 2009.

[81] Alkishriwi, N., Meinke. M., and Schroder, W., Large-eddy simulation of streamwiserotating turbulent channel flow, *Computers & Fluids*, Vol. 37, pp. 786–792, 2008.

[82] Raufeisen, A., Breuer, M., Botsch, T. and Delgado, A., LES validation of turbulent rotating buoyancy- and surface tension-driven flow against DNS, *Computers & Fluids*, Vol. 38, pp. 1549–1565, 2009.

[83] Zang, Y., Street, R. L. and Koseff, J. R., A dynamic mixed subgrid-scale model and its application to turbulent recirculating flows, *Physics of Fluids*, Vol. 5, pp. 3186–3196, 1993.

[84] Kobayashi, H., Ham, F. and Wu, X., Application of a local SGS model based on coherent structures to complex geometries, *International Journal of Heat and Fluid Flow*, Vol. 29, pp. 640–653, 2008.

[85] Kim, D., Yang. K. and Senda, M., Large eddy simulation of turbulent flow past a square cylinder confined in a channel, *Computers & Fluids*, Vol. 33, pp. 81–96, 2004.

[86] Ouvrard, H., Koobus, B., Dervieux, A. and Salvetti, V. M., Classical and variational multiscale LES of the flow around a circular cylinder on unstructured grids, *Computers & Fluids*, Vol. 39, pp. 1083–1094, 2010.

[87] Deardorff, J. W., A numerical study of three-dimensional turbulent channel flow at large Reynolds numbers, *J. Fluid Mech.*, Vol. 41, pp. 453–480, 1970.

[88] Orszag, S. A., Turbulence and Transition: A Progress Report, *Lecture Notes in Physics*, Vol. 59/1976, pp. 32–51, 1976.

[89] Ferziger, J. H. and Shaanan, S., Effect of anisotropy and rotation on turbulence production, *Phy. Fluids*, Vol. 19, p. 596, 1976.

[90] Leonard, A., Energy cascade in large-eddy simulations of turbulent fluid flows, *Adv. in Geophys.*, Vol. 18, pp. 237–248, 1974.

[91] Clark, R., Ferziger, J. H. and Reynolds, W. C., Evaluation of subgrid-scale models using an accurately simulated turbulent flow, *J. Fluid Mech.*, Vol. 91, pp. 1–16, 1979.

[92] Pope, S., *Turbulent Flows*, Cambridge Univ. Press, 2000.

[93] Lilly, D. K., The representation of small-scale turbulence in numerical simulation experiments, *NCAR MANUSCRIPT* 281, 1966.

[94] Deardorff, J. W., The use of subgrid transport equations in a three-dimensional model of atmospheric turbulence, *ASME J. Fluids Eng.*, Vol. 95, pp. 429–438, 1973.

[95] Bardina, J., Ferziger, J. H. and Reynolds, W. C., Improved turbulence models based on large-eddy simulations of homogeneous, incompressible turbulent flows, *Stanford Univ. Tech. Rept.* No. TF-19, 1983.

[96] Rogallo, R. S. and Moin, P., Numerical Simulation of Turbulent Flows, *Ann. Rev. Fluid Mech.*, Vol. 16, pp. 99–137, 1984.

[97] Schumann, U., Subgrid Scale Model for Finite Difference Simulations of Turbulent Flows in Plane Channels and Annuli, *J. Comp. Phys.*, Vol. 18, p. 376, 1975.

[98] Jimenez, J., Energy transfer and constrained simulations in isotropic turbulence, *Fluid Physics*, World Scientific, 543–557, 1995.

[99] Jimenez, J., Moisy, F., Tabeling, P., and Willaime, H., Scaling and structure in isotropic turbulence, *Intermittency in Turbulent Flows*, Cambridge Univ. Press, 193–212, 2000.

[100] Domaradzki, J. A. and Saiki, E. M., A subgrid-scale model based on the estimation of unresolved scales of turbulence, *Phys. Fluids*, Vol. 9, pp. 2148–2164, 1997.

[101] You, D. and Moin, P., A dynamic global-coefficient subgrid-scale model for large eddy simulation of turbulent scalar transport in complex geometries, *Physics of Fluids*, Vol. 21, pp. 045109, 2009.

[102] Park, N., Lee, S., Lee, J. and Choi, H., A dynamic subgrid-scale eddy viscosity model with a global model coefficient, *Physics of Fluids*, Vol. 18, pp. 2006.

[103] Anderson, J. D., *Hypersonic and High Temperature Gas Dynamics*, McGraw-Hill, New York, 1989.

[104] Brueckner, F. P., "Finite Element Analysis of High-Speed Flows with Application to the Ram Accelerator Concept," PhD Dissertation, Dept. of Aero. and Mech. Eng., Univ. of AZ, 207 p., 1991.

[105] Brueckner, F. P. and Pepper, D. W., "Parallel Finite Element Algorithm for Three-Dimensional Inviscid and Viscous Flow," *J. Thermophys. Heat Transfer*, Vol. 9, pp. 240–246, 1995.

[106] Ramachandran, P. A., *Boundary Element Methods in Transport Phenomena*, Comput. Mech. Pub., Southampton, UK, 1993.

Appendix A

SYMBOLS

Symbol	Meaning
a	width of rectangular element
A	area
$\mathbf{A}$	advection matrix
b	height of rectangular element
$\mathbf{B}$	body force
$B_{x,y}$	body force term components
c	mass fraction of material; Courant number
$\mathbf{C}$	mass (capacity) matrix
c_v	specific heat at constant volume
d	viscous damping coefficient; thickness of fluid
$\mathbf{d}$	degrees of freedom
D	mass diffusion coefficient; material derivative
$\mathbf{D'}$	anisotropic balancing diffusion
D_T	thermal diffusivity
e	internal energy per unit mass
$\|e\|$	energy norm error
f	volumetric heat source/sink; load vector; variable
$\mathbf{F}$	forcing term
Fr	Froude number
g	gravity; arbitrary variable
G	mean surface curvature
$\mathbf{G}$	mass flux vector
Gr	Grashof number
h	element length; convection (film) coefficient; height (free surface)
$I(\,)$	integral
i,j	node numbers
$\mathbf{i,j,k}$	unit vectors in the x, y, and z directions
$\mathbf{J}$	Jacobian matrix
k	turbulence kinetic energy
$\mathbf{K}$	conductance (stiffness) matrix
L	linear operator; reference length
$L_{1,2,3}$	natural coordinate system (triangles; tetrahedrals)
m	mass; total number of elements
M	Mach number; shape function for Petrov-Galerkin formulation
$\mathbf{M}$	mass matrix
n	total number of nodes
$\mathbf{n}$	outward unit vector normal to the boundary
$N(x,y,z)$	shape function
Nu	Nusselt number
p	pressure
P_{ij}	Petrov-Galerkin perturbation
Pe	Peclet number (= Pr.Re)
Pr	Prandtl number

P_s	static component of pressure
Pe	Peclet number
q	heat transfer flux
$\|\|q\|\|$	adaptation error
Q	source/sink term (heat)
R	residual; universal gas constant; outer radius
Ra	Rayleigh number
Re	Reynolds number
S	source/sink term for concentration; free surface
t	time
$\mathbf{t}$	tangent to surface
T	temperature
Ta	Taylor number
u	horizontal velocity (x); unknown variable
v	lateral velocity (y)
$\mathbf{V}$	velocity vector
w	vertical velocity (z)
$w(\mathbf{x})$	weight
x	horizontal Cartesian coordinate
$\mathbf{X}$	unknown variable vector
y	lateral Cartesian coordinate
y_c	characteristic length
z	vertical Cartesian coordinate
α	Petrov-Galerkin optimal value; x, y, z value for trilinear element
β	Petrov-Galerkin stability parameter (1)
γ	Petrov-Galerkin stability parameter (2); specific heat ratio; cell Peclet number
Γ	boundary
Δ	interval (time or space)
δ_{ij}	Kronecker delta
ε	turbulence dissipation rate
ζ	natural (nondimensional) coordinate (z)
η	natural (nondimensional) coordinate (y); maximum value for mesh regeneration error
θ	relaxation parameter
κ	thermal conductivity; von Karman's constant
λ	second coefficient of viscosity; penalty parameter; eigenvalue
μ	dynamic viscosity
μ_t	turbulent eddy viscosity
ν	kinematic viscosity
ϕ	variable; viscous heat dissipation
τ_{ij}	Reynolds stress term
ρ	density
σ_{ij}	stress tensor
ξ	nondimensional distance (x)
ξ_e	mesh enrichment error parameter
ψ	streamfunction
Ω	domain (area or volume)
∇	divergence operator

SUBSCRIPTS

i,j	node numbers; column-row reference in vectors
i,j,k	unit vectors in the x, y, and z directions
u,v,w	velocity components in the x,y,z directions
x,y,z	x,y,z coordinate directions
o	reference (free stream)
h	approximate value
t	time

SUPERSCRIPTS

e	element
n	previous time level
n+1	new time level
p	order of shape function
′	dimensional value; perturbation
.	time dependence
*	exact value

Appendix B

Applications of the finite element method to second order partial differential equations employ the use of Green's theorem to produce the weak formulation of the governing residual equation. Green's theorem allows for the reduction of a second order equation integrated over a volume or area to a first order equation integrated over an area or line, respectively.

Error and convergence estimations of the inner product projections are accomplished with the development of the Sobolev space. This is a space of order m consisting of those functions which are square integrable (belonging to L^2) including their partial derivatives up to and including those of order m and is equipped with an inner product. Therefore, Sobolev spaces are a subset of Hilbert spaces, that is, $H^m \subset L_2$ [B1]. These spaces are defined mathematically by

$$H^m(\Omega) = \left\{ u \in L^2(\Omega) : D^\alpha(u) \in L^2(\Omega) \forall \alpha s.t. |\alpha| \le m \right\}, \tag{1}$$

where the derivatives are $D^\alpha(u)$ defined in a weak (variational) sense.

The Sobolev inner product is defined as

$$(u,v)_{H^m} = \int_\Omega \sum_{|\alpha| \le m} (D^\alpha u)(D^\alpha v) dx \; for \; u,v \in H^m(\Omega), \tag{2}$$

which in turn is used to define the Sobolev norm

$$\|u\|^2_{H^m} = (u,u)_{H^m} = \int_\Omega \sum_{|\alpha| \le m} (D^\alpha u)(D^\alpha v) dx. \tag{3}$$

Also, defining a subspace of the Sobolev space, $H_o^m(\Omega)$, where certain derivatives vanish on the boundary (typical in boundary value problems for momentum, heat and mass transfer)

$$H_o^m(\Omega) = \{u \in H^m(\Omega) : u = 0 \; on \; \Gamma\}, \tag{4}$$

allows for the creation of various semi-norms, and associated bounding theorems. The bounding theorem, Poincare'-Friedrichs inequality, allows for the development of the equivalency of norms in H_o^m and H^m [B1]. This is to say that,

$$\|u\|^2_{L_2} \le C |u|^2_m. \tag{5}$$

The semi-norm $\|\bullet\|^2_m$ is defined as $\sum_{|\alpha|=m} \int_\Omega |D^\alpha u| d\Omega$ and the norm, $\|\bullet\|^2_{L_2}$ is defined as $\|u\|^2_{H^o} = \|u\|^2_{L^2} = \int_\Omega u^2 d\Omega$ with C depending on domain decretization Ω, the curvature of the exact solution, and the degree of polynomial interpolation [B2].

Determining the exactness of the approximated solution is fundamental to all approximation procedures. The robust mathematical nature of the variational methods and the associated norms helps to determine the exactness of the solution. A solution to any projection is as exact as the polynomial's interpolation allows. For example, a measure of "exactness" for temperature, T, can be defined by the norm,

$$\left\| T - \hat{T}_h \right\|_{H^o} \le ch^p, \tag{6}$$

where c is a constant dependent on the domain, 'h' is the size of an element, and 'p' is the order of the approximating polynomial [B3,B4]. This is the result of Cea's Lemma [B1]. This L_2 norm, a semi-norm of the Sobolev space, is defined by the inner product

$$< T - \hat{T}, T - \hat{T} > = \left(\int_{\Omega} \left(T - \hat{T} \right)^2 d\Omega \right)^{1/2} . \tag{7}$$

The error is said to be superconvergent for the Hilbert space inner product (the L_2 norm), since the values at the nodes at which this integral is evaluated are the exact values and the integral is identically satisfied. This is not always true for more complicated problems. This error approximation can also be developed from a Taylor series expansion of the one nodal value in terms of another.

The difficulty with determining the error with measure as currently described is of course that the exact or true solution, 'T', is not usually known. When the true solution is known there is no reason for an approximating method unless one is checking some part of a code for correctness. An approximation to the exact solution can be made if we know that the solution lies within the bounds of the approximated projection or solution [B5].

B.1 MATRIX EQUATIONS AND SOLUTION METHOD

By integrating over each element and combining the contributions from each element to nodes in common to those elements, a matrix equation is formed that can be solved for dependent nodal values [B6].

When integrating over each element and summing, the contributions of the surface flux, $\int_{d\Gamma} d\Gamma$ cancels everywhere except at the boundaries. This is an important distinction between the finite volume methods (FVM) and the classical finite element method. It also leads some to the idea that the FEM statements are not locally conservative. The FEM is conservative when properly stabilized. FVM and discontinuous Galerkin FEM forms on the other hand have truncation error associated with evaluating the surface fluxes everywhere within the domain. In general, the ability of the latter to precisely capture shocks is overshadowed by the complexity of the discretization, especially when looking to develop higher-order approximations.

B.2 TEMPORAL EVOLUTION OF THE SEMI-IMPLICIT SCHEME

B.2.1 Momentum

The FEM forms a system of equations or matrix equations when integrated (assembled) over each element for nodal values. In the following, [ ] refers to a matrix and { } refers to a column vector of the same length. Each element matrices represent a 4×4 operation in 2-D or an 8×8 in 3-D when using bilinear quadrilateral or hexahedral elements, respectively. For implicit methods these operations are assembled (added) as the entire domain is swept or integrated with the resultant matrix equation of size n × n where is n here is the total number of nodes in the domain. These matrix equations can be solved with either a direct method, for example a Cholesky or by a Krylov iterative type method, say conjugate gradient. The latter almost always requires some form of preconditioning. The Cholesky is particular handy when the left-hand side operator is not changing and only a single decomposition is required. This is when θ_2=0. Note that the explicit method is very simple to thread or vectorize and hence the algorithm is attractive for use on supercomputers. In fact, the method is able to run efficiently on chips with multiple processors for which compliers are easily capable of forming an efficient threaded executable.

The initial guess of velocity is marched in time explicitly by [B6]

$$\{\Delta\mathbf{V}_i^*\} = \Delta t\left[\mathbf{M}_v^{-1}\right]\left[\left[\mathbf{A}_u\right]\{\mathbf{V}_i\} + [\mathbf{K}_\tau]\{\mathbf{V}_i\} - \{\mathbf{F}_{v_i}\} - \frac{\Delta t}{2}([\mathbf{K}_{char}]\{\mathbf{V}_i\} - \{\mathbf{F}_{char_i}\})\right]^n, \tag{8}$$

where

$$\{\Delta\mathbf{V}_i^*\} = \{\mathbf{V}_i^*\} - \{\mathbf{V}_i^n\}. \tag{9}$$

Introduction of higher order terms in Eq. (8) is from an integration along the characteristic. Pressure (incompressible flow) is then determined from solving an implicit Poisson equation

$$\left(\left[\mathbf{M}_p\right] + \Delta t^2\theta_1\theta_2\mathbf{H}\right)\{\Delta p_i\} = \Delta t\left[[\mathbf{G}]\left\{\mathbf{V}_i^n\right\} + \theta_1[\mathbf{G}]\left\{\mathbf{V}_i^*\right\}\Delta t\theta_1[\mathbf{H}]\left\{p_i^n\right\} - \left\{\mathbf{F}_{p_i}\right\}\right]. \tag{10}$$

Note, the tensor H, is a weakened second derivative term giving rise to pressure boundary conditions. The values of θ_1 and θ_2 range for ½ to 1 depending on the amount of semi-implicitness desired. If θ_2=0, the system represents an explicit time marching – for compressible form only. The final velocity is obtained with the explicit corrector

$$\{\Delta\mathbf{V}_i\} = \left\{\Delta\mathbf{V}_i^n\right\} - \Delta t\left[\mathbf{M}_u^{-1}\right]\left[\left[\mathbf{G}\right]\left(\left\{p_i^n\right\} + \theta_2\{\Delta p_i\}\right) - \frac{\Delta t}{2}\mathbf{P}\left\{p_i^n\right\}\right]. \tag{11}$$

and again Δ refers to time level differences. Writing the equations to form concise left hand side results in the usual predictor for velocity,

$$\{\mathbf{V}_i^*\} = \{\mathbf{V}_i^n\} - \Delta t[\mathbf{M}_v^{-1}]\left[[\mathbf{A}_u]\{\mathbf{V}_i\} + [\mathbf{A}_\tau]\{\mathbf{V}_i\} - \{\mathbf{F}_{v_i}\} - \frac{\Delta t}{2}\left([\mathbf{K}_{char}]\{\mathbf{V}_i\} - \{\mathbf{F}_{char_i}\}\right)\right]^n \tag{12}$$

But at this point it is more convenient to simply solve for the change or $\left\{\Delta\mathbf{V}_i^*\right\}$ that is employed directly. Using this projection term or Δp, the pressure (or density) is updated,

$$p^{n+1} = p^n + \Delta p. \tag{13}$$

The final velocity is given by corrector

$$\{\Delta\mathbf{V}_i\} = \left\{\mathbf{V}_i^{n+1}\right\} - \left[\mathbf{V}_i^n\right] = \left\{\Delta\mathbf{V}_i^*\right\} - \Delta t\left[\mathbf{M}_u^{-1}\right]\left[[\mathbf{G}]\left(\left\{p_i^n\right\} + \theta_2\{\Delta p_i\}\right) - \frac{\Delta t}{2}\mathbf{P}\left\{p_i^n\right\}\right] \tag{14}$$

The velocity at the n+1st time step in a concise a left hand side is simply

$$\left\{\mathbf{V}_i^{n+1}\right\} = \left\{\mathbf{V}_i^*\right\} - \Delta t\left[\mathbf{M}_u^{-1}\right]\left[\left[\mathbf{G}\right]\left(\left\{p_i^n\right\} + \theta_2\{\Delta p_i\}\right) - \frac{\Delta t}{2}\mathbf{P}\left\{p_i^n\right\}\right] \tag{15}$$

B.2.2 Continuity

For compressible flow mass is advected with the velocity prediction. From Eq. (14) density is then determined,

$$\Delta\rho_i\left(\frac{1}{c^2}\right)^n \Delta p_i = -\left[\mathrm{M}_\rho^{-1}\right]\left(\frac{\partial U^n}{\partial x} + \theta_1\frac{\partial \Delta U^*}{\partial x} - \Delta t\theta_1\left(\frac{\partial^2 p^n}{\partial x\partial x} + \theta_2\frac{\partial^2 \Delta p}{\partial x\partial x}\right)\right). \tag{16}$$

For a thermally perfect gas, enthalpy and internal energy are a function of temperature only and after solving for the energy equation and the temperature is evaluated. Also, γ is the ratio of specific heats which are constant for a calorically perfect gas. Sound speed, 'c' is calculated by

$$c = \sqrt{\gamma RT} \tag{17}$$

where R is the specific gas constant, for air 287 J/kg °K or can be determined by a mass-averaging process. Continuity is maintained from the advection and calculation of the density Poisson equation. Finally, total energy is evolved with source terms explicitly.

B.2.3 Energy

Transport of energy is performed or modeled by the solution of

$$\Delta E_i = \Delta t\left[\mathbf{M}_e^{-1}\right]\begin{bmatrix}\left[\mathbf{A}_e\right]\{\mathbf{E}_\mathbf{i}\}+\left[\mathbf{C}_P\right]\{p_\mathbf{i}\}+\left[\mathbf{K}_\mathbf{T}\right]\{\mathbf{T}_\mathbf{i}\}+\{\mathbf{F}_{e\mathbf{i}}\} \\ \Delta t\left(\left[\mathbf{K}_{uE}\right]\{\mathbf{E}_\mathbf{i}\}+\left[\mathbf{K}_{up}\right]\{p_\mathbf{i}\}+\{\mathbf{F}_{es\mathbf{i}}\}\right)\end{bmatrix}^n. \tag{18}$$

Time step size must be a consideration on this explicit statement. The time scale of most engineering and environmental problems are governed by the faster time scales of turbulence and momentum transport.

B.2.4 Turbulent Kinetic Energy and Specific Dissipation Rate (k-ω)

Scalar transport for turbulent kinetic energy and species is performed and represented by scalar transport. For incompressible flow, time or Reynolds averaging processes are employed. Choosing a two-equation turbulent closure model given by,

$$\Delta k_i = \Delta \mathbf{t}\left[\mathbf{M}^{-1}\right]\begin{bmatrix}\{\mathbf{F}_k\}+\left[\mathbf{P}_k\right]+\{\beta^*\mathbf{kw}\}- \\ \left[\mathbf{K}_k\right]\{k\}-\left[\mathbf{A}_v\right]\{k\}\end{bmatrix}^n, \tag{19}$$

$$\Delta\omega_\mathbf{i} = \Delta \mathbf{t}\left[\mathbf{M}^{-1}\right]\begin{bmatrix}\{\mathbf{F}_\varpi\}+\left[\alpha \mathbf{k}/\mathbf{w}\right]\{\mathbf{P}_k\}+\{\beta\mathbf{w}^2\}- \\ \left[\mathbf{K}_\varpi\right]\{\omega\}-\left[\mathbf{A}_v\right]\{\varpi\}\end{bmatrix}^n. \tag{20}$$

the equation set is capable of producing realistic engineering results for many cases, even when some assumptions about their derivations are violated.

For compressible flow, a density averaged or Farve averaging processes is employed. When a two-equation turbulent closure model is utilized the addition of terms relating to density fluctuations is needed. Specifically since the flow is not divergence free, and has time averaged density variations, the model requires a pressure-dilatation and pressure work term, respectively. These equations are then given as

$$\Delta k_i = \Delta \mathbf{t}\left[\mathbf{M}^{-1}\right]\begin{bmatrix}\{\mathbf{F}_k\}+\left[\mathbf{P}_k\right]+\{\beta^*\mathbf{kw}\}- \\ \left[\mathbf{K}_k\right]\{k\}-\left[\mathbf{A}_v\right]\{k\}+ \\ \{p\}\left[\mathbf{C}\right]\{u\}-\{u\}\left[\mathbf{C}\right]\{p\}\end{bmatrix}^n, \tag{21}$$

$$\Delta\omega_\mathbf{i} = \Delta \mathbf{t}\left[\mathbf{M}^{-1}\right]\begin{bmatrix}\{\mathbf{F}_\varpi\}+\left[\alpha \mathbf{k}/\mathbf{w}\right]\{\mathbf{P}_k\}+\{\beta\mathbf{w}^2\}- \\ \left[\mathbf{K}_\varpi\right]\{\omega\}-\left[\mathbf{A}_v\right]\{\varpi\}\end{bmatrix}^n. \tag{22}$$

The explicit and implicit equations for velocity and pressure are always solved to the boundaries, based on the latest update to the boundary conditions, i.e., those boundaries which are changing with the flow. These are the turbulent closure model boundary conditions k, ω and μ_t, When using the law of the wall, the k−ω equations are solved to the point next to the solid boundary because the boundary for these points is determined by a wall law or function. Otherwise the model can be solved to all boundaries provided the grid resolution is sufficient enough to provide for accurate solution in the boundary layer.

B.2.5 Matrix Formulation

The individual matrices for these equations are defined as

$$\mathbf{M}_v = \int_\Omega \{N_k\} \sum_{l=1}^{nodes} [N_l] d\Omega, \tag{23}$$

$$\mathbf{M}_p = \int_\Omega \{N_k\} \left(\frac{1}{c^2}\right)^n \sum_{l=1}^{nodes} [N_l] d\Omega, \tag{24}$$

$$\mathbf{M}_e = \int_\Omega \{N_k\} \sum_{l=1}^{nodes} [N_l] d\Omega \tag{25}$$

$$\mathbf{M}_e = \int_\Omega \{N_k\} \sum_{l=1}^{nodes} [N_l] d\Omega \tag{26}$$

$$\mathbf{K}_\tau = \int_\Omega \{N_i\} \left(\nabla \left(\mathbf{V}_k [N_k]\right)\right) d\Omega + \int_\Omega \left([\mu + \mu_t] \left(\frac{\partial N_j}{\partial x_i} \frac{\partial N_i}{\partial x_j} - \frac{2}{3} \delta_{ij} \frac{\partial N_j}{\partial x_i} \right) - \frac{2}{3} \delta_{ij} \rho k \right) d\Omega, \tag{27}$$

$$\mathbf{K}_{char} = \left(\int_\Omega \left[\frac{\partial N_k}{\partial x_l} \right] \{u_l\} \left(\nabla \left(\mathbf{V}_k [N_k]\right)\right) d\Omega d\Omega \right) \tag{28}$$

$$\mathbf{C} = \int_\Omega \left[\frac{\partial N_j}{\partial x_i} \right] N_i d\Omega, \tag{29}$$

$$\mathbf{F}_v = \int_\Omega N_j f(x_i) d\Omega + \int_\Gamma [\mu + \mu_t] N_j \mathbf{n}_j \frac{\partial \mathbf{V}_j}{\partial x_j} d\Gamma, \tag{30}$$

$$\mathbf{F}_{char} = \int_\Omega \left[\frac{\partial \mathbf{V}_k}{\partial x_l} \right] (u_l) \{\rho g_l\} d\Omega \tag{31}$$

$$\mathbf{K}_T = -\left(\int_\Omega \{N_i\} \left[\frac{\partial N_j}{\partial x_i} \right] \left\{ \frac{\mu_t}{\Pr_t} \right\} \left[\frac{\partial N_j}{\partial x_i} \right] d\Omega \right) + \int_\Omega \left(\left[\kappa + \frac{\mu_t}{\Pr_t} \right] \frac{\partial N_i}{\partial x_j} \frac{\partial N_i}{\partial x_j} \right) d\Omega, \tag{32}$$

$$\mathbf{K}_k = -\left(\int_{\Omega} \{N_i\} \left[\frac{\partial N_j}{\partial x_i} \right] \{\sigma^* \mu_t\} \left[\frac{\partial N_j}{\partial x_i} \right] d\Omega \right) \tag{33}$$

$$+ \int_{\Omega} \left(\left[\mu + \sigma^* \mu_t \right] \frac{\partial N_i}{\partial x_j} \frac{\partial N_i}{\partial x_j} \right) d\Omega$$

$$\mathbf{K}_{\varpi} = -\left(\int_{\Omega} \{N_i\} \left[\frac{\partial N_j}{\partial x_i} \right] \{\sigma \mu_t\} \left[\frac{\partial N_j}{\partial x_i} \right] d\Omega \right) \tag{34}$$

$$+ \int_{\Omega} \left(\left[\mu + \sigma \mu_t \right] \frac{\partial N_i}{\partial x_j} \frac{\partial N_i}{\partial x_j} \right) d\Omega$$

$$\mathbf{F}_T = -\left(\rho \int_{\Omega} \{N_i\}\{Q_i\} d\Omega \right) \left(\int_{\Gamma} \{N_i\}[N_j]\{q_i\} d\Gamma \right), \tag{35}$$

$$\mathbf{P}_k = -\int_{\Omega} \left[\{N_j\}\{\mu_{t_i}\} \left(\left[\frac{\partial N_j}{\partial x_i} \right] \{u_i\} + \left[\frac{\partial N_i}{\partial x_j} \right] \{u_i\} - \frac{2}{3} \frac{\partial N_k}{\partial x_k} \{u_k\} d_{ij} \right) - \delta_{ij} \frac{2}{3} \rho [N_j] \{k_i\} \right] d\Omega. \tag{36}$$

REFERENCES

[B1] Reddy, D. B., *Introductory Functional Analysis with Applications to Boundary Value Problems and Finite Elements*, Springer-Verlag, N.Y., NY, 1998.

[B2] Eriksson, K., Estep, D., Hansbo, P., and Johnson, C., *Computational Differential Equations*, Cambridge University Press, 1996.

[B3] Hughes, Thomas. J. R., *The Finite Element Method – Linear Static and Dynamic Finite Element Analysis*, Prentice-hall, Inc, Englewood Cliffs, New Jersey, pp. 26–27, 1987.

[B4] Carey, Graham F., *Computational Grids – Generation, Adaptation, and Solution Strategies*, Taylor & Francis, Washington, D.C., pp. 161–168, 1997.

[B5] Zienkiewicz, O. C., Zhu, J. Z., and Gong, N.G., "Effective and Practical h-p Version Adaptive Analysis Procedures for the Finite Element Method," *International Journal for Numerical Methods in Engineering*, Vol. 28, John Wiley & Sons, Ltd., pp. 879–891, 1989.

[B6] Carrington, D. B., "An h-adaptive k-w Finite Element Model for Turbulent Thermal Flow," Los Alamos National Laboratory, report LA-UR-06-8432, Los Alamos, NM, 2007.

PART II
THE BOUNDARY ELEMENT METHOD

Chapter 1

INTRODUCTION

The boundary element method (BEM) is an integral-equation-based numerical technique that in many cases offers several advantages over Finite Difference Methods (FDM), Finite Volume Methods (FVM), or Finite Element Methods (FEM). The BEM relies on the formulation of a boundary integral equation (BIE) for the field problem to be analyzed, and specifically, this is predicated on the availability of the Green's free-space solution for the problem of interest. One of the most striking features of BEM is that, for many field problems of engineering, a boundary integral equation is discretized to solve the field problem of interest. Consequently, only the bounding surface of the domain is discretized, thereby reducing the dimension of the problem by one. For instance, in the analysis of linear and non-linear isotropic steady-state heat conduction without heat generation, a boundary discretization is only required to resolve the temperature field.

The unknowns appearing in the discretized problem are the field variable (temperature) and its normal derivative (related to the heat flux via Fourier's law of heat conduction) values at the boundary. Thus, for a certain class of problems, for which Green's free-space solutions are available, the BEM solution can be expressed in terms of boundary integrals only. Theoretical background and numerical implementation of the BEM can be found in the monographs by Brebbia and Walker [1], Brebbia et al. [2], Gipson [3], Liggett and Liu [4], Banerjee [5], Divo and Kassab [6], Brebbia and Dominguez [7], and Wrobel and Aliabadi [8]. Recent reviews of the BEM in heat transfer can be found in Kassab and Wrobel [9] and Kassab et al. [10].

This section of the book presents the basic principles of heat transfer modeling using the boundary element method (BEM). The foundations of the method are presented for steady-state heat conduction in homogeneous, isotropic media, with no internal heat generation. The mathematical model is simply the Laplace equation, and the presentation follows classical concepts of potential theory. A detailed formulation is provided. The Numerical implementation of the BEM is treated in detail in Chapter 2. The following chapters address specific issues in BEM application to heat transfer; for instance, modeling energy generation while retaining the boundary-only feature of the BEM. This is followed by application of the BEM to transient heat conduction by the Laplace transform method coupled with numerical inversion of the Laplace transform, and by the dual-reciprocity method coupled to a time-marching scheme. Then, several types of non-linearities which are common in heat transfer problems are discussed and their BEM formulation described in some detail.

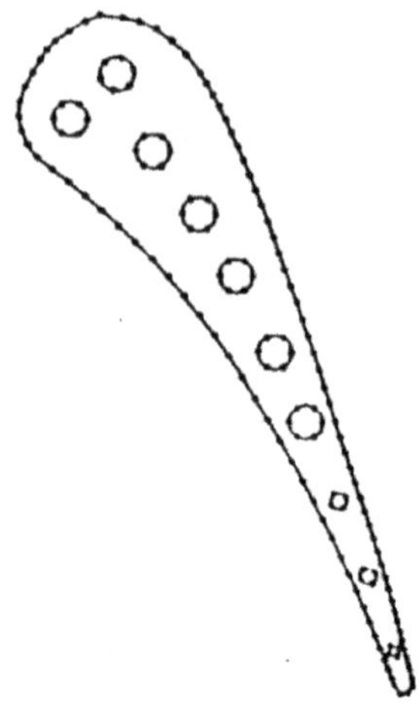

Figure 1. A BEM discretization of an internally cooled turbine blade suitable for heat conduction analysis.

Chapter 2

BEM FUNDAMENTALS

The BEM came of age in the early 1960's when its indirect formulation and implementation was developed in the aerospace industry in the pioneering work of Hess and Smith [11] into the panel method and in the work of Jaswon [12,13] in potential theory. The indirect BEM refers to the fact that the unknowns appearing in the discretized integral equations are indirectly related to the field variable. For instance, in panel methods, the magnitudes strengths of points sources, sinks, vortices or doublets at sought on the boundary to determine the potential, and these quantities have no physical meaning. As a matter of fact, the concepts underlying the indirect BEM actually may be thought to date back to the developments of the Prandtl lifting line theory where horseshoe vortices of appropriate strengths were superimposed along the span of a wing to provide the effective flow field and lift.

The direct BEM saw its early developments in the works of Cruze [14–17], Rizzo [18–21] and Shaw [22,23] in mechanics, heat transfer, and acoustics in what was at the time called the Boundary Integral Equation Method (BIEM). In this case, the direct BEM indicates that the variables that appear in the discretized integral equations indeed do have a physical meaning, for instance in mechanics the nodal displacement and the traction are sought on the boundary while in heat transfer, the nodal values of the temperature and heat flux are sought at the boundary.

The method adopted its current name, the Boundary Element Method (BEM), after the 1978 publication of the first textbook on the BEM *The Boundary Element Method for Engineers* by Brebbia [24] and subsequent publications from Brebbia and colleagues [1,2,7]. Brebbia provided an interpretation of the BEM through the method of weighted residuals (MWR) framing the method in a familiar context for engineers trained in the finite element method (FEM). An excellent and comprehensive historical review of the BEM is provided in Cheng and Cheng [25]. The direct BEM is closely related to Green's function methods and can be thought of as a numerical Green's function method. As a matter of fact, all BEM formulations rely on the existence and knowledge of what is called the Green's free space solution for the problem under consideration.

We will focus on the direct BEM that is now introduced and related to the Green's Function method for solving partial differential equations. The BEM is considered as a numerical approach to the Green's function method. To this end, we specifically examine a 2D steady heat conduction problem.

2.1 A FAMILIAR EXAMPLE: GREEN'S THIRD IDENTITY FOR POTENTIAL PROBLEMS

Let us first consider a familiar example of Green's third identity for problems that are governed by the Laplace equation for the potential, φ,

$$\nabla^2\varphi(\xi) = 0 \tag{1}$$

that is the following integral relation in 2D, see Kellog [26]

$$\varphi(\xi)=\frac{1}{2\pi}\oint_{\Gamma(x)}\left[\varphi(x)\frac{\partial\, ln(r(x,\xi))}{\partial n}-ln\, r(x,\xi)\frac{\partial\varphi(x)}{\partial n}\right]d\Gamma(x) \tag{2}$$

states that the potential φ at any interior point ξ to the domain $\Omega(x)$ is given by the oriented contour integral taken in the counterclockwise sense around the domain boundary $\Gamma(x)$ of the value of the normal derivative of the potential $\partial\varphi/\partial n$ and the potential itself φ weighted by $(-1/2\pi)\, ln\,(r(x,\xi))$ and by its derivative respectively, see Fig. 2. In Eq. (2), the Euclidan distance between the boundary point x and the field point ξ is denoted by $r(x,\xi)$, a notation utilized throughout the BEM chapters of part II of this book. The weight appearing in this equation, $(-1/2\pi)\, ln\,(r(x,\xi))$, is what is technically called the **Green's free-space solution** or **fundamental solution** for the Laplace equation in 2D. There is a corresponding fundamental solution in 3D for this problem, and in the sequel, we will find how this expression is derived and how it comes to play such a role in the relation we just considered.

Although this is an exact result, it is not immediately useful in determining the potential. This is due to the fact that in a well-posed problem, we specify the potential at any point on the boundary (Dirichlet condition), its derivative (Neumann condition), or the ratio of the two (Robin condition). In any case, we never know both the potential and its derivative at every point on the boundary for us to utilize Green's third identity. In order to resolve this problem there are two approaches that can be taken:

1. *Analytical Approach:* Green's function method for PDE's [27–30]
   Here a boundary integral formulation is obtained by constructing the Green's function for the problem whereby the weighting term (Green's free space solution) is augmented in such a way that the contour integral vanishes on portions of the boundary where we do not know the particular value needed in Eq. (2). This modified weighting term is called **the Green's function** for the problem and is associated with a particular geometry and specification of boundary conditions. The Green's function can be found by several methods, typically by solving an appropriate auxiliary problem that lends itself to analytical solution. As such, the analytical Green's function method is limited by geometry, for example, separable geometries, half spaces and such geometries that can be tackled by the method of images.

2. *Numerical Approach:* the BEM [1–25]
   Here, a singular boundary integral equation is formulated relating the values of the dependent variable and its derivative at the boundary. For instance, Eq. (2)

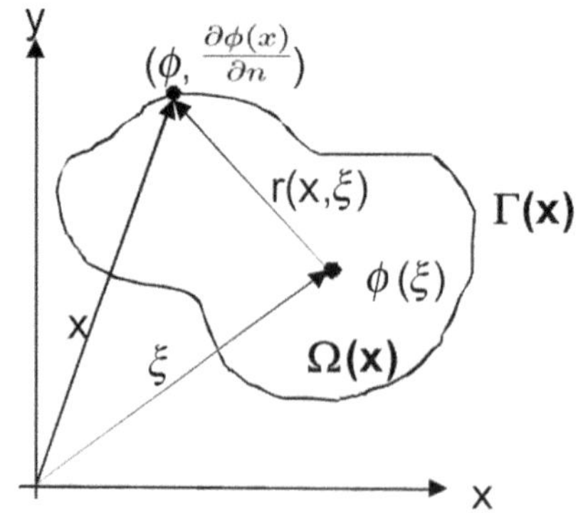

Figure 2. Green's third identity for potential problems.

is applied in a limiting fashion by taking the point ξ to the boundary. The integral equation is discretized by parametrizing the geometry, the variable and its derivative using shape functions in the same fashion developed for FEM. The discrete boundary value problem is solved, and the discrete version of Eq. (2) can then be applied to any interior point ξ. This method has no geometric limitation, only the requirement to know the **Green's free space solution** for the differential equation.

The two methods are theoretically connected, and we now look into the details of the BEM and its relation to the analytical Green's function method.

2.2 THE 2D HEAT CONDUCTION PROBLEM

Let us consider a 2D steady state heat conduction problem with a known energy generation term $u_G(x,y)$ in a rectangular $[L_x \times L_y]$ region with constant thermal conductivity, k, whose side walls are imposed with a heat flux and with temperatures prescribed at the upper and lower walls as illustrated in Fig. 3. The governing equation and boundary conditions for this problem are:

$$\nabla^2 T(x,y) + \frac{1}{k} u_G(x,y) = 0 \tag{3}$$

$$T\big|_{\Gamma_1} = T_1 \tag{4}$$

$$q\big|_{\Gamma_2} = q_2 \tag{5}$$

$$T\big|_{\Gamma_3} = T_3 \tag{6}$$

$$q\big|_{\Gamma_4} = q_4 \tag{7}$$

The boundary, $\Gamma = \Gamma_1 \cup \Gamma_2 \cup \Gamma_3 \cup \Gamma_4$, encloses the domain, Ω. We now proceed from the partial differential equation formulation of the problem to an integral equation formulation and its subsequent solution by an analytical approach (Green's function method) and by a numerical approach (BEM).

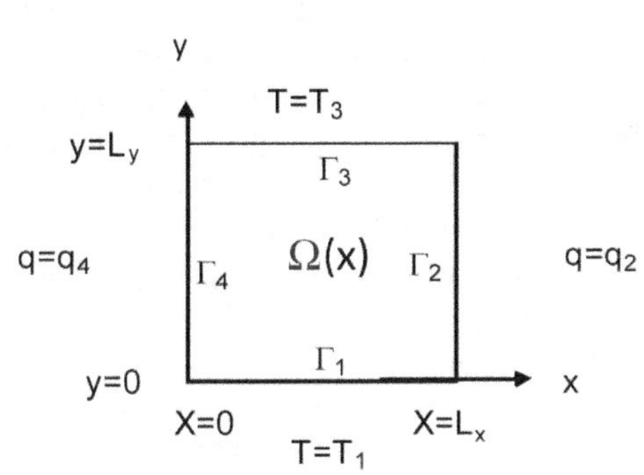

Figure 3. Steady state heat conduction problem in a rectangular region.

2.3 GENERATING THE INTEGRAL EQUATION: *WEIGHTING FUNCTION AND GREEN'S SECOND IDENTITY*

The first step in the general procedure to generate the integral equation for the problem under consideration is to introduce a weighting function that we hereon identify as T^* and to integrate the product of the differential equation and the test function over the domain of the problem (space in our case and space and time for time-dependent problems),

$$\oiint_{\Omega(x)} \left[\nabla^2 T(x) + \frac{1}{k} u_G(x) \right] T^*(x,\xi) d\Omega(x) = 0 \tag{8}$$

here the independent variable x generically denotes space dependence, for instance $x = (x, y)$ in 2D and $x = (x, y, z)$ in 3D. Notice that the weighting function, T^*, for our steady problem specifically depends on what is called the field point x and source point ξ located at a fixed position within the problem domain $\Omega(x)$, see Fig. 4. We will soon see why that is the case. Technically, mathematicians refer to this procedure as taking the inner product of the differential equation and the weighting function.

The second step involves utilizing Green's second identity (which amounts to integration by parts twice) to switch the differential linear operator ($L = \nabla^2$) from T to T^*. This identity applied to our 2D problem states that

$$\begin{aligned} \oiint_{\Omega(x)} \nabla^2 T(x) T^*(x,\xi) d\Omega(x) &= \oiint_{\Omega(x)} \nabla^2 T^*(x,\xi) T(x) d\Omega(x) \\ &+ \oint_{\Gamma(x)} \left[T^*(x,\xi) \frac{\partial T(x)}{\partial n} - T(x) \frac{\partial T^*(x,\xi)}{\partial n} \right] d\Gamma(x) \end{aligned} \tag{9}$$

where $\partial/\partial n$ denotes the normal derivative with respect to the outward-drawn normal to the boundary, $\Gamma(x)$. So that we now have the exact relation

$$\begin{aligned} \oiint_{\Omega(x)} \nabla^2 T^*(x,\xi) T(x) d\Omega(x) + \oint_{\Gamma(x)} \left[T^*(x,\xi) \frac{\partial T(x)}{\partial n} - T(x) \frac{\partial T^*(x,\xi)}{\partial n} \right] d\Gamma(x) \\ + \oiint_{\Omega(x)} T^*(x,\xi) \frac{1}{k} u_G(x) d\Omega(x) = 0 \end{aligned} \tag{10}$$

Mathematically, the differential operator acting on the weight function T^* as a result of this manipulation is called the **adjoint operator** ($L^* = \nabla^2$), and in our problem it is the same operator that acts on the temperature ($L = \nabla^2$). In such a case, where $L = L^*$, the differential operator of the governing equation is said to be **self-adjoint**. When $L \neq L^*$ then the operator of the governing equation is said to be **non-adjoint**. Self adjoint operators arise in differential equations that have only even derivatives, while non-adjoint operators arise in differential equations that have odd derivatives that are associated with irreversibilities.

The third step is now to judiciously choose the test function such that the relation in Eq. (10) involves only boundary integration of unknown quantities. To this end, the key step is to choose the weight function T^* a solution of the adjoint equation perturbed by Dirac Delta function acting at the source point, ξ, located at the interior of $\Omega(x)$, that is in our case

$$\nabla^2 T^*(x,\xi) = -\delta(x-\xi) \tag{11}$$

The reason why this choice is made is that the Dirac Delta function is a generalized function [31,32] defined by how it operates on other functions. In particular, the Dirac Delta function is uniquely defined by its sifting property

$$\int_{-\infty}^{+\infty} \delta(x-\xi) f(x)dx = f(\xi) \tag{12}$$

and also by the requirement that when, $f(x) = 1$, the above integrates to a value of one. So that making by making the choice for the weighting function satisfying Eq. (11)

$$\oiint_{\Omega(x)} \nabla^2 T^*(x,\xi)T(x)d\Omega(x) = \oiint_{\Omega(x)} -\delta(x,\xi)T(x)d\Omega(x) = -T(\xi) \tag{13}$$

then Eq. (10) reduces to,

$$T(\xi) = \oint_{\Gamma(x)} \left[\frac{\partial T(x)}{\partial n} T^*(x,\xi) - T(x)\frac{\partial T^*(x,\xi)}{\partial n} \right] d\Gamma(x) + \oiint_{\Omega(x)} T^*(x,\xi)\frac{1}{k}u_G(x)d\Omega(x) \tag{14}$$

Notice that with $u_G(x) = 0$ and the choice

$$T^*(x,\xi) = -\frac{1}{2\pi} ln(r(x,\xi)) \tag{15}$$

the above equation is Green's third identity (see Eq. (2)) for the potential problem corresponding to

$$\nabla^2 T(x) = 0 \tag{16}$$

Indeed, the $T^*(x,\xi)$ given in Eq. (15) is the solution to the adjoint equation Eq. (11), and it is called the **Green's free space solution** or **fundamental solution** for the Laplace equation. A derivation of the fundamental solution for the Laplace equations in 2D and 3D is provided in the Appendix of this chapter. The development so far then explains how Green's third identity is obtained. Expanding Eq. (14) over the rectangular geometry of our problem, we find that the temperature at any interior point is related to the weighted integrals of temperature and its normal derivative at the boundary and the weighted integral of the energy source over the domain.

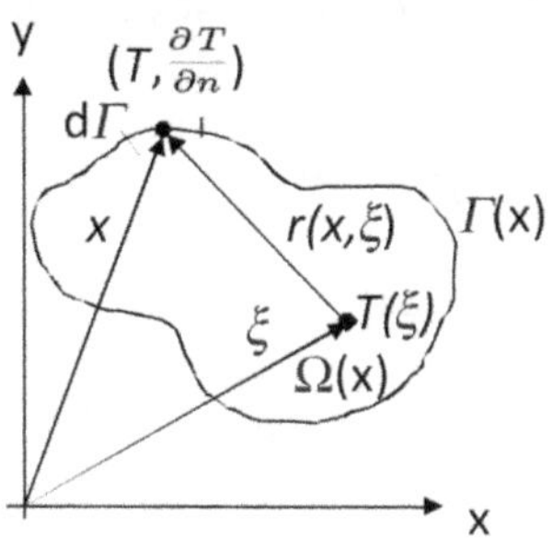

ξ – source point located anywhere in the interior domain Ω(x) and not on the boundary Γ(x)
X – field point located on the boundary Γ(x)

Figure 4. The boundary integral relationships for heat conduction.

$$
\begin{aligned}
T(\xi) = & \int_{\Gamma_1} \frac{\partial T(x)}{\partial n} T^*(x,\xi) d\Gamma(x) + \int_{\Gamma_2} \frac{\partial T(x)}{\partial n} T^*(x,\xi) d\Gamma(x) \\
& + \int_{\Gamma_3} \frac{\partial T(x)}{\partial n} T^*(x,\xi) d\Gamma(x) + \int_{\Gamma_4} \frac{\partial T(x)}{\partial n} T^*(x,\xi) d\Gamma(x) \\
& - \int_{\Gamma_1} T(x) \frac{\partial T^*(x,\xi)}{\partial n} d\Gamma(x) - \int_{\Gamma_2} T(x) \frac{\partial T^*(x,\xi)}{\partial n} d\Gamma(x) \\
& - \int_{\Gamma_3} T(x) \frac{\partial T^*(x,\xi)}{\partial n} d\Gamma(x) - \int_{\Gamma_4} T(x) \frac{\partial T^*(x,\xi)}{\partial n} d\Gamma(x) \\
& + \oiint_{\Omega(x)} \frac{1}{k} u_G(x) T^*(x,\xi) d\Omega(x)
\end{aligned} \tag{17}
$$

The issue raised in applying the above relation is that, at any point on the boundary for a well-posed heat conduction problem, we specify only one of the following typical conditions

$$T(x) = \bar{T} \quad \text{on} \quad \Gamma_T \tag{18}$$

$$\frac{\partial T}{\partial n}(x) = \bar{q} \quad \text{on} \quad \Gamma_q \tag{19}$$

$$\frac{\partial T(x)}{\partial n} + HT(x) = HT_\infty \quad \text{on} \quad \Gamma_h \tag{20}$$

Where, $\bar{q} = -q_s/k$, with q_s as surface heat flux and k the thermal conductivity, and $H = h/k$, with h as the convective heat transfer coefficient and T_∞ as the ambient temperature. Consequently, four of the eight boundary integrals cannot be evaluated. We now consider two approaches to resolve this problem.

2.4 ANALYTICAL SOLUTION: *GREEN'S FUNCTION METHOD AND THE AUXILIARY PROBLEM*

An obvious way to resolve the problem is to adjust the weighting function such that not only does it satisfy the adjoint equation Eq. (11) but also certain boundary conditions that eliminate the integrals we cannot evaluate in Eq. (17). So that if we require that T^* satisfies

$$\nabla^2 T^*_{GF}(x,\xi) = -\delta(x-\xi) \tag{21}$$

$$\left.\frac{\partial T^*_{GF}}{\partial n}(x,\xi)\right|_{\Gamma_1} = 0 \tag{22}$$

$$\left.T^*_{GF}(x,\xi)\right|_{\Gamma_2} = 0 \tag{23}$$

$$\left.\frac{\partial T^*_{GF}}{\partial n}(x,\xi)\right|_{\Gamma_3} = 0 \tag{24}$$

$$T^*_{GF}(x,\xi)\Big|_{\Gamma_4} = 0 \tag{25}$$

where at this point T^* is identified as **the Green's function**, T^*_{GF} for the problem, then temperature is explicitly evaluated as

$$\begin{aligned} T(\xi) = & -\int_{\Gamma_1} T(x,\xi)\frac{\partial T^*_{GF}(x)}{\partial n}d\Gamma(x) + \int_{\Gamma_2} \frac{\partial T(x,\xi)}{\partial n}T^*_{GF}(x)d\Gamma(x) \\ & -\int_{\Gamma_3} T(x,\xi)\frac{\partial T^*_{GF}(x)}{\partial n}d\Gamma(x) + \int_{\Gamma_4} \frac{\partial T(x,\xi)}{\partial n}T^*_{GF}(x)d\Gamma(x) + \oiint_{\Omega(x)} \frac{1}{k}u_G(x)T^*_{GF}(x,\xi)d\Omega(x) \end{aligned} \tag{26}$$

The problem now is relegated to finding the problem Green's function, T^*_{GF}. The usual manner in which the Green's function is determined is by induction. That is, we set-up and solve an *auxiliary problem* where the boundary conditions of original problem under consideration are rendered homogeneous, but, the governing equation is kept non-homogeneous if a source term exists originally or is rendered non-homogeneous if the original problem did not have a source term. In our case, the auxiliary problem corresponding to Eq. (3)–(7) is

$$\nabla^2 T_{aux}(x,y) + \frac{1}{k}u_G(x,y) = 0 \tag{27}$$

$$T_{aux}\Big|_{\Gamma_1} = 0 \tag{28}$$

$$q_{aux}\Big|_{\Gamma_2} = 0 \tag{29}$$

$$T_{aux}\Big|_{\Gamma_3} = 0 \tag{30}$$

$$q_{aux}\Big|_{\Gamma_4} = 0 \tag{31}$$

whose solution by variation of parameters can be written in the form (see Appendix A of this chapter)

$$T_{aux}(\xi) = \oiint_{\Omega(x)} \frac{1}{k}u_G(x)\left(\frac{1}{N(\delta_n)N(\beta_m)}\sum_n\sum_m W_{nm}(x)W_{nm}(\xi)\right)d\Omega(x) \tag{32}$$

where $W_{nm}(x)$ is the problem eigenfunction associated with the eigenvalues δ_n and β_m and $N(\delta_n)$ and $N(\beta_m)$ are normalizing integrals, see Ozisik [27]. By comparison to the solution of the auxilliary problem given by Eq. (26) with homogeneous conditions,

$$T_{aux}(\xi) = \oiint_{\Omega(x)} T^*_{GF}(x,\xi)\frac{1}{k}u_G(x)d\Omega(x) \tag{33}$$

the Green's function for the problem is identified by inspection as

$$T^*_{GF}(x,\xi) = \frac{1}{N(\delta_n)N(\beta_m)}\sum_n\sum_m W_{nm}(x)W_{nm}(\xi) \tag{34}$$

and can subsequently be introduced into Eq. (26) to find the temperature for our problem with non-homogeneous boundary conditions.

Example 2.1: Green's function method in for conduction in a square. Let us consider a specific set of boundary conditions for the example problem. Specifically consider no generation, adiabatic side walls and imposed temperature of $T_1 = 0$ and $T_3 = 100$, in a square region $L_x = L_y = 1$, that is

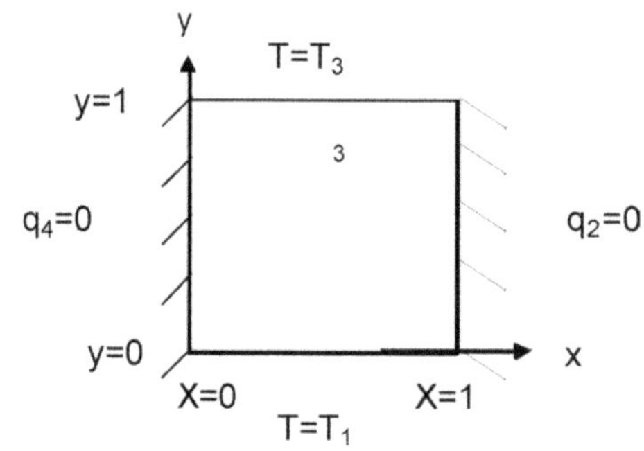

$$\nabla^2 T(x,y) = 0 \tag{35}$$

$$T\big|_{\Gamma_1} = T_1 \tag{36}$$

$$q\big|_{\Gamma_2} = 0 \tag{37}$$

$$T\big|_{\Gamma_3} = T_3 \tag{38}$$

$$q\big|_{\Gamma_4} = 0 \tag{39}$$

The analytical solution of the problem is

$$T(x,y) = T_1 + (T_3 - T_1)\frac{y}{L_y} \tag{40}$$

The Green's function for the problem is

$$\begin{aligned} T^*_{GF}(x,y,x',y') &= \sum_{m=1}^{\infty} \frac{2}{(\lambda_{0,m}L)^2} \sin(\delta_m y)\sin(\delta_m y') \\ &+ \sum_{n=1}^{\infty}\sum_{m=1}^{\infty} \frac{4}{(\lambda_{n,m}L)^2} \cos(\beta_n x)\cos(\beta_n x')\sin(\delta_m y)\sin(\delta_m y') \end{aligned} \tag{41}$$

Where $\beta_n = \dfrac{n\pi}{L}$, $\delta_m = \dfrac{m\pi}{L}$, and $\lambda^2_{n,m} = \beta_n^2 + \delta_m^2$. A plot of the Green's function for the problem evaluated at (x',y') = (0.5,0.5) and at (x',y') = (0.25,0.75) clearly reveals the singularity at source point, see Fig. 5.

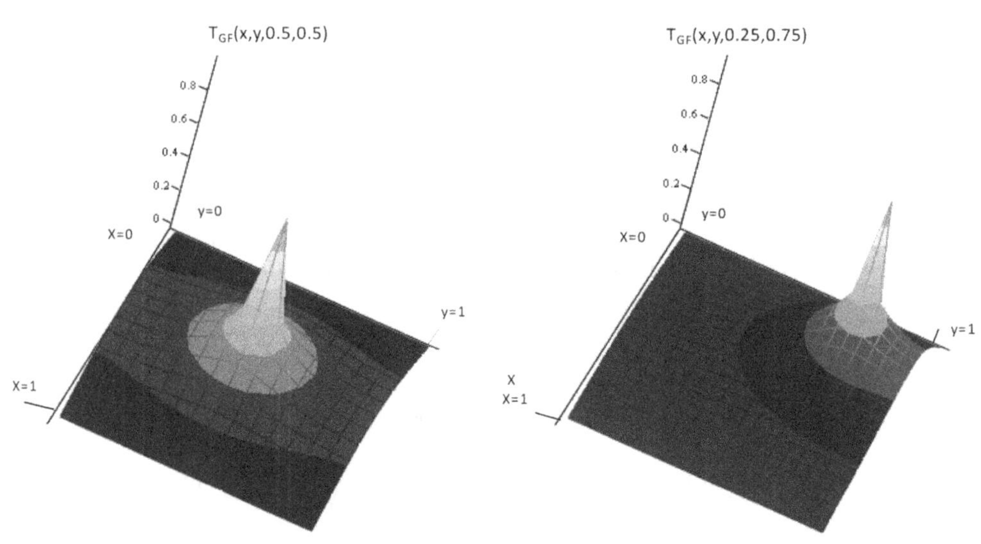

Figure 5. Plot of the Green's function, Eq. (41), at (x',y') = (0.25,0.5) and (x',y') = (0.5,0.75).

The Green's function solution is from Eq. (26), with the imposed boundary conditions of this problem,

$$T(x,y) = -\int_0^L T_1\left(-\frac{\partial T_{GF}^*(x,y,x',y'=0)}{\partial y}\right)dx' - \int_0^L T_3\left(\frac{\partial T_{GF}^*(x,y,x',y'=L)}{\partial y}\right)dx' \quad (42)$$

and, evaluating the integrals we find that the Green's function solution is

$$\begin{aligned} T(x,y) = T_1 \sum_{m=1}^{\infty} \frac{2\beta_m}{(\lambda_{0,m}L)^2} \sin(\beta_m y)\cos(\beta_m y')\Big|_{y'=0} \\ - T_3 \sum_{m=1}^{\infty} \frac{2\beta_m}{(\lambda_{0,m}L)^2} \sin(\beta_m y)\cos(\beta_m y')\Big|_{y'=L} \end{aligned} \quad (43)$$

These series converge slowly, and taking 2000 terms, the first and second series can be evaluated easily to sum to the following values in the table below. These values can be used to compute the resulting temperature for different values at T_1 and T_3.

y-location	First series at y′ = 0	Second series at y′ = L	Temperature ($T_1 = 0$, $T_3 = 100$)
0.25	0.74962	–0.24993	24.993
0.5	0.49984	–0.49984	49.984
0.75	0.24993	–0.74962	74.962

It is the need to solve the auxiliary problem that limits the analytical Green's function method to simple geometries. That is, any problem posed over a domain whose geometric boundaries cannot be framed in a separable coordinate system, for instance, Cartesian, cylindrical, spherical, or whose geometry does not lend itself to alternative method of images (semi-infinite media with first or second kind BC's) or similar such construction see Greenberg [28], cannot be solved by this method. Essentially, the analytical Green's function method is superposition carried out in one integral relation and is limited by the same constraints as the method of separation of variables and superposition. An excellent reference on the Green's function method applied to heat conduction is Beck et al. [29]. We now turn to a more general method that relies on a numerical approach and is free of geometric constraints.

2.5 NUMERICAL SOLUTION: *THE BEM AND THE BOUNDARY INTEGRAL EQUATION*

Alternatively to the analytical Green's function method that is constrained to relatively simple geometries due to the need to find the Green's function for the problem, we can form a boundary value problem for the unknown quantities at the boundary by taking the source point ξ from the interior to the boundary and by evaluating the resulting boundary integral equation which is now singular as $r(x,\xi) \to 0$ over the contour integral. In a limiting process [2,3,7,24], it can be shown that the boundary integral equation (BIE) is

$$\begin{aligned} C(\xi)T(\xi) = \oint_{\Gamma(x)} \left[\frac{\partial T(x)}{\partial n} T^*(x,\xi) - T(x)\frac{\partial T^*(x,\xi)}{\partial n}\right] d\Gamma(x) \\ + \iint_{\Omega(x)} T^*(x,\xi)\frac{1}{k} u_G(x)\, d\Omega(x) \end{aligned} \quad (44)$$

Because of the singularity of the fundamental solution, the singular contour integrals are evaluated in the Cauchy Principal Value (CPV) sense and the jump term, $C(\xi)$, is

$$C(\xi) = \frac{\theta_{int}}{2\pi} \quad 2D$$
$$= \frac{\theta_{int}}{4\pi} \quad 3D \tag{45}$$

Where θ_{int} is the internal angle subtended by the boundary at the source point, ξ (in radians in 2-D and steradians in 3-D). As illustrated in 2D in Fig. 6, the jump term $C(\xi) = 1/2$ for smooth boundaries. Numerically, this term can always be evaluated using a uniform temperature field argument as we will show. Moreover, we can use Eq. (44) at the interior with the jump term simply set to one in such a case.

Let us set the generation term to zero ($u_G = 0$). We can deal with cases where u_G is non-zero as special cases using particular solutions, when these are available, or the dual-reciprocity method (DRM) [33], which can be thought of as a generalized manner to generate particular solutions for the BEM. The BIE is valid everywhere on the boundary and can be used to formulate a boundary value problem for the unknown temperature or its normal derivative.

The singular boundary integral equation can be solved numerically using the boundary element method. Application of the method requires two types of approximations:

(1) spatial approximation of the boundary geometry, $\Gamma(x)$.
(2) functional approximation of the boundary temperature and its normal derivative.

In the BEM, a pattern of nodes is laid out on the boundary, and the BIE is discretized to yield an algebraic relation between the nodal temperatures and their normal derivatives. First, the boundary geometry is approximated in a piecewise fashion, that is the boundary Γ is subdivided into N_e elements, Γ_j, such that

$$C(\xi)T(\xi) = \sum_{j=1}^{N_e} \int_{\Gamma_j} \left[T^*(x,\xi)\frac{\partial T(x)}{\partial n} - T(x)\frac{\partial T^*(x,\xi)}{\partial n} \right] d\Gamma_j(x) \tag{46}$$

as illustrated in Fig. 7. Typically linear, quadratic or cubic interpolating polynomials shape functions such as those utilized in FEM (see Part I of this book) are used for this purpose, Brebbia et al. [1,2] although other approximations are possible, for instance cubic splines Butler et al. [34] or B-spline models proposed by Cabral et al. [35,36].

Next, the functional dependence of the temperature and its normal derivative on the boundary is approximated in terms of their values over a set of nodal points using

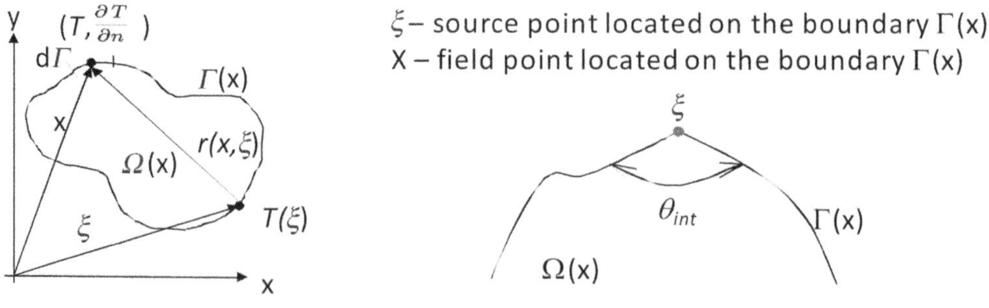

Figure 6. Taking the source point to the boundary to form a boundary integral equation.

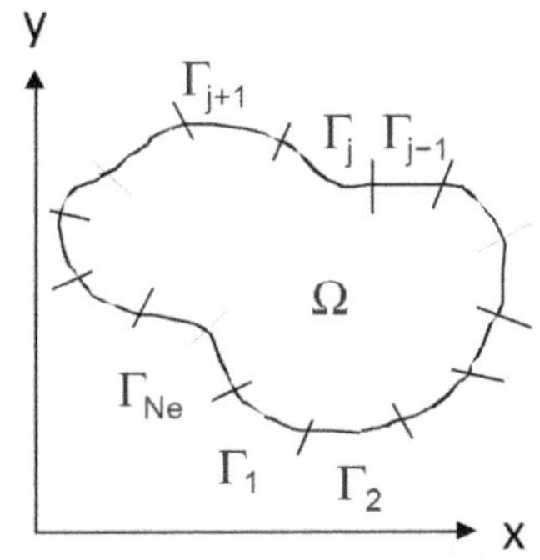

Figure 7. The discretization of the boundary Γ into N_e elements, Γ_j.

suitable interpolation functions. Typically, constant, linear or quadratic polynomial interpolation is utilized for this purpose, although trigonometric approximation and various other schemes have been proposed, see for instance cubic Hermite interpolators proposed by Watson [37–39] and Durodola and Fenner [40]. Boundary Elements much like Finite Elements, can be,

(a) *iso-parametric:* if both the boundary geometry and temperature and its normal derivative are approximated using the same level of interpolation.
(b) *sub-parametric:* if the interpolation for the geometry is of higher level than temperature and its normal derivative.
(c) *super-parametric:* if the interpolation for the geometry is of lower level than temperature and its normal derivative.

Moreover, the nodes that are used to model the geometry and the dependent variables can coincide leading to continuous elements [2,5,8], see Fig. 8, or they may not coincide leading to discontinuous elements [2,42], see Fig. 9. An advantage of discontinuous elements is that the discontinuity of the outward-drawn normal at the corner between two adjacent boundary elements is naturally incorporated in the formulation, while special treatment must be given to such cases when using continuous elements [43,44]. In addition, discontinuous boundary conditions where they are imposed are naturally enforced when using discontinuous elements. However, such elements lead to a heavier computational burden as more unknowns result for a similar order of approximation in comparison to continuous elements.

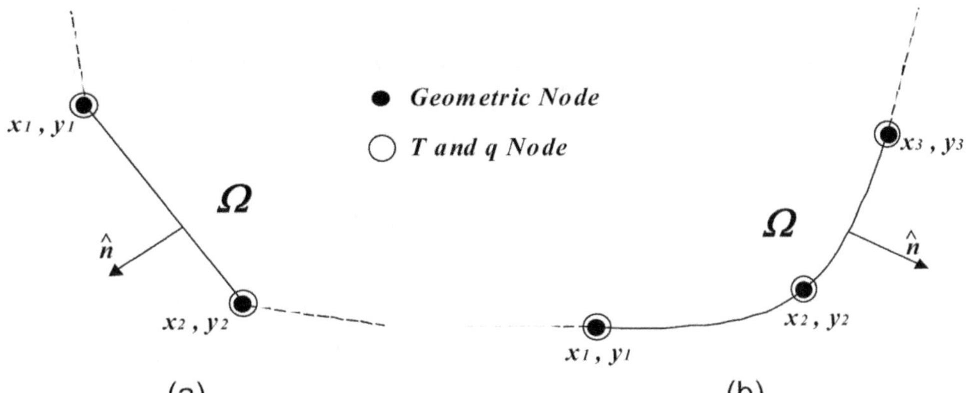

Figure 8. Examples of two-dimensional continuous (a) linear and (b) quadratic isoparametric boundary elements.

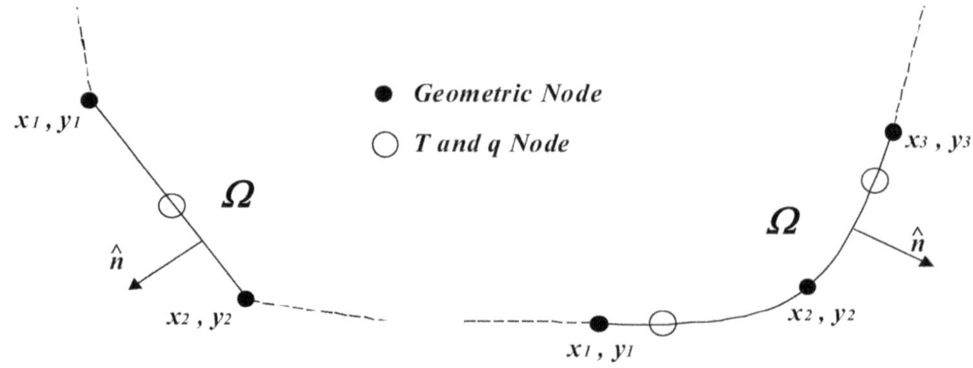

Figure 9. Examples of two-dimensional (a) linear and (b) quadratic subparametric discontinuous boundary elements with the element in (a) called a constant element.

The simplest possible approximation is the linear sub-parametric (constant) boundary elements which model the geometry as piecewise linear and the dependent variables as constant over the element, see Fig. 10. In such a case, the BIE becomes

$$C(\xi)T(\xi) = \sum_{j=1}^{N_e} q_j \int_{\Gamma_j} T^*(x,\xi)d\Gamma_j(x) - \sum_{j=1}^{N_e} T_j \int_{\Gamma_j} q^*(x,\xi)d\Gamma_j(x) \tag{47}$$

where the notation $q = \partial T/\partial n$ and $q^* = \partial T^*/\partial n$ has been used to denote the normal derivatives and will be adhered to in the rest of part II of the book. This notation is consistent with the BEM literature and should not be confused with the heat flux. We will specifically utilize q'' to denote the heat flux. Also, T_j and q_j denote the values of T and q at node j that is located at the midpoint of the *j-th* boundary element, $\Gamma_j(x)$. In the case of constant elements, the number of nodal unknowns is equal to the number of elements. There are N_e-nodal unknowns T_j or q_j whichever is not specified by the boundary condition at the node.

In order to arrive at an algebraic set that will be solved for these N_e-nodal unknowns, a collocation procedure may be adopted whereby the source point, ξ, is located at each $i = 1,2\ldots N_e$ boundary node to yield the algebraic analog to the BIE,

$$C(\xi_i)T(\xi_i) = \sum_{j=1}^{N_e} G_{ij}q_j - \sum_{j=1}^{N_e} \hat{H}_{ij}T_j \qquad i = 1,2\ldots N_e \tag{48}$$

where,

$$G_{ij} = \int_{\Gamma_j} T^*(x,\xi_i)d\Gamma_j(x), \qquad \hat{H}_{ij} = \int_{\Gamma_j} q^*(x,\xi_i)d\Gamma_j(x) \tag{49}$$

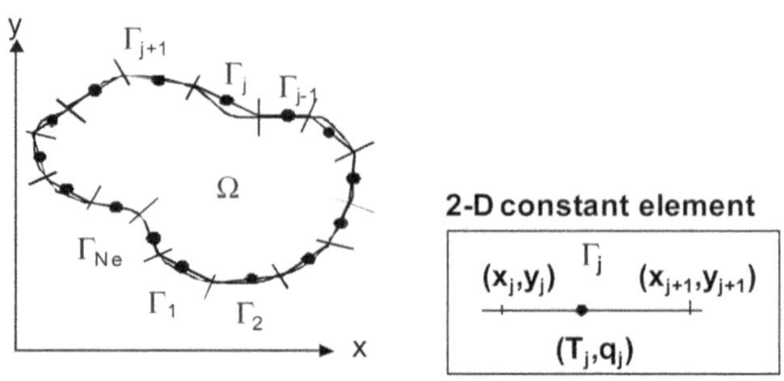

Figure 10. A constant boundary element discretization.

The coefficients can be analytically evaluated or can be computed automatically via Gauss-Legendre quadratures. Further defining,

$$H_{ij} = C(\xi_i)\delta_{ij} + \hat{H}_{ij}, \qquad \text{here } \delta_{ij} = \begin{cases} 0 & \text{if} \quad i \neq j \\ 1 & \text{if} \quad i = j \end{cases} \tag{50}$$

we then arrive at the standard BEM algebraic form

$$[H]\{T\} = [G]\{q\} \tag{51}$$

The square $N_e \times N_e$ matrices [H] and [G] contain the influence-coefficients, H_{ij} and G_{ij}, and {T} and {q} are vectors whose elements are nodal values of boundary temperature, T_j, and its normal derivative, q_j. Once the boundary conditions are applied, the above can be re-arranged in the standard algebraic form

$$[A]\{x\} = \{b\} \tag{52}$$

in which all unknowns have been collected into the vector {x}, while the right-hand side vector, {b}, is the "load" vector. As every point on the boundary is related to every other point, the matrix [A] is fully-populated, and it is generally un-symmetric and non-diagonally dominant. A Galerkin approach can be taken to formulate the BEM [45] and produce a symmetric coefficient matrix, however, the process is very computationally intensive requiring double quadratures and, as such, it is not a commonly used BEM formulation.

This system can be solved by standard direct schemes such as Gauss elimination with pivoting and equilibration for most 2-D problems. In 3-D applications, as the number of unknowns increases, one must often resort to iterative methods. Given the nature of the coefficient matrix, Jacobi, Gauss-Seidel or SOR fail to converge and as such, symmetrization and pre-conditioning can be used to solve the system by minimization techniques such as the bi-conjugate gradient method [46] or directly by non-symmetric iterative equation solvers such as the general minimization of residuals method (GMRES) [47,48]. Alternatively, the domain may be sub-sectioned in a domain decomposition approach, particularly in 3D and for large scale problems and the resulting equations solved either using block methods [49,50] or iteratively [51].

Some comments regarding the numerics of the BEM are in order. It is common practice to evaluate the influence-coefficients numerically, via Gauss-type quadrature. However, care must be given to the singular coefficients. The G_{ii} terms are weakly singular and can be integrated analytically or numerically using adaptive quadratures, for example Gauss-Kronrod rules in 2D [52]. However, the H_{ii} integrals are strongly singular and can often be evaluated by an equi-potential argument. For instance, if the temperature were uniform throughout the field, then the flux would be zero everywhere, and consequently Eq. (48) yields the relation

$$H_{ii} = -\sum_{\substack{j=1 \\ j \neq i}}^{N_e} H_{ij} \tag{53}$$

It is noted that this relationship is true for all closed domains, i.e. interior problems. When writing the BEM equations for an infinite or semi-infinite domain, that is for an exterior problem a shown in, Fig. 11, the equipotential argument must be modified as

$$H_{ii} = 1 - \sum_{\substack{j=1 \\ j \neq i}}^{N_e} H_{ij} \tag{54}$$

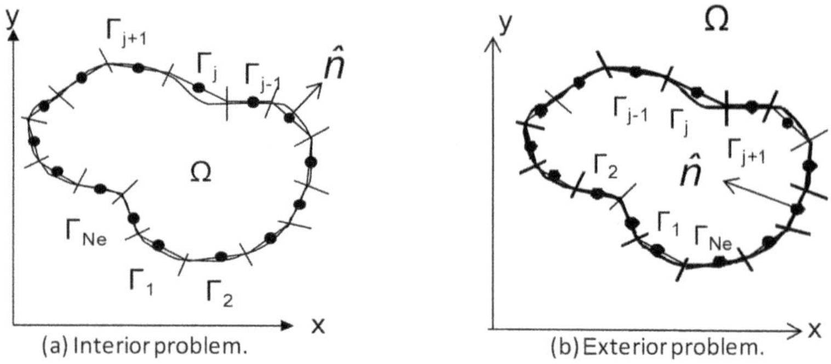

Figure 11. Exterior and interior BEM problem.

to account for the fact that a unit temperature prescribed in a boundless domain is due to a unit source that produces a value of −1 for the integral over Γ_∞

$$\oint_{\Gamma_\infty} \frac{\partial T^*}{\partial n} d\Gamma = -1 \tag{55}$$

Higher-order models for T and q are often used in BEM, with quadratic interpolation the most common higher-order approximation in use. We will address the details of the numerics and corresponding expressions for $H_{i,j}$ and $G_{i,j}$, address the generation term, address transient as well as applications to heat conduction in orthotropic or anisotropic regions, non-linear heat conduction, heat conduction in non-homogeneous media, and applications to inverse problems and convection heat transfer in the coming chapters. We now focus on a simple 2D example in order to illustrate the essentials and the BEM.

Example 2.2: Constant element BEM for conduction in a square. Consider the heat conduction problem we previously solved by the Green's function method, and which we now solve by the BEM. Specifically, we will utilize the simplest boundary element called a constant element. Here, the geometry is taken as piecewise linear and the temperature and its normal derivative are taken as constant over the element. Moreover, the placement of the location of T and q nodes is taken at the midpoint of the constant element as shown in Fig. 12. We will evaluate the integrals for H_{ij} and G_{ij} analytically, relegating the numerical implementation to the next chapter.

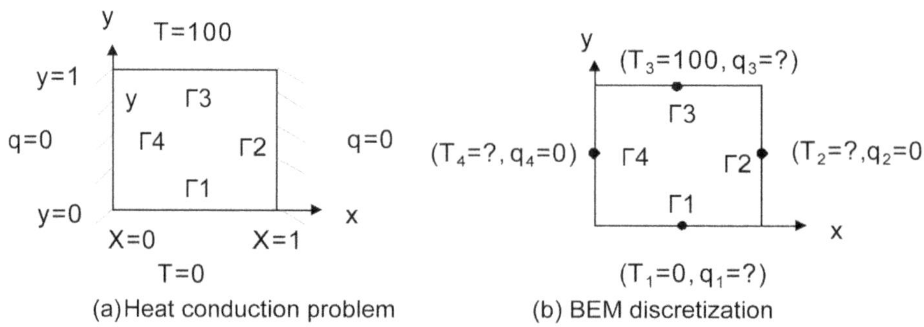

Figure 12. Heat conduction problem in a square and 4-noded BEM discretization.

We now consider evaluation of the coefficients in the BEM algebraic equation. Eq. (51). To this end, we identify two types of coefficients:

(a) off-diagonal coefficients: those account for the effect of other elements on the element of interest, and they are readily evaluated.
(b) diagonal elements: those account for the self influence of an elements, and they are evaluated using the CVP definition of a singular integral and the potential field argument.

Let us first consider the off-diagonal integrals. Here, when evaluating, $\partial T^*/\partial n$, the chain rule is utilized, that is

$$\frac{\partial T^*}{\partial n} = \frac{\partial T^*}{\partial r}\frac{\partial r}{\partial n} = \frac{\partial T^*}{\partial r}\nabla r \bullet \hat{n} \tag{56}$$

with $\hat{n}$ denoting the **outward-drawn** normal, and r being the position vector. So that in 2D, we have

$$\frac{\partial T^*}{\partial n} = \frac{-1}{2\pi}\frac{\vec{r}(x,\xi)\bullet\hat{n}}{|\vec{r}(x,\xi)|^2} \tag{57}$$

where $r(x,\xi) = |\vec{r}(x,\xi)|$ is Euclidean distance between the source point, ξ, and the field point, x. Moreover, as we develop the problem, it will be apparent that in this case due to the geometric symmetry there results symmetry in the H_{ij} and G_{ij} coefficients. This will save much work as we can then only generate a few of the sixteen coefficients.

Evaluating typical off-diagonal coefficients:
Let us consider $\hat{H}_{23}$ and G_{23} for instance. The source point is located at, $\xi = (1,0.5)$ at the midpoint of element, Γ_2, the field point, x, ranges over, $\Gamma_3(x)$, and the position vector $\vec{r}(x,\xi)$ points from the source to the field point. The associated geometry and integrals as provided below

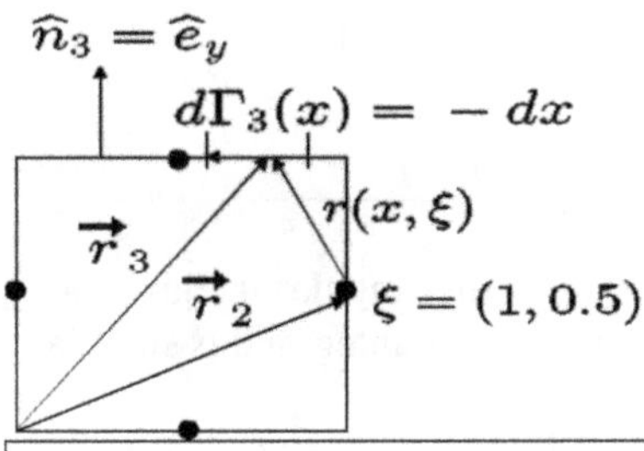

$$\begin{aligned}
\vec{r}_2 &= \hat{e}_x + 0.5\hat{e}_y\\
\vec{r}_3 &= x\hat{e}_x + \hat{e}_y\\
\hat{n}_3 &= \hat{e}_y\\
d\Gamma_3(x) &= -dx\\
\vec{r}_{23} &= (x-1)\hat{e}_x + 0.5\hat{e}_y\\
|\vec{r}_{23}| &= \sqrt{(x-1)^2 + (0.5)^2}\\
\vec{r}_{23}\cdot\hat{n}_3 &= 0.5
\end{aligned}$$

$$\begin{aligned}
\hat{H}_{2,3} &= -\frac{1}{2\pi}\int_{\Gamma_3}\frac{\vec{r}_{23}\cdot\hat{n}_3}{|\vec{r}_{23}|^2}d\Gamma_3(x)\\
&= -\frac{1}{2\pi}\int_1^0\frac{0.5}{(x-1)^2+(0.5)^2}(-dx)\\
&= \frac{1}{2\pi}tan^{-1}\left(\frac{x-1}{0.5}\right)\Big|_1^0\\
&= -0.176
\end{aligned}$$

$$\begin{aligned}
G_{2,3} &= -\frac{1}{2\pi}\int_{\Gamma_3} ln|\vec{r}_{23}|d\Gamma_3(x)\\
&= -\frac{1}{2\pi}\int_1^0\frac{1}{2}ln[(x-1)^2+(0.5)^2]dx\\
&= -\frac{1}{4\pi}\Big\{ln[(x-1)^2+(0.5)^2]\\
&\qquad -2(x-1)+tan^{-1}\left(\frac{x-1}{0.5}\right)\Big\}\Big|_1^0\\
&= 0.0533
\end{aligned}$$

Due to symmetry, it is easy to show that such coefficients relating adjacent faces evaluate to the same numerical value

$$\hat{H}_{23} = \hat{H}_{32} = \hat{H}_{12} = \hat{H}_{21}\ldots \quad \text{and} \quad G_{23} = G_{32} = G_{12} = G_{21}\ldots \tag{58}$$

Similary, the other off-diagonal coefficients may be readily evaluated, for instance, coefficients corresponding to influences of elements facing each other, say $\hat{H}_{13}$ and G_{13}

$$\hat{H}_{13} = -0.1476 \quad \text{and} \quad G_{13} = -0.0062 \tag{59}$$

and, similar symmetries hold in this case due to the geometric symmetry of the problem, that is

$$\hat{H}_{13} = \hat{H}_{31} = \hat{H}_{24} = \hat{H}_{42} \qquad \text{and} \qquad G_{13} = G_{31} = G_{24} = G_{42} \tag{60}$$

Evaluating typical diagonal coefficients:
Let us consider $\hat{H}_{11}$ and G_{11} for instance. In this case, we are evaluating the effect of the boundary element on itself. The source point is located at, $\xi = (0.5,0)$ at the midpoint of element, Γ_1, the field point, x, ranges over, $\Gamma_1(x)$ and the position vector $\vec{r}(x,\xi)$ points from the source to the field point. The associated geometry and integrals as provided below

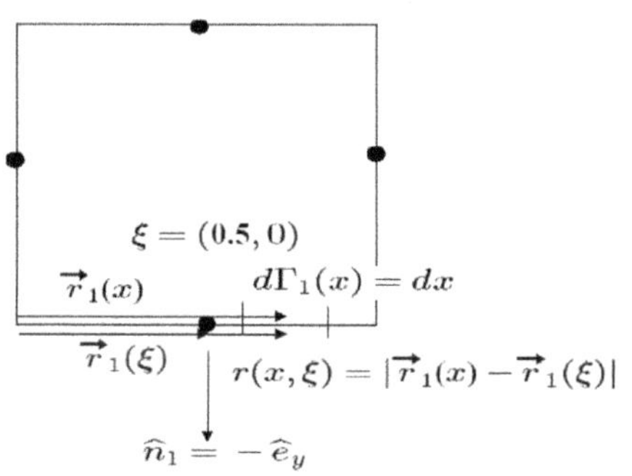

$$\begin{aligned}
\vec{r}_1(x) &= x\hat{e}_x \\
\vec{r}_1(\xi) &= 0.5\hat{e}_x \\
\hat{n}_1 &= -\hat{e}_y \\
d\Gamma_1(x) &= dx \\
\vec{r}_{11} &= x\hat{e}_x \\
|\vec{r}_{11}| &= x \\
\vec{r}_{11}\cdot\hat{n}_1 &= 0
\end{aligned}$$

$$\hat{\boldsymbol{H}}_{1,1} = -\frac{1}{2\pi}\int_{\Gamma_1}\frac{\vec{\boldsymbol{r}}_{11}\cdot\hat{\boldsymbol{n}}_1}{|\vec{\boldsymbol{r}}_{11}|^2}d\Gamma_1(\boldsymbol{x}) = 0$$

$$\begin{aligned}
G_{1,1} &= -\frac{1}{2\pi}\int_{\Gamma_1} ln|x-0.5|d\Gamma_1(x) \\
&= -\frac{1}{2\pi}\lim_{\epsilon\to 0}\left[\int_0^{0.5-\epsilon}\frac{1}{2}ln(0.5-x)dx + \int_{0.5+\epsilon}^{1}\frac{1}{2}ln(x-0.5)\right]dx \\
&= 0.2695
\end{aligned}$$

where the Cauchy Principal Value (CPV) of the singular integral for G_{11} has been used to evaluate the integral, that is for an integral whose integrand is singular at x_o

$$CPV\int_a^b f(x)dx = \lim_{\varepsilon\to 0}\left[\int_a^{x_o-\varepsilon} f(x)dx + \int_{x_o+\varepsilon}^{b} f(x)dx\right] \tag{61}$$

It is interesting to note that for the constant element, the strongly singular integral for $\hat{H}_{11}$ vanishes as the position vector and outward-drawn normal are orthogonal to each other. This is also the case for the linear boundary element, but no longer the case when there is a curvature over the element as in quadratic and higher order elements. Here again due to the symmetry of our problem

$$\hat{H}_{11} = \hat{H}_{22} = \hat{H}_{33} = \hat{H}_{44} \qquad \text{and} \qquad G_{11} = G_{22} = G_{33} = G_{44} \tag{62}$$

Now, given that the boundary is smooth, the jump term, $C(\xi_i)$, is readily evaluated as 1/2, and in light of the results from Eq. (62), the diagonal coefficients of the [H] matrix are all 0.5 that is $H_{ii} = 0.5$ $i = 1,2\ldots4$. With all these coefficients in hand, we can now assemble the BEM equations. These are displayed below for

$$\begin{bmatrix} 0.5 & -0.1762 & -0.1476 & -0.1762 \\ -0.1762 & 0.5 & -0.1762 & -0.1476 \\ -0.1476 & -0.1762 & 0.5 & -0.1762 \\ -0.1762 & -0.1476 & -0.1762 & 0.5 \end{bmatrix} \begin{Bmatrix} T_1 \\ T_2 \\ T_3 \\ T_4 \end{Bmatrix} = \begin{bmatrix} 0.2695 & 0.0533 & -0.0062 & 0.0533 \\ 0.0533 & 0.2695 & 0.0533 & -0.0062 \\ -0.0062 & 0.0533 & 0.2695 & 0.0533 \\ 0.0533 & -0.0062 & 0.0533 & 0.2695 \end{bmatrix} \begin{Bmatrix} q_1 \\ q_2 \\ q_3 \\ q_4 \end{Bmatrix} \tag{63}$$

Notice that although we did not use the equi-potential argument, we find indeed that

$$H_{ii} = -\sum_{\substack{i=1 \\ i\neq j}}^{4} H_{ij} \quad i = 1,2\ldots4 \tag{64}$$

$$0.5 = -(-0.1762 - 0.1476 - 0.1762)$$

This equi-potential argument is typically invoked and carried out in the build [H] and [G] matrix phase of a BEM code to generate the H_{ii} diagonal coefficients.

Introducing boundary conditions, re-arranging, and solving:
Next boundary conditions are introduced, and given that, in our problem, the boundary conditions specify T_1,q_2,T_3,q_4 and that q_1,T_2,q_3,T_4 are unknowns, the above equations are rearranged as:

$$\begin{bmatrix} -0.2695 & -0.1762 & 0.0062 & -0.1762 \\ -0.0533 & 0.5 & -0.0533 & -0.1476 \\ 0.0062 & -0.1762 & -0.2695 & -0.1762 \\ -0.0533 & -0.1476 & -0.0533 & 0.5 \end{bmatrix} \begin{Bmatrix} q_1 \\ T_2 \\ q_3 \\ T_4 \end{Bmatrix} = \begin{bmatrix} -0.5 & 0.0533 & 0.1476 & 0.0533 \\ 0.1762 & 0.2695 & 0.1762 & -0.0062 \\ 0.1476 & 0.0533 & -0.5 & 0.0533 \\ 0.1762 & -0.0062 & 0.1762 & 0.2695 \end{bmatrix} \begin{Bmatrix} T_1 \\ q_2 \\ T_3 \\ q_4 \end{Bmatrix} \tag{65}$$

This was accomplished by multiplying the column of the vector element that was moved by −1 and switching the corresponding columns. For instance as q_1 and T_1 switched positions, then so did negatives of columns 1 of the [H] and [G] matrices take place of each other's location in the respective matrix column location, that is

$$\text{if} \quad T_j \leftrightarrow q_j \quad \text{then} \quad -H_{ij} \leftrightarrow -G_{ij} \quad i = 1,2\ldots N \tag{66}$$

Imposing the boundary conditions, namely that $T_1 = 0$, $q_2 = 0$, $T_3 = 100$ and $q_4 = 0$, the right-hand-side matrix-vector product yields the load vector for our problem and we now have the following set of linear equations to solve,

$$\begin{bmatrix} -0.2695 & -0.1762 & 0.0062 & -0.1762 \\ -0.0533 & 0.5 & -0.0533 & -0.1476 \\ 0.0062 & -0.1762 & -0.2695 & -0.1762 \\ -0.0533 & -0.1476 & -0.0533 & 0.5 \end{bmatrix} \begin{Bmatrix} q_1 \\ T_2 \\ q_3 \\ T_4 \end{Bmatrix} = \begin{Bmatrix} 14.578 \\ 17.62 \\ -50 \\ 17.62 \end{Bmatrix} \tag{67}$$

Once the boundary conditions are imposed, the matrix equations are no longer symmetric, even for a symmetric problem, and the coefficient matrix is not diagonally dominant. This poses a problem with standard Jacobi, Gauss-Seidel or SOR solvers. The conditioning number of the coefficient matrix is 2.557 in the L_2-norm. Solving using a direct solver with partial pivoting and equilibration is thus a necessity for any size BEM problem. The above are solved to yield

$$\begin{Bmatrix} q_1 \\ T_2 \\ q_3 \\ T_4 \end{Bmatrix} = \begin{Bmatrix} -117.4 \\ 49.99 \\ 117.4 \\ 49.99 \end{Bmatrix} \tag{68}$$

While the exact solution is

$$\begin{Bmatrix} q_1 \\ T_2 \\ q_3 \\ T_4 \end{Bmatrix} = \begin{Bmatrix} -100 \\ 50 \\ 100 \\ 50 \end{Bmatrix} \tag{69}$$

So that the 4-noded BEM model produce almost exact nodal temperatures and 17% error in the normal derivative. That is the typical trend that the temperature is more accurately computed than the normal derivative in a BEM solution.

The procedure of solving the boundary value problem for the unknowns at the boundary then consists of three steps:

1. Assemble the [H] and [G] matrix system corresponding to the algebraic analog of the boundary integral equation for the problem generated by the discretization of the boundary into elements and the distribution of the field variable and its derivative on the boundary by a chosen interpolation. This step generates the classical BEM form:

$$[H]\{T\} = [G]\{q\}$$

2. Introduce the boundary conditions and re-arrange the BEM equations into the standard algebraic form:

$$[A]\{x\} = \{b\}$$

3. Solve the fully-populated algebraic system for the boundary unknowns.

Post processing the interior temperature:

Finding the temperature at any interior point is now a matter of applying the discrete version of the integral equation for the temperature, Eq. (14), or

$$T(\xi) = \sum_{j=1}^{N_e} q_j \int_{\Gamma_j} T^*(x,\xi)d\Gamma_j(x) - \sum_{j=1}^{N_e} T_j \int_{\Gamma_j} q^*(x,\xi)d\Gamma_j(x) \tag{70}$$

And following our notation, we have in discrete form

$$T(\xi) = \sum_{j=1}^{N_e} G_{ij}q_j - \sum_{j=1}^{N_e} \hat{H}_{ij}T_j \qquad \xi \in \Omega(x) \tag{71}$$

Choosing an interior point say, $\xi = (0{,}5{,}0.5)$, we now consider the source point, ξ, fixed and the field point, x, traverses the boundary, $\Gamma(x)$, see Fig. 13. In this case, there are no singular integrals. There must be care taken when evaluating an interior temperature close to the boundary as the integrand can become very nearly singular, however, this is readily handled with adaptive quadratures. Carrying out the such integration, there results

$$T(0.5,\ 0.5) = 49.93 \tag{72}$$

In comparison, the exact value is $T(0.5,0.5) = 50$.

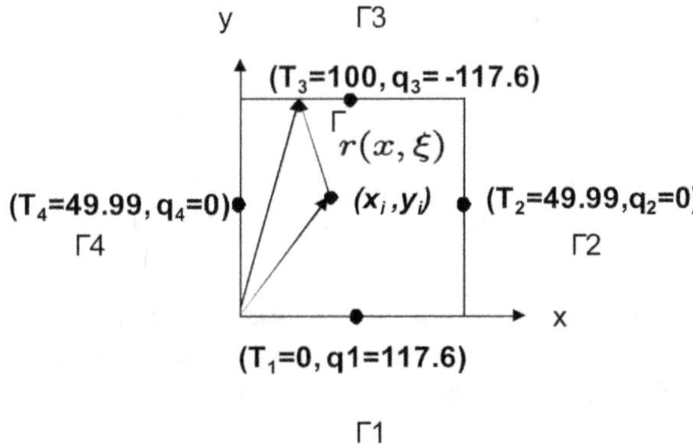

Figure 13. Post-processing temperature at internal point $(x_i, y_i) = (0.5.0.5)$.

The rapid convergence of the BEM under nodal refinement (p-refinement) is readily evident investigated by considering further refinement and examining interior solutions at various interior points. Results for convergence under p-refinement are provided by examining the temperature values at the 9 interior points shown in Fig. 14 for 4, 8, 12, and 16 constant elements. Rapid convergence under p- and h-refinement is illustrated in Fig. 15 by introducing higher order discontinuous boundary elements and evaluating the maximum error in temperature at the same 9 interior points displayed in Fig. 14.

The question arises now as how to treat other types of boundary conditions commonly encountered in heat transfer. It is clear that if a non-adiabatic boundary condition is imposed at a given node, say a heat flux q_s'' is imposed, then

$$q = -q_s''/k \tag{73}$$

Point	x-coord	y-coord	*T*(4)	*T*(8)	*T*(12)	*T*(16)	*T*(exact)
1	0.25	0.25	28.92	24.80	24.82	24.89	25.00
2	0.50	0.25	26.79	24.80	24.93	24.95	25.00
3	0.75	0.25	28.92	24.80	24.82	24.89	25.00
4	0.25	0.50	49.39	49.99	50.04	50.00	50.00
5	0.50	0.50	49.93	50.00	50.03	50.00	50.00
6	0.75	0.50	49.39	49.99	50.04	50.00	50.00
7	0.25	0.75	72.04	75.14	75.26	75.12	75.00
8	0.50	0.75	71.99	75.19	75.12	75.05	75.00
9	0.75	0.75	72.04	75.14	75.25	75.12	75.00

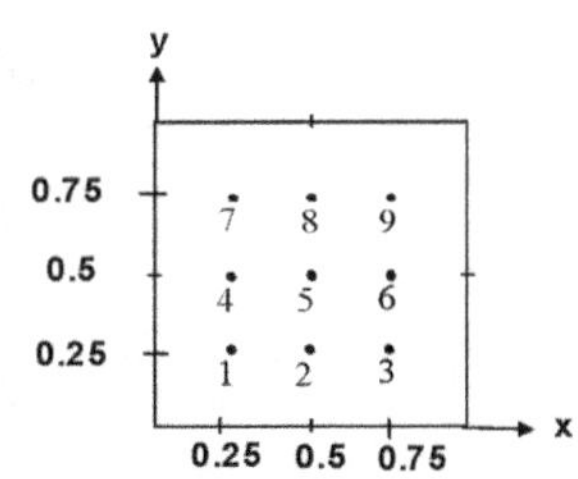

Figure 14. Location of the nine points where the temperature is computed in post-processing and the resulting values at these points.

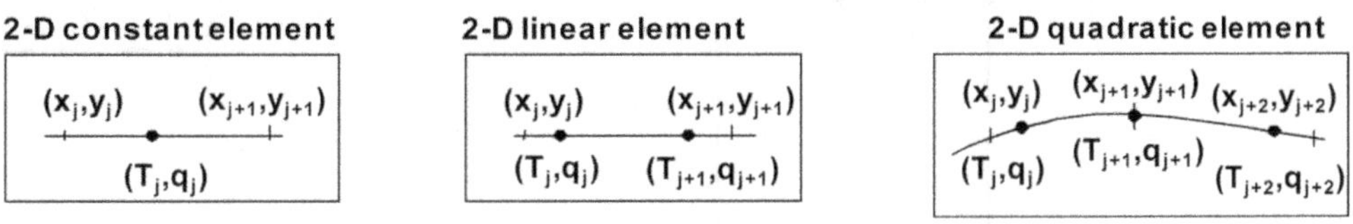

	Constant	Linear	Quadratic
4 Elements	4.83 %	1.17 %	0.004 %
8 Elements	0.276 %	0.05 %	0 %
12 Elements	0.289 %	0.002 %	0 %
16 Elements	0.148 %	0 %	0 %

Figure 15. Convergence of the constant BEM model under p- and h-refinement.

Similarly, if a convective boundary condition is imposed, then

$$q = -\frac{h}{k}(T - T_\infty) \tag{74}$$

Where, h, is the convective heat transfer coefficient and T_∞ is the ambient temperature. The BEM form is then re-arranged accordingly. This is illustrated in the next example.

Example 2.3: Convective boundary condition. Let us consider the same problem and geometry, however, we now impose a convection boundary condition at $y = 1$, as depicted in Fig. 16. The exact solution for this problem is

$$T(y) = T_1 + \left[\frac{\left(\frac{y}{L}\right)}{1+\left(\frac{hL}{k}\right)^{-1}}\right] T_\infty \tag{75}$$

Let us take the values: $T_1 = 0$, $h = 25$, $k = 5$, and $T_\infty = 100$. Implementing the BEM procedure leads to BEM equations that are unchanged from the previous example. However, at the re-arrangement stage, a BEM equations at any node, i, reads

$$H_{i1}T_1 + H_{i2}T_2 + H_{i3}T_3 + H_{i4}T_4 = G_{i1}q_1 + G_{i2}q_2 + G_{i3}q_3 + G_{i4}q_4 \tag{76}$$

with the convective boundary condition

$$q_3 = -\frac{h}{k}(T_3 - T_\infty) \tag{77}$$

is consequently re-arranged as

$$H_{i1}T_1 + H_{i2}T_2 + (H_{i3} + \frac{h}{k}G_{i3})T_3 + H_{i4}T_4 = G_{i1}q_1 + G_{i2}q_2 + G_{i3}\left(\frac{h}{k}T_\infty\right) + G_{i4}q_4 \tag{78}$$

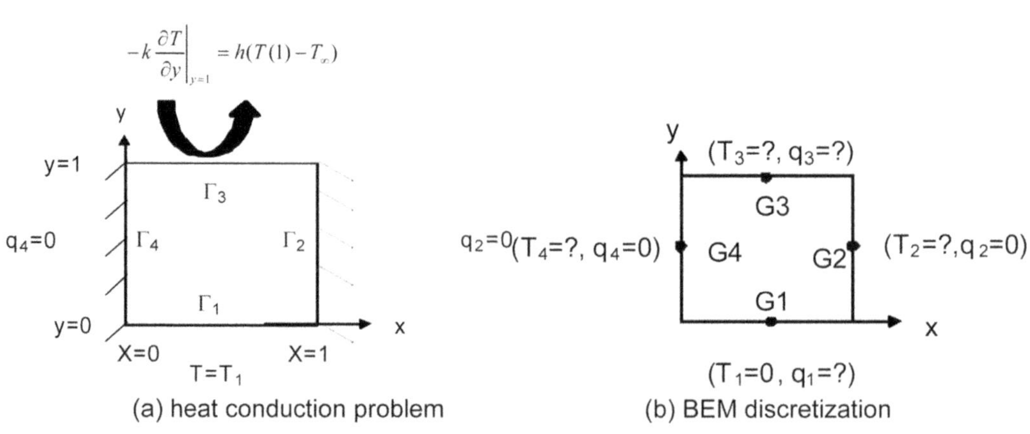

Figure 16. Conduction in a square with a convective boundary condition at $y = 1$.

This is reflected in the matrix system as

$$\begin{bmatrix} -0.2695 & -0.1762 & 0.1476+\frac{h}{k}(0.0062) & -0.1762 \\ -0.0533 & 0.5 & 0.1762+\frac{h}{k}(0.0533) & -0.1476 \\ 0.0062 & -0.1762 & 0.5+\frac{h}{k}(0.2965) & -0.1762 \\ -0.0533 & -0.1476 & 0.1762+\frac{h}{k}(0.0533) & 0.5 \end{bmatrix} \begin{Bmatrix} q_1 \\ T_2 \\ T_3 \\ T_4 \end{Bmatrix} = \begin{bmatrix} -0.5 & 0.0533 & 0.0062 & 0.0533 \\ 0.1762 & 0.2695 & 0.0533 & -0.0062 \\ 0.1476 & 0.0533 & 0.2965 & 0.0533 \\ 0.1762 & -0.0062 & 0.0533 & 0.2695 \end{bmatrix} \begin{Bmatrix} T_1 \\ q_2 \\ \frac{h}{k}T_\infty \\ q_4 \end{Bmatrix} \tag{79}$$

where the re-arrangement has been left in symbolic form for emphasis. Thus in this case the re-arrangement accommodating the boundary conditions entailed:

$$\begin{aligned} &1.\quad \text{switching} \quad T_1 \leftrightarrow q_1 \quad \text{so that for } j=1 \quad -H_{ij} \leftrightarrow -G_{ij} \quad i=1,2...N \\ &2.\quad \text{replacing} \quad H_{i3} \rightarrow H_{i3} + \frac{h}{k}G_{i3} \quad \text{for } i=1,2...N \quad \text{and replacing} \quad q_3 \rightarrow \frac{h}{k}T_\infty \end{aligned} \tag{80}$$

Choosing the given values for our problem, there results the following algebraic system:

$$\begin{bmatrix} -0.2695 & -0.1762 & -0.1662 & -0.1762 \\ -0.0533 & 0.5 & -0.0163 & -0.1476 \\ 0.0062 & -0.1762 & 1.3084 & -0.1762 \\ -0.0533 & -0.1476 & -0.0163 & 0.5 \end{bmatrix} \begin{Bmatrix} q_1 \\ T_2 \\ T_3 \\ T_4 \end{Bmatrix} = \begin{Bmatrix} -3.096 \\ 26.65 \\ 134.73 \\ 26.65 \end{Bmatrix} \tag{81}$$

whose solution produces

$$\begin{Bmatrix} q_1 \\ T_2 \\ T_3 \\ T_4 \end{Bmatrix} = \begin{Bmatrix} -95.11 \\ 40.49 \\ 80.98 \\ 40.49 \end{Bmatrix} \tag{82}$$

While the exact solution is

$$\begin{Bmatrix} q_1 \\ T_2 \\ q_3 \\ T_4 \end{Bmatrix} = \begin{Bmatrix} -83.33 \\ 41.67 \\ 83.33 \\ 41.67 \end{Bmatrix} \tag{83}$$

The results are then 2.9% in error in the temperature and 14% in error in the normal derivative of the temperature. A MATHCAD spreadsheet as well as a MATLAB code for the 2D constant BEM code used to generate the results in this section are available at the accompanying website for this book: www.fbm.centercorp.com.

This concludes the introduction to the BEM and its theoretical foundations. We now consider numerical issues in the implementation of the BEM and then develop several applications in heat transfer.

APPENDIX A DERIVATION OF THE GREEN'S FUNCTION FOR THE 2D PROBLEM IN A SQUARE

The eigenfunctions utilized in the solution of the problem in the square are found by solving the associated homogeneous problem, repeated here for completeness

$$\nabla^2 T_{aux}(x,y) + \frac{1}{k} u_G(x,y) = 0 \tag{84}$$

$$T_{aux}\Big|_{\Gamma_1} = 0 \tag{85}$$

$$q_{aux}\Big|_{\Gamma_2} = 0 \tag{86}$$

$$T_{aux}\Big|_{\Gamma_3} = 0 \tag{87}$$

$$q_{aux}\Big|_{\Gamma_4} = 0 \tag{88}$$

To solve this problem, the solution is expanded in terms of the eigenfunctions of the related Helmholtz problem

$$\nabla^2 T(x,y) + \lambda^2 T(x,y) = 0 \tag{89}$$

$$T\Big|_{\Gamma_1} = 0 \tag{90}$$

$$q\Big|_{\Gamma_2} = 0 \tag{91}$$

$$T\Big|_{\Gamma_3} = 0 \tag{92}$$

$$q\Big|_{\Gamma_4} = 0 \tag{93}$$

These are found by standard methods [27] as

$$W_{nm}(x,y) = \cos(\beta_n x)\sin(\delta_m y) \quad n = 0,1,2 \ldots \infty \quad \text{and} \quad m = 1,2 \ldots \infty \tag{94}$$

where the eigenvalues are

$$\beta_n = \frac{n\pi}{L} \quad \text{and} \quad \delta_m = \frac{m\pi}{L} \tag{95}$$

Expanding the auxiliary problem temperature in terms of the eigenfunctions of the related Helmholtz problem,

$$T_{aux}(x,y) = \sum_{n=0}^{\infty}\sum_{m=1}^{\infty} C_{nm} W_{nm}(x,y) \tag{96}$$

Introducing into the governing equation, we find

$$\sum_{n=0}^{\infty}\sum_{m=1}^{\infty} C_{nm} \nabla^2 W_{nm}(x,y) + \frac{1}{k} u_G(x,y) = 0 \tag{97}$$

Now since the eigenfunctions solve the related Helmholtz problem, we have

$$\nabla^2 W_{nm}(x,y) = -\lambda_{nm}^2 W_{nm}(x,y) \tag{98}$$

and Eq. (97) consequently becomes

$$\sum_{n=0}^{\infty}\sum_{m=1}^{\infty} C_{nm}\lambda_{nm}^2 W_{nm}(x,y) = \frac{1}{k}u_G(x,y) \tag{99}$$

Utilizing the orthogonality property of the eigenfunctions, we find that

$$C_{nm} = \frac{1}{N(\beta_n)N(\delta_m)}\int_0^L\int_0^L \frac{1}{k}u_G(x',y')\cos(\beta_n x')\sin(\delta_m y')dx'dy' \tag{100}$$

where the normalizing integrals are

$$\begin{aligned} &N(\beta_n) = L \quad n=0 \quad \text{and} \quad N(\beta_n) = L/2 \quad n = 1,2\ldots\infty \\ &N(\delta_m) = L/2 \quad m = 1,2\ldots\infty \end{aligned} \tag{101}$$

and $\lambda_{nm}^2 = \beta_n^2 + \delta_m^2$. Finally, introducing the expansion coefficients and assembling the solution, we arrive at the final form of the solution to the auxiliary problem

$$T(x,y) = \sum_{n=0}^{\infty}\sum_{m=1}^{\infty}\left[\frac{1}{N(\beta_n)N(\delta_m)}\int_0^L\int_0^L \cos(\beta_n x')\sin(\delta_m y')\frac{1}{k}u_G(x',y')dx'dy'\right]\cos(\beta_n x)\sin(\delta_m y) \tag{102}$$

or re-arranging in the form that allows us to identify the Green's function for the problem, that is the form of Eq. (33), we find that

$$T(x,y) = \int_0^L\int_0^L\left[\sum_{n=0}^{\infty}\sum_{m=1}^{\infty}\frac{1}{N(\beta_n)N(\delta_m)}\cos(\beta_n x)\sin(\delta_m y)\cos(\beta_n x')\sin(\delta_m y')\right]\frac{1}{k}u_G(x',y')dx'dy' \tag{103}$$

So the that the Green's function for this problem is identified as the term in the brackets

$$T_{GF}(x,y,x',y') = \sum_{n=0}^{\infty}\sum_{m=1}^{\infty}\frac{1}{N(\beta_n)N(\delta_m)}\cos(\beta_n x)\sin(\delta_m y)\cos(\beta_n x')\sin(\delta_m y') \tag{104}$$

APPENDIX B DERIVATION OF THE GREEN'S FREE SPACE (FUNDAMENTAL) SOLUTION TO THE LAPLACE EQUATION

Since the Dirac Delta function is symmetric the response of its perturbation of the adjoint equation is also symmetric, thus adopting polar coordinates in 2D, we have

$$\frac{1}{r}\frac{\partial}{\partial r}\left(r\frac{\partial T^*}{\partial r}\right) = -\delta(r-\xi) \tag{105}$$

Away from the source term, the adjoint differential equation becomes

$$\frac{1}{r}\frac{\partial}{\partial r}\left(r\frac{\partial T^*}{\partial r}\right) = 0 \tag{106}$$

and it can be integrated to yield,

$$T^*(r) = Alnr + B \tag{107}$$

We can arbitrarily set $B = 0$, and we recognize the result so far is a point heat source located at $r = \xi$. Now, to determine the strength of the source, we consider a region, $\Omega(x)$ bounded by a curve $\Gamma(x)$ that encloses the source, and we carry out an integral over the region so that

$$\oiint_{\Omega(x)} \nabla^2 T^*(r) d\Omega(x) = -\oiint_{\Omega(x)} \delta(r-\xi) d\Omega(x) \tag{108}$$

From the operational property of the Dirac Delta function,

$$\oiint_{\Omega(x)} \nabla^2 T^*(r) d\Omega(x) = -1 \tag{109}$$

And using the Gauss Divergence theorem on the left-hand-side, we find

$$\oint_{\Gamma(x)} \nabla T^*(r) \bullet \hat{n}\, d\Gamma(x) = -1 \tag{110}$$

Choosing now $\Omega(x)$ to be a circle centered at ξ and $\Gamma(x)$ to be its boundary for convenience, see Fig. 10. Although it is not necessary to choose a circle as any simple connected curve enclosing the source point will yield the same result, we find

$$\int_0^{2\pi} \frac{\partial}{\partial r}\big(A ln(r)\big)\hat{e}_r \bullet \hat{e}_r\,(r d\theta) = -1 \tag{111}$$

Where, $\hat{e}_r$ is the unit vector in the radial direction, the outward-drawn normal the circle is $\hat{n} = \hat{e}_r$, and $\nabla T^*(r) = (\partial T^*/\partial r)\hat{e}_r$ in polar coordinates. Evaluating the integral results in

$$2\pi A = -1 \tag{112}$$

Or finally, in 2D the fundamental solution is

$$T^*_{2D} = -\frac{1}{2\pi} ln(r) \tag{113}$$

Following the same argument in 3D and utilizing a sphere and spherical coordinates, one readily arrives at the 3D fundamental solution for the Laplace equation,

$$T^*_{3D} = \frac{1}{4\pi r} \tag{114}$$

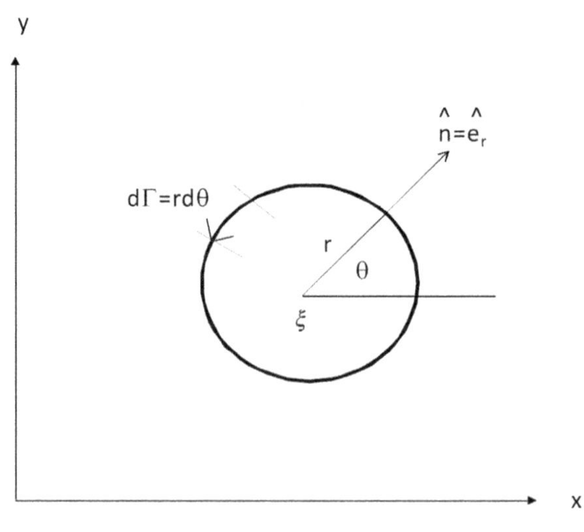

Figure 17. Circular domain useful to adopt to evaluate the strength of the source for the fundamental solution of the Laplace equation.

Chapter 3

NUMERICAL IMPLEMENTATION OF THE BEM

In typical BEM codes much of the steps we carried out in detail and analytically in the last chapter are automated in a manner similar to that in the FEM. We now consider the basic operation in the BEM code and the adoption of either classical continuous or discontinuous elements.

3.1 TWO-DIMENSIONAL BOUNDARY ELEMENTS

The generation of the BEM equations is accomplished by introducing shape functions in terms of a homogenous parameter, η, that ranges over [−1,1] in a manner suitable for Gauss-type quadratures and by parameterizing the geometry and dependent variables. That is any point (x,y) on any of the boundary element, $\Gamma_j(x)$, is parameterized as

$$
\begin{aligned}
x_j(\eta) &= \sum_{p=1}^{NGE} N^p(\eta) x_j^p \\
y_j(\eta) &= \sum_{p=1}^{NGE} N^p(\eta) y_j^p
\end{aligned}
\tag{115}
$$

Here, (x_j^p, y_j^p) denote the geometric nodes, $p = 1,2\ldots NGE$, utilized to discretize the element, $\Gamma_j(x)$, and $N^p(\eta)$ is the pth geometric shape function. Usually, linear or quadratic shape functions are used for this purpose and these, and other shape functions can be constructed by standard interpolation methods and they can be found in many FEM [53] and BEM references [1,2]. For the sake of convenience, we list in Table 1 the two most commonly utilized shape functions: linear and quadratic.

The second approximation required by the BEM is functional, necessary because although the integration along the entire boundary has been reduced to a summation of integrals over each boundary element, we do not know how the temperature and its normal derivative vary within each element, that is as we have seen, one or the other function or their ratio, in the case of a convective boundary condition, is usually known from the boundary conditions of the problem, but not both. Thus, we approximate the variation of T and q within each element by writing them in terms of their values at some nodal

TABLE 1. Linear and quadratic shape functions utilized for 2D boundary elements

Linear Shape Functions	Quadratic Shape Function
$N^1(\eta) = \frac{1}{2}(1-\eta)$	$N^1(\eta) = -\frac{1}{2}(1-\eta)\eta$
$N^2(\eta) = \frac{1}{2}(\eta+1)$	$N^2(\eta) = (\eta+1)(1-\eta)$
	$N^3(\eta) = \frac{1}{2}(\eta+1)\eta$

points using suitable interpolating shape functions again in terms of the homogeneous variable, η,

$$T_j(\eta) = \sum_{k=1}^{NPE} M^k(\eta) T_j^k$$
$$q_j(\eta) = \sum_{k=1}^{NPE} M^k(\eta) q_j^k \tag{116}$$

Here, T_j^k and q_j^k denote the nodal values at the, $k = 1,2\ldots NPE$, nodes utilized on each element to discretize the temperature and its normal derivative on each element, $\Gamma_j(x)$, and $M^k(\eta)$ is the kth field variable shape function.

There are two types of boundary elements that are typically used in the BEM: continuous and discontinuous elements. Such elements are displayed in Figs. 18 and 19, where isoparametric and sub-parametric elements are illustrated. The constant element we have used so far is a discontinuous subparametric element since the geometry is modeled as linear and T and q are modeled as constant. The distinctions between such elements are:

1. *Continuous elements*:
   a. The shape functions, $N^p(n)$, for the geometry and the shape functions, $M^k(n)$, for T and q, are the same; that is $N^k(\eta) = M^k(\eta)$.
   b. Consequently the geometric nodes and the T and q nodes coincide.
   c. Both T and q are continuously interpolated over the element.
   d. These elements pose issues at corner nodes where 1st kind BC's are imposed and when discontinuous boundary conditions are imposed. This can be resolved by developing additional independent equation(s) at a corner which can become onerous in 3D [43,44,54].
   e. Require a connectivity matrix in 3D.
   f. No proliferation of nodes.
2. *Discontinuous elements*:
   a. The shape functions, $N^p(\eta)$, for the geometry and the shape functions, $M^k(\eta)$, for T and q, are different; that is $N^k(\eta) \neq M^k(\eta)$ with the nodes for $M^k(\eta)$ offset to the interior of the geometric nodes within the boundary element. Consequently the geometric nodes and the T and q nodes do not coincide.
   b. Both T and q are discontinuous: interpolated between the variable nodes and extrapolated to the geometric end-nodes.
   c. These elements pose no issues at corner nodes where 1st kind BC's are imposed and naturally accommodate discontinuous boundary conditions. There is no need to develop additional independent equation(s) at a corner or a star point in 3D [42].
   d. Do not require a connectivity matrix in 3D.
   e. More accurate normal derivatives than continuous elements.
   f. Increase the number of variable nodes in the model over continuous elements. However, there are effective approaches to resolve such issues [49,51].

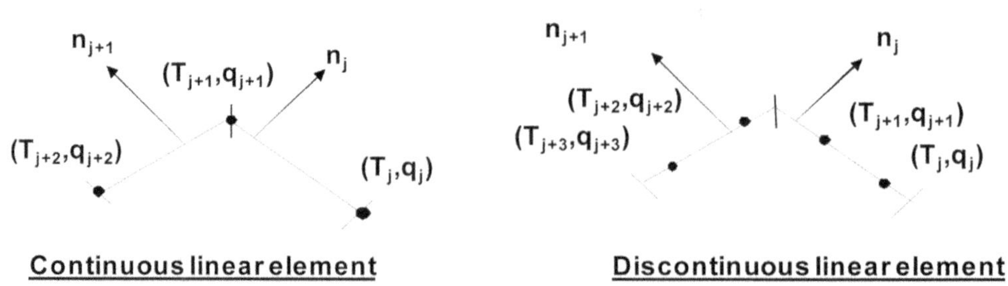

Figure 18. Contiguous continuous and discontinuous iso-parametric linear elements.

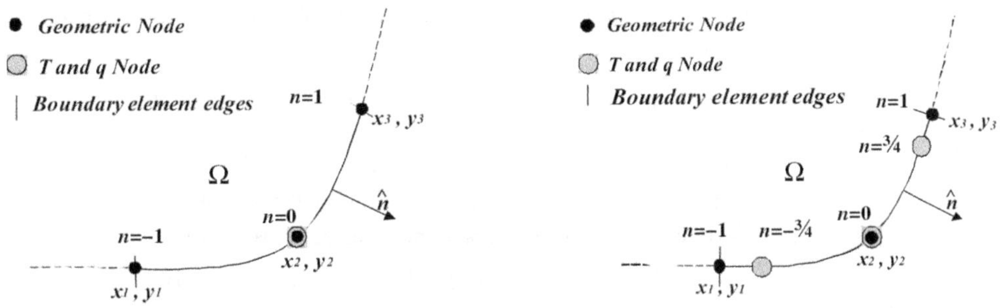

(a) discontinuous sub-parametric quadratic boundary element.

(b) discontinuous iso-parametric boundary element.

Figure 19. Discontinuous sub-parametric and iso-parametric boundary elements.

There are advantages and disadvantages in utilizing both types of elements. Both continuous and discontinuous elements have been successfully utilized in the course of the development of the BEM, and they both provide accurate results if they are implemented properly. The issue of proliferation of unknowns arising from discontinuous elements can readily be resolved by methods of domain decomposition as will be discussed in the following chapters. From the point of view of ease of coding and implementation of discontinuous boundary conditions often encountered in heat transfer, discontinuous elements are a natural choice, and they carry much less overhead than continuous elements. Regardless of whether continuous of discontinuous elements are chosen, it is important to use higher order elements, typically quadratic in 2D and at least bilinear in 3D, to take advantage of the high level of accuracy provided by the BEM.

Introducing these discretizations into the boundary integral equation, the following results

$$C(\xi_i)T(\xi_i)+\sum_{j=1}^{N_e}\sum_{k=1}^{NPE}\hat{H}_{ij}^k T_j^k=\sum_{j=1}^{N_e}\sum_{k=1}^{NPE}G_{ij}^k q_j^k \qquad i=1,2\ldots N_e \tag{117}$$

where

$$\begin{aligned}\hat{H}_{ij}^k &= \int_{-1}^{+1} q^*(\eta,\xi_i)M^k(\eta)d\Gamma_j(\eta)\\ G_{ij}^k &= \int_{-1}^{+1} T^*(\eta,\xi_i)M^k(\eta)d\Gamma_j(\eta)\end{aligned} \tag{118}$$

and

$$d\Gamma_j(\eta)=J_j(\eta)d\eta, \quad \text{where} \quad J_j(\eta)=\sqrt{\left(\frac{dx(\eta)}{d\eta}\right)^2+\left(\frac{dy(\eta)}{d\eta}\right)^2} \tag{119}$$

The BEM coefficients are evaluated numerically via Gauss-Legendre quadratures with special adaption when evaluating the CPV self-influence integrals, when evaluating integrals over elements that are very close to each other due to discretization of very thin regions such as airfoil trailing edges, and when post-processing interior points that are very close to the boundary. An example of a very robust adaptive quadrature is the DQUAGS routine in QUADPACK [52] that is based on the Gauss-Kronrod pairs, G_7K_{15}. The assembly of the coefficient matrices follow the same concepts developed for the FEM in part I of this book and, this is not repeated here.

Example 3.1: Sub-parametric quadratic boundary element. In such a case the geometry is modeled as quadratic so that, $NGE = 3$, and

$$\begin{aligned} x(\eta) &= N^1(\eta)x_1 + N^2(\eta)x_2 + N^3(\eta)x_3 \\ y(\eta) &= N^1(\eta)y_1 + N^2(\eta)y_2 + N^3(\eta)y_3 \end{aligned} \tag{120}$$

with the geometric shape functions given in Table 1, and with the temperature and its normal derivative are taken as constant. Consequently, with

$$T(\eta) = T_j \quad \text{and} \quad q(\eta) = q_j \quad \text{on} \quad \Gamma_j \tag{121}$$

A typical integral over an element, Γ_j, is then,

$$\begin{aligned} \int_{\Gamma_j} T(\eta) q^*(\eta, \xi_i) d\Gamma_j &= T_j \int_{-1}^{+1} q^*(\eta, \xi_i) J_j(\eta) d\eta \\ &= T_j \underbrace{\sum_{n=1}^{NGP} w_n q^*(\eta_n, \xi_i) J_j(\eta_n)}_{=\hat{H}_{ij}} \end{aligned} \tag{122}$$

where we used a NGP (number of Gauss points) Gauss-type rule where w_k and η_k are the Gauss-weights and points. The metrics needed to compute the Jacobian from Eq. (119) are then

$$\frac{dx(\eta)}{d\eta} = \sum_{k=1}^{3} \frac{dN^k(\eta)}{d\eta} x^k \quad \text{and} \quad \frac{dy(\eta)}{d\eta} = \sum_{k=1}^{3} \frac{dN^k(\eta)}{d\eta} y^k \tag{123}$$

Example 3.2: Iso-parametric discontinuous quadratic boundary element. In such a case the geometry is modeled as quadratic as in the previous example, however, the temperature and heat flux are modeled utilizing the discontinuous quadratic shape functions in columns 2 of Table 2 and with $NPE = NGE = 3$,

$$\begin{aligned} T(\eta) &= M^1(\eta)T_j^1 + M^2(\eta)T_j^2 + M^3(\eta)T_j^3 \\ q(\eta) &= M^1(\eta)q_j^1 + M^2(\eta)q_j^2 + M^3(\eta)q_j^3 \end{aligned} \tag{124}$$

A typical integral over an element, Γ_j, is then,

$$\begin{aligned} \int_{\Gamma_j} T(\eta) q^*(\eta, \xi_i) d\Gamma_j &= \sum_{k=1}^{3} T_j^k \int_{-1}^{+1} q^*(\eta, \xi_i) M^k(\eta) J(\eta) d\eta \\ &= \sum_{k=1}^{3} T_j^k \underbrace{\left[\sum_{n=1}^{NGP} w_n q^*(\eta_n, \xi_i) M^k(\eta_n) J(\eta_n) \right]}_{=\hat{H}_{ij}^k} \end{aligned} \tag{125}$$

where again, w_k and η_k are the Gauss-weights and points.

Table 2. Linear and quadratic discontinuous shape functions utilized for 2D boundary elements (variable end-nodes offset at $\eta = \pm 3/4$)

Linear Discontinuous Shape Functions	Quadratic Discontinuous Shape Function
$M^1(\eta) = \frac{1}{6}(3 - 4\eta)$	$M^1(\eta) = \frac{2}{9}(4\eta - 3)\eta$
$M^2(\eta) = \frac{1}{6}(3 + 4\eta)$	$M^2(\eta) = \frac{1}{9}(3 - 4\eta)(3 + 4\eta)$
	$M^3(\eta) = \frac{2}{9}(4\eta + 3)\eta$

Example 3.3: Continuous quadratic boundary element and the corner node problem. We illustrate an issue that concerns continuous elements when these are utilized to model corner points, especially where a Dirichlet condition is imposed. At a corner, the heat flux is double valued because of the non-uniqueness of the outward-drawn normal. This fact poses a numerical problem for linear, quadratic, and higher order continuous element formulations as flux and temperature nodes are located at geometric corners for these elements. Discontinuous prescription of boundary conditions poses a numerical problem as well. Improper treatment of this problem can severely degrade the solution around the corner region where high accuracy is often desired. Poor resolution of corner heat flux values can also degrade the BEM computed interior temperatures.

Most BEM codes that utilize continuous elements adopt a so-called double noded BEM solution to address this issue. Considering continuous elements, the BEM equations are written as

$$\sum_{j=1}^{L} H_{ij}T_j = \sum_{j=1}^{M} G_{ij}q_j \quad i = 1,2 \ldots N \tag{126}$$

Here, N is the number of nodes used to discretize the domain, $L = N$ and $M = 2N$ for linear elements, and $L = N/2$ and $M = 3N/2$ for quadratic elements (N has to even in this case). The coefficients H_{ij} and G_{ij} are numerically evaluated as before, and the components of q have been separated allowing for discontinuous boundary conditions and double valued normal derivatives of the temperature to be treated explicitly. The tangent to the boundary possesses a discontinuity at each node which is common to two elements, and the normal derivative of the temperature will be doubled valued there. Such node is termed a corner. There are three variables at such nodes: the temperature and two normal derivatives of the temperature.

Boundary conditions provide N of the $2N$ unknowns in Eq. (126) which allows the formation of a square system [H]{T} = [G]{q} that can be re-arranged in standard algebraic forma and solved. This is true except when a Dirichlet (temperature) condition is prescribed at the corner node. In this case an additional equation must be provided otherwise the two fluxes must be set arbitrarily equal to each other.

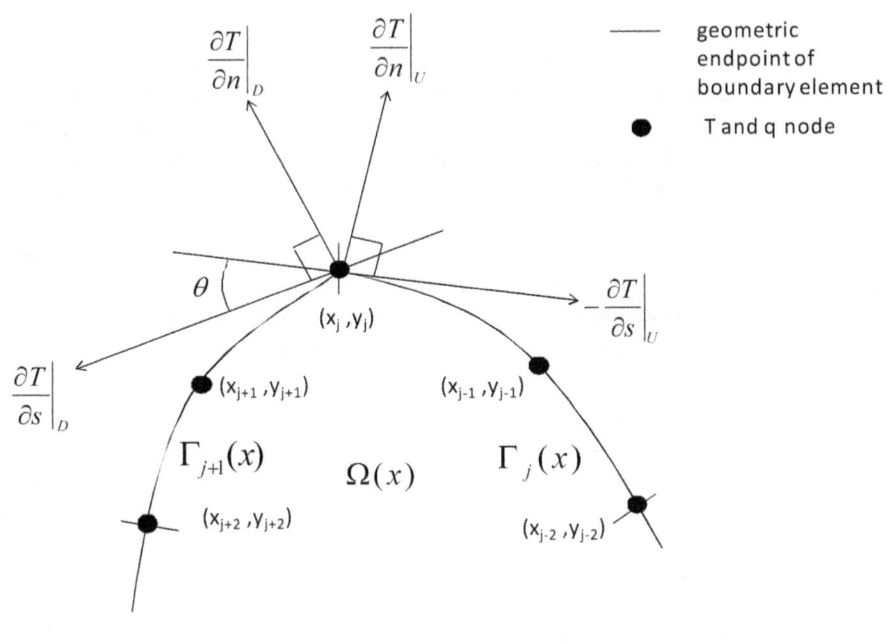

Figure 20. Two continuous quadratic elements Γ_j and Γ_{j+1} meet at the common corner node j.

Referring to Fig. 20, the upstream and downstream normal gradients can be written in terms of the tangential and normal gradients of the adjoining element as

$$\left.\frac{\partial T}{\partial n}\right|_U = \left[\left.\frac{\partial T}{\partial n}\right|_D \cos(\theta) - \left.\frac{\partial T}{\partial s}\right|_D \sin(\theta)\right]$$
$$\left.\frac{\partial T}{\partial n}\right|_D = \left[\left.\frac{\partial T}{\partial n}\right|_U \cos(\theta) + \left.\frac{\partial T}{\partial s}\right|_U \sin(\theta)\right] \tag{127}$$

The normal fluxes can be found independently of the BEM analysis in terms of tangential gradients that are calculated by differentiating the parametrized temperature along the element of interest. This introduces error in the flux calculation and computing the fluxes independently of the BEM does not capitalize on the smoothing effect which would be gleaned by incorporating the heat flux calculation in the BEM solution. To this end, re-arranging the above, we find the normal derivatives expressed in terms of the tangential derivatives

$$\left.\frac{\partial T}{\partial n}\right|_U = \frac{1}{\sin(\theta)}\left[\left.\frac{\partial T}{\partial s}\right|_U \cos(\theta) - \left.\frac{\partial T}{\partial s}\right|_D\right]$$
$$\left.\frac{\partial T}{\partial n}\right|_D = \frac{1}{\sin(\theta)}\left[\left.\frac{\partial T}{\partial s}\right|_U - \left.\frac{\partial T}{\partial s}\right|_D \cos(\theta)\right] \tag{128}$$

and subtracting these two equations there arises a relationship [43]

$$\left.\frac{\partial T}{\partial n}\right|_U = \pm\left[\left.\frac{\partial T}{\partial s}\right|_U + \left.\frac{\partial T}{\partial s}\right|_D\right]\left(\frac{\cos(\theta)-1}{\sin(\theta)}\right) + \left.\frac{\partial T}{\partial n}\right|_D \tag{129}$$

The + sign is applied for concave corners ($\theta < 180°$) and the – sign for re-entrant corners ($\theta > 180°$). Re-entrant corners nodes require finer discretization near the node than do concave corners. Since the boundary downstream of the node turns into the domain for concave corners, the temperatures used in calculating the tangential gradients are more representative of the temperatures near the node.

To implement this formulation, expressions for the tangential temperature gradients are determined by differentiating the parametrized temperature along the boundary element of interest with respect to the intrinsic coordinate used to parametrize the temperature along that element. The resulting expressions relate the tangential gradients to linear combinations of nodal temperatures. The turn angle θ is readily calculated from the local geometry.

Given the boundary conditions, Eq. (129) can now be substituted for the either the upstream or downstream corresponding flux component of the vector {q} at any corner node at which only the temperature is prescribed. This equation is expressed in terms of upstream and downstream nodal temperatures (arising from the approximation for the upstream and downstream tangential heat fluxes) plus the yet unsolved for normal heat flux (upstream or downstream). The structure of Eq. (129) allows for two ways in which to adjust the matrices to accommodate its inclusion. One way is to simply compute the tangential gradients prior to introducing Eq. (129). The [G] matrix and {q} vector are then appropriately adjusted to render the system solvable. The other way is to absorb the coefficients multiplying the nodal temperatures in Eq. (129) in the [H] matrix after multiplying by the appropriate G_{ij} coefficients and to adjust the matrix such that the unsolved heat flux is left as an unknown in the right hand side vector. For either case, once all corner nodes have been accounted for, all the known and unknown variables in the BEM equations are separated to the right and left-hand sides,

Node number	q-exact	% Error in q 26 elements	% Error in q 32 elements
1u	-0.1097	-2.726	0.738
1d	0.9889	0.147	-0.237
11u	0.9789	16.610	4.812
11d	1.2939	-10.233	3.231
21u	3.5791	0.861	0.081
21d	-0.5118	4.265	0.733
27u	-3.2979	0.425	0.431
27d	-3.4390	-0.349	-0.343

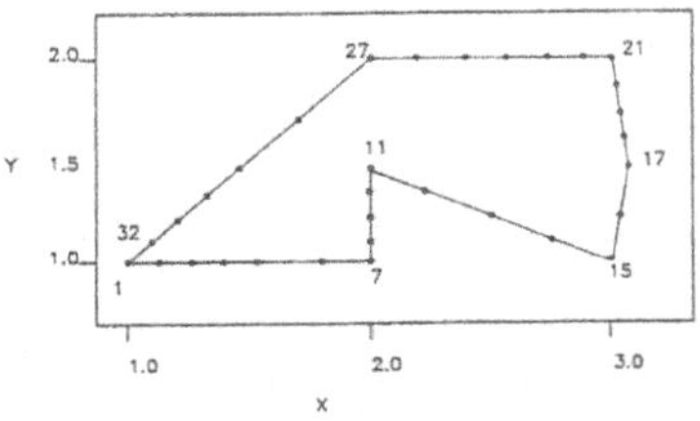

Figure 21. Test problem for the continuous element corner problem [43].

respectively, as before. The second of each pair of corner heat fluxes is then calculated after the BEM analysis using Eq. (129). Although slightly more complicated, the latter approach more fully integrates Eq. (129) into the BEM solution and yields better results.

An exact solution to the Laplace equation for the temperature is taken as T(x,y) = sin(x) cosh(y) to impose Dirichlet boundary conditions on all portions of the boundary displayed to the right in Fig. 21. This problem provides a severe test of the method by coupling an increased number of corner nodes with coarse nodal spacing. There are six concave corner nodes, with turn angles of 20, 45, 80, 90, 107, and 135°, and one re-entrant corner node of 117°. Two nodal distributions, one using twenty-six nodes and one using thirty-two nodes, were successively used to discretize the boundary. Quadratic elements were used to model the temperature and heat flux. Results for this problem are given in Fig. 21.

The corresponding problem in 3D leads to so-called star-points and overhead associated with the double node approach and Dirichlet conditions are such points can become quickly onerous.

3.2 THREE-DIMENSIONAL BOUNDARY ELEMENTS

In three dimensional BEM modeling the surface is represented using 2D shape functions so that any point (x,y,z) on any boundary element, $\Gamma_j(x)$, is parametrized utilizing homogeneous coordinates (η,ξ) as

$$x_j(\eta,\zeta) = \sum_{k=1}^{NGE} N^k(\eta,\zeta)x_j^k$$
$$y_j(\eta,\zeta) = \sum_{k=1}^{NGE} N^k(\eta,\zeta)y_j^k \qquad (130)$$
$$z_j(\eta,\zeta) = \sum_{k=1}^{NGE} N^k(\eta,\zeta)z_j^k$$

Here, (x_j^k, y_j^k, z_j^k) denote the *NGE* geometric nodes, k = 1,2…NGE, utilized to discretize the element, $\Gamma_j(x)$, and $N^k(\eta,\zeta)$ is the kth geometric shape function, see Fig. 22, where NGE = 8.

Considering a specific example in detail, a 3D constant boundary element is defined based on the bilinear shape functions, with four geometric nodes, *NGE* = 4, as illustrated in Fig. 23 with the corresponding geometric continuous shape functions given in Table 3, while the temperature and heat flux are taken as constant over the element so that,

$$T_j(\eta,\zeta) = T_j \quad \text{and} \quad q_j(\eta,\zeta) = q_j \qquad (131)$$

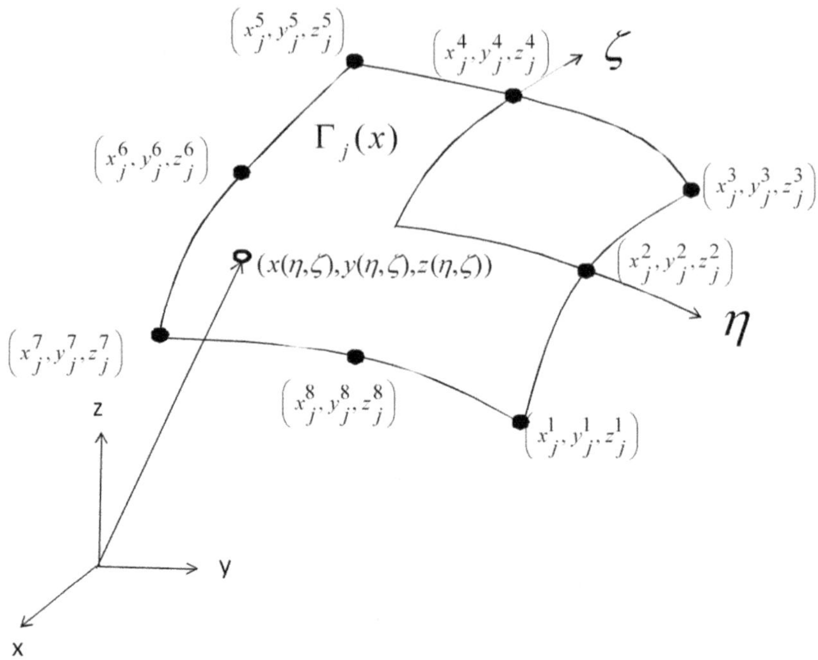

Figure 22. Geometric discretization of a 3D boundary element where *NGE* = 8.

TABLE 3. Continuous and discontinuous bilinear shape functions

Continuous Bilinear Shape Functions	Discontinuous Bilinear Shape Functions
$N^1(\eta,\zeta)=\frac{1}{4}(1-\eta)(1-\zeta)$	$M^1(\eta,\zeta)=\frac{1}{36}(3-4\eta)(3-4\zeta)$
$N^2(\eta,\zeta)=\frac{1}{4}(1+\eta)(1-\zeta)$	$M^2(\eta,\zeta)=\frac{1}{36}(3+4\eta)(3-4\zeta)$
$N^3(\eta,\zeta)=\frac{1}{4}(1+\eta)(1+\zeta)$	$M^3(\eta,\zeta)=\frac{1}{36}(3+4\eta)(3+4\zeta)$
$N^4(\eta,\zeta)=\frac{1}{4}(1-\eta)(1+\zeta)$	$M^4(\eta,\zeta)=\frac{1}{36}(3-4\eta)(3+4\zeta)$

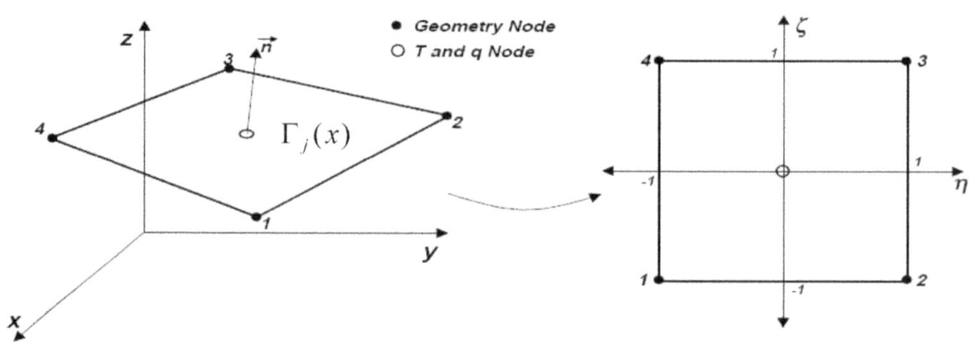

Figure 23. A constant 3D boundary element.

Consequently, a typical integral over the boundary element, $\Gamma_j(x)$, becomes using a *NGP* point Gauss-type quadrature rule,

$$\iint_{\Gamma_j} T(x)q^*(x,\xi_i)d\Gamma_j(x) = T_j \int_{-1}^{+1}\int_{-1}^{+1} q^*(\eta,\zeta,\xi_i)J_j(\eta,\zeta)d\eta d\zeta$$

$$= T_j \underbrace{\sum_{n=1}^{NGP}\sum_{m=1}^{NGP} w_n w_m q^*(\eta_n,\zeta_m,\xi_i)J_j(\eta_n,\zeta_m)}_{=\hat{H}_{ij}} \tag{132}$$

The Jacobian is evaluated in terms of the components of the metric tensor as,

$$J_j(\eta,\zeta) = \sqrt{g_{\eta\eta}g_{\zeta\zeta} - g_{\eta\zeta}^2} \tag{133}$$

with the components of the metric tensor, g_{ij}, are

$$g_{\eta\eta} = \left(\frac{\partial x}{\partial \eta}\right)^2 + \left(\frac{\partial y}{\partial \eta}\right)^2 + \left(\frac{\partial z}{\partial \eta}\right)^2$$

$$g_{\zeta\zeta} = \left(\frac{\partial x}{\partial \zeta}\right)^2 + \left(\frac{\partial y}{\partial \zeta}\right)^2 + \left(\frac{\partial z}{\partial \zeta}\right)^2 \tag{134}$$

$$g_{\eta\zeta} = \left(\frac{\partial x}{\partial \eta}\frac{\partial x}{\partial \zeta}\right) + \left(\frac{\partial y}{\partial \eta}\frac{\partial y}{\partial \zeta}\right) + \left(\frac{\partial z}{\partial \eta}\frac{\partial z}{\partial \zeta}\right)$$

while the components of the outward-drawn unit normal are evaluated as

$$n_x = \left(\frac{\partial y}{\partial \eta}\frac{\partial z}{\partial \zeta} - \frac{\partial z}{\partial \eta}\frac{\partial y}{\partial \zeta}\right)\Big/ J_j(\eta,\zeta)$$

$$n_y = \left(\frac{\partial z}{\partial \eta}\frac{\partial x}{\partial \zeta} - \frac{\partial x}{\partial \eta}\frac{\partial z}{\partial \zeta}\right)\Big/ J_j(\eta,\zeta) \tag{135}$$

$$n_z = \left(\frac{\partial x}{\partial \eta}\frac{\partial y}{\partial \zeta} - \frac{\partial y}{\partial \eta}\frac{\partial x}{\partial \zeta}\right)\Big/ J_j(\eta,\zeta)$$

The metrics are given as

$$\frac{\partial x(\eta,\zeta)}{\partial \eta} = \sum_{k=1}^{4}\frac{dN^k(\eta,\zeta)}{d\eta}x^k, \quad \frac{\partial x(\eta,\zeta)}{\partial \zeta} = \sum_{k=1}^{4}\frac{dN^k(\eta,\zeta)}{d\zeta}x^k, \quad \frac{\partial y(\eta,\zeta)}{\partial \eta} = \sum_{k=1}^{4}\frac{dN^k(\eta,\zeta)}{d\eta}y^k, \ldots \tag{136}$$

A bilinear 3D discontinuous boundary element, shown in Fig. 24, utilizes the bilinear continuous shape functions, $N^k(\eta,\zeta)$, in Table 3 for the spatial discretizations and bilinear discontinuous functions, $M^k(\eta,\zeta)$, for T and q. For such an element a typical integral over the element, becomes

$$\iint_{\Gamma_j} T(x)q^*(x,\xi_i)d\Gamma_j(x) = \sum_{k=1}^{4} T_j^k \int_{-1}^{+1}\int_{-1}^{+1} M^k(\eta,\zeta)q^*(\eta,\zeta,\xi_i)J_j(\eta,\zeta)d\eta d\zeta$$

$$= \sum_{k=1}^{4} T_j^k \underbrace{\sum_{n=1}^{NGP}\sum_{m=1}^{NGP} w_n w_m M^k(\eta_n,\zeta_m)q^*(\eta_n,\zeta_m,\xi_i)J_j(\eta_n,\zeta_m)}_{=\hat{H}_{ij}^k} \tag{137}$$

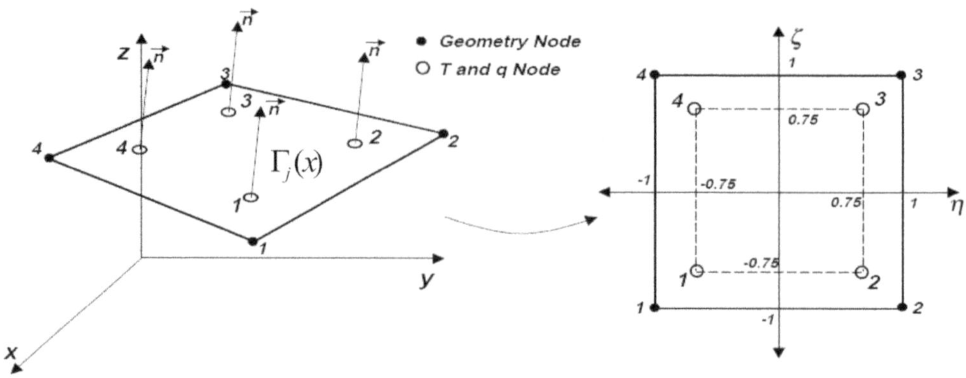

Figure 24. A bilinear discontinuous boundary element.

To better capture the geometry without increasing the number of unknowns, a discontinuous bi-quadratic sub-parametric boundary element is constituted by utilizing continuous bi-quadratic shape functions given in Table 4 for the geometry and bilinear discontinuous bilinear shape functions for T and q.

In practice constant elements should not be used in 3D as they provide low accuracy and slow convergence under refinement. They can effectively be used to provide an initial condition for an iterative solution such as in the case of non-linear heart conduction with convective boundary conditions or during iterative solution coupled with domain decomposition for large-scale problems. At least bilinear (continuous or discontinuous) elements or sub-parametric bi-quadratic elements should be used, see Fig. 25 and Table 4.

TABLE 4. Bi-quadratic continuous shape functions

$N^1(\eta,\zeta)=\frac{1}{4}(1-\eta)(1-\zeta)(-1-\eta-\zeta)$	$N^5(\eta,\zeta)=\frac{1}{2}(1-\eta^2)(1-\zeta)$
$N^2(\eta,\zeta)=\frac{1}{4}(1+\eta)(1-\zeta)(-1+\eta-\zeta)$	$N^6(\eta,\zeta)=\frac{1}{2}(1+\eta)(1-\zeta^2)$
$N^3(\eta,\zeta)=\frac{1}{4}(1+\eta)(1+\zeta)(-1+\eta+\zeta)$	$N^7(\eta,\zeta)=\frac{1}{2}(1-\eta^2)(1+\zeta)$
$N^4(\eta,\zeta)=\frac{1}{4}(1-\eta)(1+\zeta)(-1-\eta+\zeta)$	$N^8(\eta,\zeta)=\frac{1}{2}(1+\eta)(1-\zeta^2)$

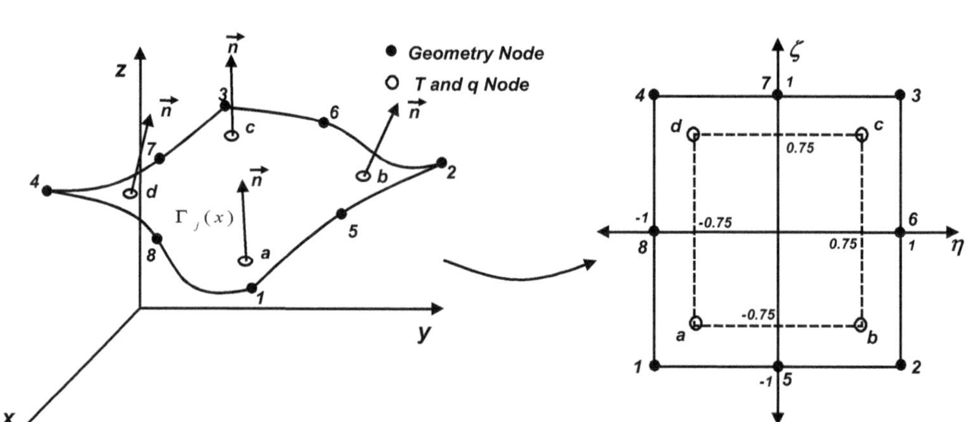

Figure 25. A discontinuous sub-parametric bi-quadratic boundary element.

We do not discuss continuous 3D elements in this book. However, the reader should be aware of the issues raised by such elements when a Dirichlet condition is imposed at a node and at so-called star-point nodes. There are several approaches that have been proposed to address these issued and the literature should be consulted.

3.3 ADAPTIVE QUADRATURE IN 3D

Just as in 2D, there arises the need to perform adaptive quadrature in 3D. Such quadratures are necessary for computation of self-influence integrals, for integration over thin regions where boundary elements can be very close to each other, and when evaluating an interior point close to a boundary element. Unfortunately, unlike in 2D, there are no general rules to perform adaptive quadrature in 3D. However, we can adapt the number of Gauss points in the quadrature rules to every integral depending of the variability of the integrand depending on the case at hand.

Non-Singular Elements: The influence coefficients G_{ij} are inversely proportional to the Euclidean distance between the field or integration point and the collocation point ξ_i, and the influence factors H_{ij} are inversely proportional to the square of the Euclidean distance between the field or integration point and the collocation point. Therefore, as the collocation point ξ_i is positioned closer to the integration element, the variability of the integrand increases, requiring an increase in the number of Gaussian points, *NGP*, for the integral approximation to provide a similar level of accuracy. Hence, a simple rule can be heuristically employed to change the number of Gaussian points depending on how far the collocation point is to the integration element. Basing our adaptation on a normalized measure of the Euclidean distance from the collocation point ξ_i to the element, $\Gamma_j(x)$, defined as

$$r_c = \frac{r_{ij}\big|_{\max}}{r_{ij}\big|_{\min}} \tag{138}$$

A heuristic rule that has proved effective in 3D BEM codes is provided in Table 5. Gauss rules are increased until $r_c > 10$ then the boundary element is subdivided and Gauss rules are applied to each sub-domain of the element.

Singular Elements: When computing self influence integrals, the collocation point ξ_i lies on the integration element. The integrand becomes singular and cannot be accurately

TABLE 5. Heuristic integration rules for non-singular 3D boundary elements

r_c Range	# of Integration Cells	Gaussian Points
$1 < r_c \le 1{,}2$	1	9 (3 × 3)
$1.2 < r_c \le 1.5$	1	25 (5 × 5)
$1.5 < r_c \le 2.0$	1	49 (7 × 7)
$2.0 < r_c \le 3.0$	1	100 (10 × 10)
$3.0 < r_c \le 5.0$	1	225 (15 × 15)
$5.0 < r_c \le 7.5$	1	400 (20 × 20)
$7.5 < r_c \le 10$	1	625 (25 × 25)
$10 < r_c \le 25$	4 (2 × 2)	4 × 900 (30 × 30)
$25 < r_c \le 50$	9 (3 × 3)	9 × 900 (30 × 30)
$50 < r_c \le 100$	16 (4 × 4)	16 × 900 (30 × 30)
$100 < r_c \le 200$	25 (5 × 5)	25 × 900 (30 × 30)
$200 < r_c \le 500$	36 (6 × 6)	36 × 900 (30 × 30)
$r_c > 500$	64 (8 × 8)	64 × 900 (30 × 30)

approximated by a standard Gaussian rule. Although the integral for the H_{ii} coefficient is strongly singular because its integrand is inversely proportional to the distance squared ($1/r^2$), this integral need not be computed directly as it can be instead evaluated using the equi-potential argument by summing the off-diagonal terms in the row of the [H] matrix, see Eqs. (53) and (54). The integral for the G_{ii} coefficient is weakly singular because its integrand is inversely proportional to the distance ($1/r$). This singularity may be avoided through an extra transformation of the local coordinate system. Fig. 26 shows such a primary segmentation of the singular discontinuous boundary elements.

For the constant boundary elements, twelve quadrilateral subdivisions have been made to the singular boundary element in addition to the shaded area that surround the source point, while, the discontinuous bilinear element the element has been subdivided into 10 quadrilateral subdivisions plus the shaded area that surrounds the source point. The adaptive Gauss quadrature procedure already discussed is implemented over the non-shaded segments.

On the shaded region that surrounds the collocation point, a local polar coordinate system becomes useful for our purposes, as illustrated in Fig. 27. Relating the homogeneous coordinates (η,ζ) to the local polar coordinates (ρ,θ) as

$$\begin{aligned} \eta &= \rho\cos(\theta) \\ \zeta &= \rho\sin(\theta) \end{aligned} \tag{139}$$

and the contribution over the shaded region of to the G_{ii} coefficient that we denote by G_{ii}^0 is comprised of the integrals over the 4 triangular sub-segments $\Gamma_{01}+\Gamma_{02}+\Gamma_{03}+\Gamma_{04}$, so that

$$\begin{aligned} G_{ii}^0 = &\int_{-\frac{\pi}{4}}^{\frac{\pi}{4}}\int_0^{1/\cos(\theta)} q^*(\eta,\zeta,\xi_i)J(\eta,\zeta)\rho_1 d\rho_1 d\theta_1 \\ &+\int_{\frac{\pi}{4}}^{\frac{3\pi}{4}}\int_0^{1/\sin(\theta)} q^*(\eta,\zeta,\xi_i)J(\eta,\zeta)\rho_2 d\rho_2 d\theta_2 \\ &+\int_{\frac{3\pi}{4}}^{\frac{5\pi}{4}}\int_0^{-1/\cos(\theta)} q^*(\eta,\zeta,\xi_i)J(\eta,\zeta)\rho_3 d\rho_3 d\theta_3 \\ &+\int_{\frac{5\pi}{4}}^{\frac{7\pi}{4}}\int_0^{-1/\sin(\theta)} q^*(\eta,\zeta,\xi_i)J(\eta,\zeta)\rho_4 d\rho_4 d\theta_4 \end{aligned} \tag{140}$$

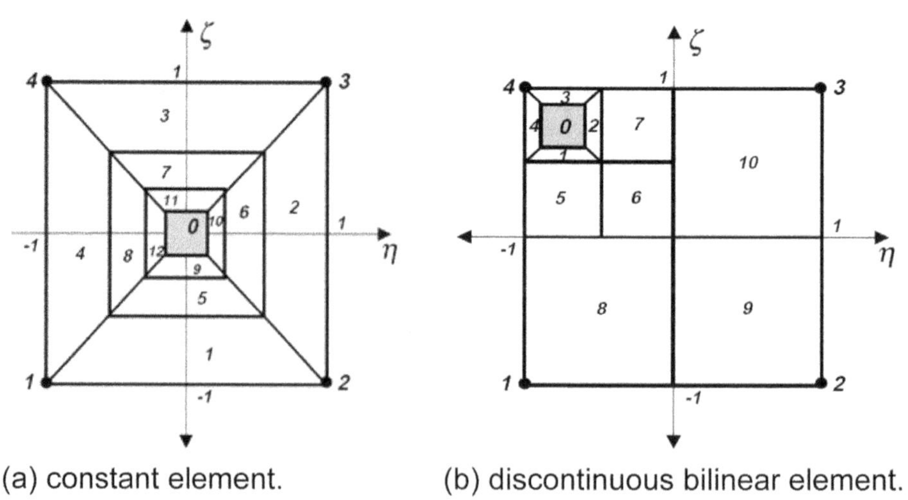

Figure 26. Segmentation for quadrature of a singular boundary element.

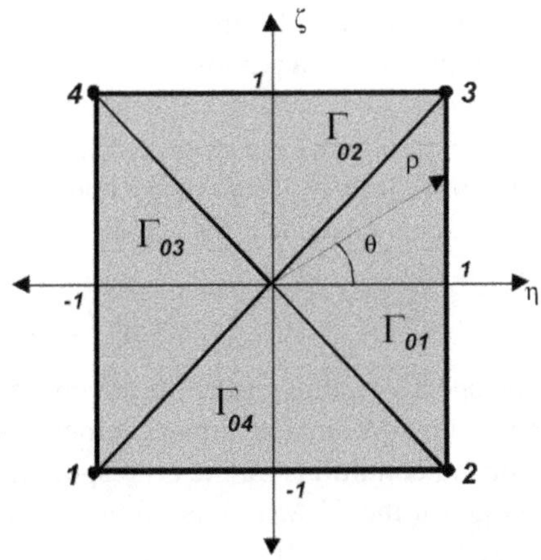

Figure 27. Sub-segmentation about the singular point and local polar coordinate system about the source point.

TABLE 6. Transformation of local polar coordinates (ρ,θ) onto $(-1,1)$

Γ_{01}	$\rho_1 = \frac{s+1}{2\cos(\theta_1)}$	$\theta_1 = \frac{\pi}{4}t$
Γ_{02}	$\rho_2 = \frac{s+1}{2\sin(\theta_2)}$	$\theta_2 = \frac{\pi}{4}(t+2)$
Γ_{03}	$\rho_3 = -\frac{s+1}{2\cos(\theta_3)}$	$\theta_3 = \frac{\pi}{4}(t+4)$
Γ_{04}	$\rho_4 = -\frac{s+1}{2\sin(\theta_4)}$	$\theta_4 = \frac{\pi}{4}(t+6)$

The Jacobian of the transformation weakens the (1/r) singularity and effectively concentrates the Gauss points close to the source point. These integrals are evaluated by Gauss quadrature by applying an additional transformation into (s,t) homogeneous coordinates provided in Table 6.

Finally, the G_{ii}^0 contribution to the G_{ii} coefficient is computed by numerically evaluating via Gauss quadratures, the four integrals

$$\begin{aligned} G_{ii}^0 = &\int_{-1}^{1}\int_{-1}^{1} q^*(\eta,\zeta,\xi_i)J(\eta,\zeta)\frac{\pi\rho_1}{8\cos(\theta_1)}dsdt \\ &+\int_{-1}^{1}\int_{-1}^{1} q^*(\eta,\zeta,\xi_i)J(\eta,\zeta)\frac{\pi\rho_2}{8\sin(\theta_2)}dsdt \\ &+\int_{-1}^{1}\int_{-1}^{1} q^*(\eta,\zeta,\xi_i)J(\eta,\zeta)\frac{-\pi\rho_3}{8\cos(\theta_3)}dsdt \\ &+\int_{-1}^{1}\int_{-1}^{1} q^*(\eta,\zeta,\xi_i)J(\eta,\zeta)\frac{-\pi\rho_4}{8\sin(\theta_4)}dsdt \end{aligned} \tag{141}$$

3.4 NUMERICAL SOLUTION OF THE BEM EQUATIONS

Upon re-arrangement of the BEM equations, the resulting algebraic system is fully-populated, non-symmetric and not diagonally dominant. For moderate number of unknowns encountered in most 2D models, these equations are effectively solved by direct methods such as Gauss elimination with partial pivoting and equilibration of the matrix or LU decomposition if a parametric study is undertaken varying the boundary conditions leading to several load vectors {b} while the coefficient matrix remains constant.

For larger problems, encountered in 3D, direct methods become prohibitive due to round-off. Consequently, iterative methods become a necessity. However, classical stationary methods such a Jacobi iteration, Gauss-Seidel or SOR, that solve the standard linear system by the general fixed-point iteration

$$\{x\}^{n+1} = [T]\{x\}^{n} + \{c\} \tag{142}$$

require the spectral radius of the iterative matrix [T] to be less than one. Unlike the Finite Difference, Finite Volume or Finite Element Method, where the Scarsborough criteria provides sufficient conditions for the corresponding iterative matrix to possess a spectral radius less than one, in the BEM this is often not the case, and the spectral radius of the corresponding stationary iterative matrix can often be very close to one leading to extremely slow convergence or greater than one leading to divergence.

The iterative solutions of the BEM equations should be undertaken by either more rapidly converging symmetric solvers such at conjugate gradient or bi-conjugate gradient methods with the BEM equations having been symmetrized and pre-conditioned or directly by a non-symmetric iterative solvers such at the generalized minimization of residuals (GMRES) method of Saad and Schultz [55] with restart and appropriate pre-conditioning.

Example 3.4: Comparison of Gauss-Seidel and Conjugate Gradient for the solution of the BEM equations. For the case of the 4 noded BEM example problem, the matrix system of Eq. (67) is solved using Gauss-Seidel (GS) and the conjugate gradient method (CGM), with a initial guess of {x} = {0} and a convergence criteria for the residual of 10^{-5}. The CGM converged in 4 iterations while the GS method took 13 iterations. The plot of the L_2 norm of the residuals is provided in Fig. 28. The GMRES algorithm converged in 3 iterations for this problem. The MATHCAD pseudo-codes for the conjugate gradient algorithm for a non-symmetric algebraic set of simultaneous equations as well as a basic the GMRES algorithm are provided in the Appendix to this chapter.

It should be noted that there are specialized approaches to solve large-scale BEM problems that arise in 3D. We will leave this topic to a later chapter and conclude this chapter on the basics of the numerical implementation of the BEM. Attention is now given to specific concerns of the application of the BEM to heat transfer.

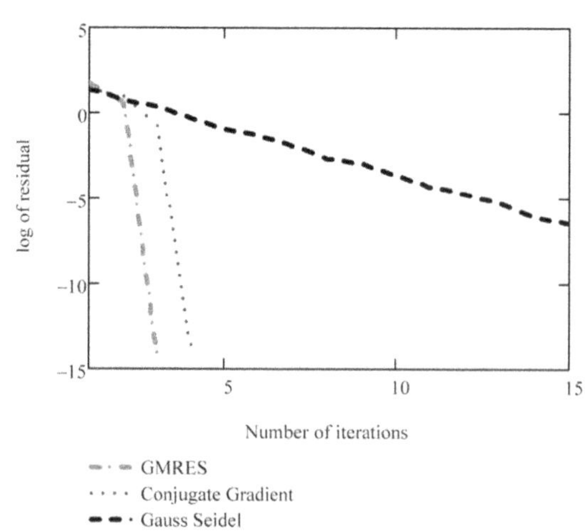

Iteration	GMRES	CG	GS
1	58.404	21.978	24.102
2	3.541	9.702	5.959
3	$5.024 \cdot 10^{-15}$	0.153	2.355
4		$1.297 \cdot 10^{-14}$	0.501
5			0.112
6			0.048
7			0.011
8			$2.178 \cdot 10^{-3}$
9			$9.573 \cdot 10^{-4}$
10			$2.424 \cdot 10^{-4}$
11			$4.345 \cdot 10^{-5}$
12			$1.892 \cdot 10^{-5}$
13			$5.246 \cdot 10^{-6}$
...			...

Figure 28. Comparison of the L_2 norm of the residual for the GS and CG iterative solution of the 4 noded BEM example problem.

APPENDIX A CONJUGATE GRADIENT AND GMRES MATHCAD PSEUDO-CODES

The Conjugate Gradient (CG) iterative solver: The CG-based iterative solver for the solution of the simultaneous equations,

$$[A]\{x\} = \{b\} \tag{143}$$

is based on utilizing the CG to minimize the quadratic functional

$$f(\{x\}) = \frac{1}{2}\{x\}^T[A]\{x\} - \{x\}^T\{b\} \tag{144}$$

that has a minimum at [A]{x} = {b} **if [A] is symmetric** as was observed by G. Temple [56]. When [A] is un-symmetric, then the symmetrized form

$$[\hat{A}]\{x\} = \{\hat{b}\}, \quad \text{where} \quad [\hat{A}] = [A]^T[A] \tag{145}$$

$$\{b\}^T = [A]^T\{b\}$$

is solved. In the CG, there are basic matrix vector operations,

$$[A]\{v\} \tag{146}$$

In the case of the BEM where the equations are symmetrized, these operations are carried out in two steps, that is

$$\{w\} = [\hat{A}]\{p\} \quad \rightarrow \quad \{v\} = [A]\{p\} \quad \text{then} \quad \{w\} = [A]^T\{v\} \tag{147}$$

As the conditioning number of $[\hat{A}]$ is the square of the conditioning number of [A], this procedure is often accompanied by a preconditioning of the symmetrized set of equations. Effective pre-conditioning includes diagonal pre-conditioning and incomplete LU preconditioning.

The MATHCAD pseudo-code for the CGM-iterative method accommodating a non-symmetric coefficient matrix is provided in Fig. 29. The input is the coefficient matrix [A], the right-hand side vector {b} and the initial guess for the solution {xo}. The procedure iterates until the L_2-norm of the residual vector is reduced to a value less than or equal to 10^{-5}. The code returns a vector containing the number of iterations to convergence, K, the solution vector, {x}, and the vector of the L_2-norm of the residual at each iteration.

Below are the MATHCAD results for the 4 node BEM problem in Example 2.2

Linear system generated by BEM and initial guess

$$A = \begin{pmatrix} -0.2695 & -0.1762 & 0.0062 & -0.1762 \\ -0.0533 & 0.5 & -0.0533 & -0.1476 \\ 0.0062 & -0.1762 & -0.2695 & -0.1762 \\ -0.0533 & -0.1476 & -0.0533 & 0.5 \end{pmatrix} \quad b = \begin{pmatrix} 14.758 \\ 17.62 \\ -50 \\ 17.62 \end{pmatrix} \quad xo = \begin{pmatrix} 0 \\ 0 \\ 0 \\ 0 \end{pmatrix}$$

Call the conjugate gradient non-symmetric solver

$$\begin{pmatrix} K\_CG \\ x\_CG \\ r\_CG \end{pmatrix} := \text{Conjugate\_gradiant\_non\_symm}\ (A, b, xo, k\max)$$

Number of iterations to convergence for the CG non-symmetric solver

$$K\_CG = 4$$

$$
\text{Conjugate\_gradient\_non\_symm}(A,b,x,kmax) := \left|
\begin{array}{l}
x_1 \leftarrow x \\
b \leftarrow A^T \cdot b \\
v \leftarrow A \cdot x_1 \\
r_1 \leftarrow b - (A)^T \cdot x_1 \\
res_1 \leftarrow |r_1| \\
\text{for } k \in 2..kmax \\
\quad \left|
\begin{array}{l}
p_2 \leftarrow r_1 \quad \text{if } k = 2 \\
\beta \leftarrow \dfrac{(r_{k-1})^T \cdot r_{k-1}}{(r_{k-2})^T \cdot r_{k-2}} \quad \text{if } k > 2 \\
p_k \leftarrow r_{k-1} + \beta \cdot p_{k-1} \quad \text{if } k > 2 \\
v \leftarrow A \cdot p_k \\
w \leftarrow (A)^T \cdot v \\
\alpha \leftarrow \dfrac{(r_{k-1})^T \cdot r_{k-1}}{(p_k)^T \cdot w} \\
x_k \leftarrow x_{k-1} + \alpha \cdot p_k \\
r_k \leftarrow r_{k-1} - \alpha \cdot w \\
K \leftarrow k \\
res_k \leftarrow |r_k| \\
\text{break if } res_k \le 10^{-5}
\end{array}
\right. \\
\begin{pmatrix} K \\ x \\ res \end{pmatrix}
\end{array}
\right.
$$

Figure 29. MATHCAD pseudo-code for the CG-iterative solution for a set of algebraic simultaneous equations.

Solution as a function of iterations

$$
x\_CG_1 = \begin{pmatrix} 0 \\ 0 \\ 0 \\ 0 \end{pmatrix} \quad x\_CG_2 = \begin{pmatrix} -30.534 \\ 61.511 \\ 57.881 \\ 61.511 \end{pmatrix} \quad x\_CG_3 \begin{pmatrix} -116.006 \\ 49.515 \\ 119.23 \\ 49.515 \end{pmatrix} \quad x\_CG_4 \begin{pmatrix} -117.458 \\ 49.999 \\ 117.466 \\ 49.999 \end{pmatrix}
$$

Residual norm as a function of iteration:

$$r\_CG = \begin{pmatrix} 21.978 \\ 9.702 \\ 0.153 \\ 1.297\times10^{-14} \end{pmatrix}$$

Non-symmetric iterative GMRES solver: The non-symmetric GMRES iterative solver, minimizes the residual of the algebraic set of equations over the same Krylov subspace that the CG utilizes in its search for a minimum. The notion is that since the Krylov subspace was optimal for CG application to solving the simultaneous symmetric system of equations, then it might be an appropriate choice for the non-symmetric case.

The GMRES iteration generates the k-th correction to the initial guess, $\{x_o\}$ as

$$\{x_k\} = \{x_0\} + \{z_k\} \tag{148}$$

Where the k-th correction is expressed as a series in terms of orthonormal bases, $\{q_i\}$, of the Kylov subspace, generated by the Arnoldi algorithm,

$$\{z_k\} = \sum_{i=1}^{k} y_i\{q_i\} = [Q_k]\{y_k\} \tag{149}$$

Where the columns of $[Q_k]$ are the orthonormal bases, $\{q_i\}$. The coefficients y_i are found by minimizing the residual in a least squares sense.

The residual can be expressed

$$\begin{aligned}\{r_k\} &= \{b\} - [A]\{x_k\} \\ &= \{r_0\} - [A][Q_k]\{y_k\}\end{aligned} \tag{150}$$

From the Arnoldi algorithm which is a modified Gram-Schmidt applied to recursively generate the orthonormal basis of the Krylov subspace with the first basis vector as the normalized initial residual $\{q_1\} = \{r_0\}/\|\{r_0\}\|$

$$h_{k+1,k}\{q_{k+1}\} = [A]\{q_k\} - \sum_{i=1}^{k} h_{i,k}\{q_i\} \tag{151}$$

the following result can be shown

$$\begin{aligned}[A][Q_k] &= [Q_k][H_k] + h_{k+1,k}\{q_k\}\{e_k\}^T \\ &= [Q_{k+1}][H_k^*]\end{aligned} \tag{152}$$

Here $\{e_k\}^T$ is the transpose of the k-th column of the identity matrix and $[H_k^*]$ is an upper Hessenberg matrix formed with the coefficients generated by the Arnoldi algorithm. For instance, for k = 4,

$$[H_4^*] = \begin{bmatrix} h_{1,1} & h_{1,2} & h_{1,3} \\ h_{2,1} & h_{2,2} & h_{2,3} \\ 0 & h_{3,2} & h_{3,3} \\ 0 & 0 & h_{4,3} \end{bmatrix} \tag{153}$$

It can then be shown that the k-th residual is expressed as

$$\{r_k\} = [Q_{k+1}](\beta\{e_1\}^T - [H_k^*]\{y_k\}) \tag{154}$$

where the norm of the initial residual is denoted by $\beta = \|\{r_0\}\|$.

The GMRES statement is then, to find the coefficients $\{y_k\}$ that minimize the k-th residual in a least-squares sense. Using the orthogonality property of $[Q_{k+1}]$ leads to the following least-squares problem

$$\text{find} \quad \{y_k\} \quad \min \rightarrow \left\| \beta\{e_1\} - [H_k^*]\{y_k\} \right\| \tag{155}$$

The QR method is used to solve this least-squares problem, with Givens rotations as the most effective means of generating the QR factors of the modified upper Hessenberg matrix $[H_k^*]$. That is, given that we find by Givens rotations

$$[H_k^*] = [Q_H^*][R_H^*] \tag{156}$$

the QR factors of $[H_k^*]$, then solving the least-squares problem of Eq. (155) requires the solution by back-substitution of

$$[R_H^*]\{y_k\} = [Q_H^*]^T \beta\{e_1\} \tag{157}$$

In practice, although there are no general rules as to when to restart, a good practice with BEM equations is to carry out the GMRES process over a number of iterations that is 10% of the dimension of [A] and to restart the GMRES iterative process with the current updated value of the solution as an initial guess and with its corresponding residual. The restart process is then repeated until convergence. This restarted implementation of GMRES is commonly denoted as GMRES(m) with m denoting the dimension of the Krylov subspace (Kmax in the MATHCAD routine below) that is utilized at each GMRES solution before each restart. Kmax can at most be the size of the NxN matrix [A]. Kmax is set arbitrarily to 50 in the displayed MATCHAD code and should be adjusted accordingly for each problem under consideration. Moreover, as with the CG method, this procedure is often accompanied by a pre-conditioning of the linear system of equations. Effective pre-conditioning includes diagonal pre-conditioning and incomplete LU pre-conditioning.

The GMRES requires two supporting routines, a Givens rotation routine and a routine to solve Eq. (157) by back-substitution. The MATHCAD pseudo-code and sample results are provided below for a non-preconditioned and no-restart GMRES iterative solution of the BEM equations.

$$
\text{Hessenberg\_back}(H, g) := \left|
\begin{array}{l}
N \leftarrow \text{length}(g) - 1 \\
y_N \leftarrow \dfrac{g_N}{H_{N,N}} \\
\text{for } i \in 1, 2 .. N - 1 \\
\quad \left| \begin{array}{l}
k \leftarrow N - i \\
y_k \leftarrow \dfrac{\left[g_k - \displaystyle\sum_{j=k+1}^{i+1} (H_{k,j} \cdot y_j) \right]}{H_{k,k}}
\end{array} \right. \\
y
\end{array}
\right.
$$

Below are the MATHCAD results for the 4 node BEM problem in Example 2.2.

```
Givens(H, g) := | j ← length(g)
                | h ← H
                | for k ∈ 1, 2..j − 1
                |   | c_k ← h_{k,k} / [(h_{k,k})^2 + (h_{k+1,k})^2]^(1/2)
                |   | s_k ← h_{k+1,k} / [(h_{k,k})^2 + (h_{k+1,k})^2]^(1/2)
                |   | for ii ∈ 1, 2..j
                |   |   for jj ∈ 1, 2..j
                |   |     | G_{ii,jj} ← 0  if ii ≠ jj
                |   |     | G_{ii,ii} ← 1
                |   | G_{k,k} ← c_k
                |   | G_{k,k+1} ← s_k
                |   | G_{k+1,k} ← −s_k
                |   | G_{k+1,k+1} ← c_k
                |   | h ← G·h
                |   | g ← G·g
                | (h)
                | (g)
```

```
GMRES(A, b, xo) := | K_max ← 50
                   | ro ← b − A·xo
                   | g_1 ← |ro|
                   | break  if g_1 < 10^-5
                   | q_1 ← ro / g_1
                   | q_1 ← q_1
                   | h_{1,1} ← q_1^T·A·q_1
                   | q_2 ← A·q_1 − h_{1,1}·q_1
                   | h_{2,1} ← |q_2|
                   | q_2 ← q_2 / h_{2,1}
                   | y_1 ← g_1 / h_{1,1}
                   | x_1 ← xo + q_1·y_1
                   | for j ∈ 2, 3..K_max
                   |   | q_j ← q_j
                   |   | for i ∈ 1, 2..j
                   |   |   h_{i,j} ← q_i^T·A·q_j
                   |   | sum ← A·q_j
                   |   | for i ∈ 1, 2..j
                   |   |   sum ← sum − h_{i,j}·q_i
                   |   | q_{j+1} ← sum
                   |   | h_{j+1,j} ← |q_{j+1}|
                   |   | q_{j+1} ← q_{j+1} / h_{j+1,j}
                   |   | g_{j+1} ← 0
                   |   | R ← Givens(h, g)
                   |   | y ← Hessenberg_back(R_1, R_2)
                   |   | res_j ← (R_2)_{j+1}
                   |   | K ← j
                   |   | (break)  if |res_j| ≤ 10^-5
                   | (K  )
                   | (y  )
                   | (q  )
                   | (res)
```

Linear system generated by BEM and initial guess
Call GMRES

$$A = \begin{pmatrix} -0.2695 & -0.1762 & 0.0062 & -0.1762 \\ -0.0533 & 0.5 & -0.0533 & -0.1476 \\ 0.0062 & -0.1762 & -0.2695 & -0.1762 \\ -0.0533 & -0.1476 & -0.0533 & 0.5 \end{pmatrix} \quad b = \begin{pmatrix} 14.758 \\ 17.62 \\ -50 \\ 17.62 \end{pmatrix} \quad xo = \begin{pmatrix} 0 \\ 0 \\ 0 \\ 0 \end{pmatrix}$$

$$\begin{pmatrix} K \\ y \\ q \\ r \end{pmatrix} := \mathrm{GMRES}(A, b, xo)$$

Reconstruct the solution at each iteration:

$$k := 1,2\ldots K \qquad K = 3$$

$$x_k := xo + \sum_{i=1}^{k} (y_i \cdot q_i)$$

The solution at each iteration:

$$x_1 = \begin{pmatrix} -25.836 \\ -30.846 \\ 87.831 \\ -30.846 \end{pmatrix} \quad x_2 = \begin{pmatrix} -110.368 \\ 52.763 \\ 121.507 \\ 52.763 \end{pmatrix} \quad x_3 = \begin{pmatrix} -117.458 \\ 49.999 \\ 117.466 \\ 49.999 \end{pmatrix}$$

The residual norm as a function of iteration:

$$r_k := b - A \cdot x_k \quad r\_GMRES_k := |r_k| \quad r\_GMRES = \begin{pmatrix} 58.404 \\ 3.541 \\ 5.024 \times 10^{-15} \end{pmatrix}$$

Chapter 4

STEADY HEAT CONDUCTION WITH VARIABLE HEAT CONDUCTIVITY

In this chapter we address several possible variation of the thermal conductivity that can addressed in the BEM formulation. Variation of the thermal conductivity with temperature when are large temperature gradients leads to non-linear heat conduction as k = K(T). In certain applications, the heat conductivity is anisotropic or orthotropic, the conductivity is a second order symmetric tensor, and the BEM methodology developed so far can readily accommodate such cases. Finally, the medium of interest may have a non-homogenous conductivity and this may be accommodated by piece-wise constant modeling of the variation of k, leading to concepts of domain decomposition that are used later in the book in the context of large-scale problems, may be accommodated in restricted cases by a transformation, or may be addressed in a more general sense through a new approach to the fundamental solution. Each of these cases are now addressed in turn.

4.1 NONLINEAR THERMAL CONDUCTIVITY

In the case of steady state non-linear heat conduction arising from the variation of the thermal conductivity as a function of temperature, leads to the following non-linear Poisson equation

$$\nabla^2 T + \frac{1}{k(T)}\frac{dK}{dT}\nabla T \bullet \nabla T = 0 \tag{158}$$

that formally leads to a domain integral in the BEM that can be treated by the Dual Reciprocity BEM [33]. However, the classical Kirchhoff transform [27] can be used to linearize the governing equation for this problem permitting solution of the resulting transformed equation by standard BEM as described in Bialecki and Nowak [57], Azevedo and Wrobel [58], and Bialecki and Nahlik [59]. Here, a new dependent variable, the Kirchhoff transform temperature, $U(T)$ is defined as the area under the $k(T)$ vs T curve

$$U(T) = \frac{1}{k_o}\int_{T_o}^{T} k(T)dT \tag{159}$$

where T_o is a reference temperature at which the reference thermal conductivity k_o is evaluated. The Kirchhoff transform is the area under the curve and, as such, is a monotonic single-valued function of temperature. The integral can be evaluated analytically or numerically via a quadrature, and the curve of $U(T)$ vs. T can readily be constructed, see Fig. 30. This curve should be made a subroutine or statement function that can be called at any time a temperature is required when operating in the Kirchhoff transform domain. The Kirchhoff transformation defines the dependent variable (with units of temperature) such that $k(T)dT = k_o dU$, and consequently

$$k(T)\nabla T = k_o \nabla U \tag{160}$$

Introducing these results into the governing heat conduction equation, leads to the Laplace equation in the Kirchhoff temperature

$$\nabla^2 U = 0 \tag{161}$$

Imposed temperature and heat flux conditions also transform linearly to give:

$$\begin{aligned} T\big|_{\vec{r}_s} = T_s \quad &\rightarrow \quad U\big|_{\vec{r}_s} = U(T_s) \\ -k\frac{\partial T}{\partial n}\bigg|_{\vec{r}_s} = q_s \quad &\rightarrow \quad -k_o\frac{\partial U}{\partial n}\bigg|_{\vec{r}_s} = q_s \end{aligned} \tag{162}$$

where $\vec{r}_s$ denotes a point on the surface. Consequently, the developments of the BEM for the Laplace equation apply directly to non-linear heat conduction with the Kirchhoff as dependent variable. In the case of convective boundary conditions, the transformation is non-linear

$$-k\frac{\partial T}{\partial n}\bigg|_{\vec{r}_s} = h(T - T_{ref}) \quad \rightarrow \quad -k_o\frac{\partial U}{\partial n}\bigg|_{\vec{r}_s} = h\left[T(U) - U(T_{ref})\right] \tag{163}$$

and iteration must be used [51, 58, and 59].

Example 4.1: If the heat conductivity varies linearly with temperature as

$$k(T) = k_o + \beta(T - T_o) \tag{164}$$

Then applying the definition of the Kirchhoff transform

$$U(T) = \left(\frac{\beta}{2k_o}\right)T^2 + \left(1 - \frac{\beta T_o}{k_o}\right)T + \left(\frac{\beta}{2k_o}T_o^2 - T_o\right) \tag{165}$$

and the inverse is found from the quadratic equation as

$$T(U) = \left(T_o - \frac{k_o}{\beta}\right) \pm \sqrt{\left(\frac{k_o}{\beta}\right)^2 + \frac{2k_o U}{\beta}} \tag{166}$$

Where the positive root is taken for $\beta > 0$ and the negative root is taken for $\beta < 0$.

Example 4.2: Consider implementation of the Kirchhoff transform in a constant BEM formulation, the discretized BEM equations are

$$[H]\{U\} = [G]\{q\} \tag{167}$$

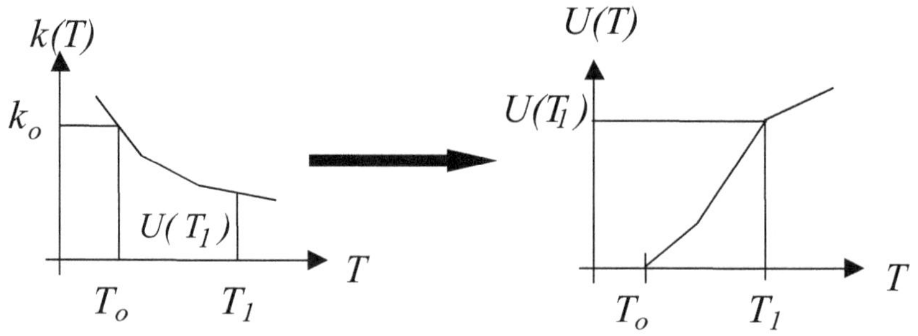

Figure 30. The Kirchhoff transform $U(T)$ is a unique monotonic increasing function of T can be found analytically or numerically.

where,

$$q = -\frac{q_s}{k_o} \qquad \text{heat flux prescribed}$$
$$q = -\frac{h}{k_o}\left[T(U) - U(T_s)\right] \qquad \text{convection prescribed} \tag{168}$$
$$q = -\frac{\sigma\varepsilon}{k_o}\left[T^4(U) - U^4(T_s)\right] \qquad \text{radiation prescribed}$$

and where σ is the Stefan-Boltzmann constant and ϵ is the emissivity. These equations can be cast in the standard non-linear form

$$\{F(\underset{\sim}{X})\} = [H]\{U\} - [G]\{q\} \tag{169}$$

with

$$\begin{aligned} X_j &= q_j \quad \text{when } T \text{ is prescribed} \\ X_j &= U_j \quad \text{otherwise} \end{aligned} \tag{170}$$

These equations are readily solved by the Newton-Raphson method

$$\{X\}^{n+1} = \{X\}^n + \{\Delta X\}^n \tag{171}$$

The correction vector, $\Delta\underset{\sim}{X}^n$, is found at each iteration by solving

$$[J(\underset{\sim}{X}^n)]\{\Delta X^n\} = -\{F(\underset{\sim}{X}^n)\} \tag{172}$$

The elements of the Jacobian matrix are found as:

(a) temperature is prescribed: $X_j = q_j \Rightarrow J_{i,j} = -G_{i,j}$

(b) any other boundary condition prescribed: $X_j = U_j \Rightarrow J_{i,j} = H_{i,j} - G_{i,j}\dfrac{\partial q_j}{\partial U_j}$ (173)

Given the relationships for q in Eq. (168), we have

$$\frac{\partial q_j}{\partial U_j} = 0 \qquad \text{heat flux prescribed}$$
$$\frac{\partial q_j}{\partial U_j} = -\frac{h}{k_o}\frac{\partial T}{\partial U_j} = -\frac{h}{k(T_j)} \qquad \text{convection prescribed} \tag{174}$$
$$\frac{\partial q_j}{\partial U_i} = -\frac{4\sigma\varepsilon}{k(T_i)}T_j^3 \qquad \text{radiation prescribed}$$

and T_j is obtained from the $U(T)$ vs. T relationship. An initial guess for this system of equations may be obtained by linearizing the convective boundary condition as $q = -\frac{h}{k_o}\left[U - U(T_s)\right]$ and solving the linear system of BEM equations.

4.2 ANISOTROPIC HEAT CONDUCTIVITY

In certain applications to laminates, composites, and crystals, the thermal conductivity is anisotropic as its value varies with orientation and it is formally a second order symmetric tensor $k_{ij} = k_{ji}$, with positive entries in the diagonal, $k_{ii} > 0$, with the off- diagonals limited by the relationship, $k_{ii}k_{jj} - k_{ij}^2 > 0$ [27]. The heat flux vector is then

$$q_i = -k_{ij}\frac{\partial T}{\partial x_j} \tag{175}$$

and the heat conduction equation becomes

$$\nabla \bullet \left(-[k] \bullet \nabla T\right) = 0 \quad \text{or} \quad \frac{\partial}{\partial x_i}\left(-k_{ij}\frac{\partial T}{\partial x_j}\right) = 0 \tag{176}$$

Given the conductivity tensor expressed in the reference coordinate system, solving the eigenvalue problem

$$\begin{vmatrix} k_{11}-\lambda & k_{12} & k_{13} \\ k_{21} & k_{22}-\lambda & k_{23} \\ k_{31} & k_{32} & k_{33}-\lambda \end{vmatrix} = 0 \tag{177}$$

produces the principal conductivities, λ_1, λ_2 and λ_3. The direction cosines L_{ij} of the principal directions, x_i', with respect to the reference coordinate system, x_i, are determined by solving

$$\begin{vmatrix} k_{11}-\lambda_i & k_{12} & k_{13} \\ k_{21} & k_{22}-\lambda_i & k_{23} \\ k_{31} & k_{32} & k_{33}-\lambda_i \end{vmatrix} \begin{vmatrix} L_{i1} \\ L_{i2} \\ L_{i3} \end{vmatrix} = 0 \tag{178}$$

$$L_{i1}^2 + L_{i2}^2 + L_{i3}^2 = 1$$

The rotated coordinate direction, x_i', that diagonalizes the conductivity matrix is related to the original Cartesian coordinate system by

$$\{x'\} = [L]\{x\} \quad \text{or} \quad x_i' = L_{ij}x_j \tag{179}$$

The governing equation becomes in reference to the principal directions, x_i',

$$\lambda_1\frac{\partial^2 T}{\partial x'^2} + \lambda_2\frac{\partial^2 T}{\partial y'^2} + \lambda_3\frac{\partial^2 T}{\partial z'^2} = 0 \tag{180}$$

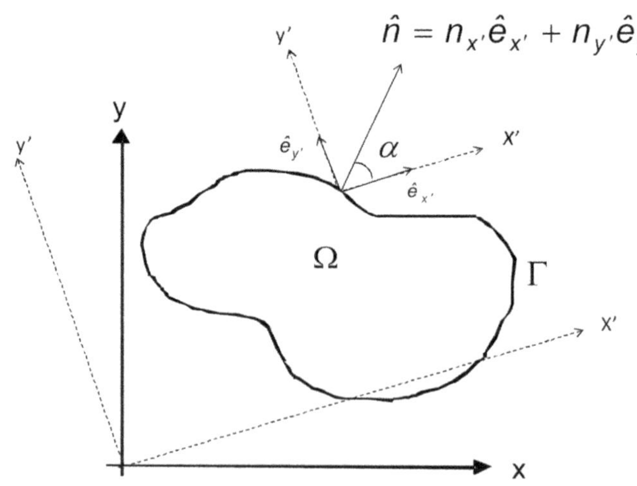

Figure 31. Rotation to principal directions (x',y').

This orthotropic form of the governing equation is also referred to as D'Arcy's equation, and it appears in applications of flow through porous media. It can be reduced to the Laplace equation by another coordinate transformation, namely the scaling

$$z_i = \frac{x_i'}{\sqrt{\lambda_i}} \tag{181}$$

So that the governing equation becomes in the scaled rotated coordinate system

$$\frac{\partial^2 T}{\partial z_1^2} + \frac{\partial^2 T}{\partial z_2^2} + \frac{\partial^2 T}{\partial z_3^2} = 0 \tag{182}$$

It should be noted that the Dirac delta function now appears in scaled coordinates, and one must account for the property of the Dirac delta function [67]

$$\delta[a(x - x_i)] = \frac{1}{|a|}\delta(x - x_i) \tag{183}$$

To take know fundamental solution in the scaled z-coordinate system to the principal coordinate system. For example, 2D and 3D fundamental solutions are

$$\begin{aligned} T^*(x',y';x_i',y_i') &= \frac{-1}{2\pi\sqrt{\lambda_1\lambda_2}} \ln\sqrt{\frac{1}{\lambda_1}(x'-x_i')^2 + \frac{1}{\lambda_2}(y'-y_i')^2} \\ T^*(x',y',z';x_i',y_i',z_i') &= \frac{1}{4\pi\sqrt{\lambda_1\lambda_2\lambda_3}} \frac{1}{\sqrt{\frac{1}{\lambda_1}(x'-x_i')^2 + \frac{1}{\lambda_2}(y'-y_i')^2 + \frac{1}{\lambda_3}(z'-z_i')^2}} \end{aligned} \tag{184}$$

In the principal coordinate frame, the normal derivative of the fundamental solution, q^*, is given by

$$q^* = \nabla_{x'} T^* \bullet \hat{n}' \tag{185}$$

Where $\nabla_{x'} T$ is the gradient taken with respect to the principal directions and $\hat{n}'$ is the outward-drawn normal defined with respect to the principal directions, see Fig. 31. For the general case of anisotropy, the heat flux is written in the principal coordinates according to tensor transformation rules as

$$\vec{q} = -[L][k][L]^T \nabla_{x'} T \quad \text{or} \quad q_i{}' = -L_{im} k_{mn} L_{jn} \frac{\partial T}{\partial x_j'} \tag{186}$$

Thus a standard BEM code can be adopted for general anisotropy by first rotating to the principal coordinate frame and using the scaled fundamental solution given above instead of the isotropic fundamental solution. Note that if the conductivity of the medium is already in orthotropic form, then there is no need for rotation to principal axes, and the formulation is simply carried out only due to the scaling.

4.3 NON-HOMOGENOUS THERMAL CONDUCTIVITY

In a non-homogeneous medium, the thermal conductivity may also vary with position and this poses a problem for BEM formulations as formally, the adjoint equation is variable coefficient and finding a fundamental solution is not possible. Shaw et al. [60–62] found transformations for linear and piecewise-linear non-homogeneous variations of the thermal conductivity that lead to solutions in terms of the axi-symmetric BEM formulation. A different approach is taken by Divo and Kassab [6,63–65] to derive a generalized boundary integral formulation valid for general heterogeneous media.

We will consider the problem of piecewise non-homogenous media as this leads to the important concept of domain decomposition that will be used later for large-scale BEM calculations and will be used in the context of meshless methods as well. This approach was first proposed by Rizzo and Shippy [66] in the context of piece-wise homogeneous elasticity. Consider a heat conducting medium constituted of two regions Ω_1 whose conductivity is k_1 and Ω_2 whose conductivity is k_2. Both regions are governed by the Laplace equation, so that the standard BEM formulation is applied to each region leading to the standard forms

$$\begin{aligned} [H_1]\{T_1\} &= [G_1]\{q_1\} \quad \text{for} \quad \Omega_1 \\ [H_2]\{T_2\} &= [G_2]\{q_2\} \quad \text{for} \quad \Omega_2 \end{aligned} \tag{187}$$

It is clear that the two domains share an interface, denoted as $\Gamma_{1,I}$ and $\Gamma_{2,I}$ whose outward-drawn normals point in opposite directions. The BEM equations are assembled into a global set of equations for the unknowns by requiring the continuity of of the temperature and heat flux at the common interface. To this end, the nodal temperatures and fluxes in each domain are identified as belonging either to the external boundary or to the interface, see Fig. 32. For region Ω_1:

$$\begin{aligned} &\{T_1^{\Gamma_1}\} \text{ and } \{q_1^{\Gamma_1}\} \quad \text{nodal temperatures and fluxes on boundary but not on interface} \\ &\{T_1^{\Gamma_{1,I}}\} \text{ and } \{q_1^{\Gamma_{1,I}}\} \quad \text{nodal temperatures and fluxes on interface but not on boundary} \end{aligned} \tag{188}$$

and for region Ω_2:

$$\begin{aligned} &\{T_2^{\Gamma_1}\} \text{ and } \{q_2^{\Gamma_1}\} \quad \text{nodal temperatures and fluxes on boundary but not on interface} \\ &\{T_2^{\Gamma_{2,I}}\} \text{ and } \{q_2^{\Gamma_{2,I}}\} \quad \text{nodal temperatures and fluxes on interface but not on boundary} \end{aligned} \tag{189}$$

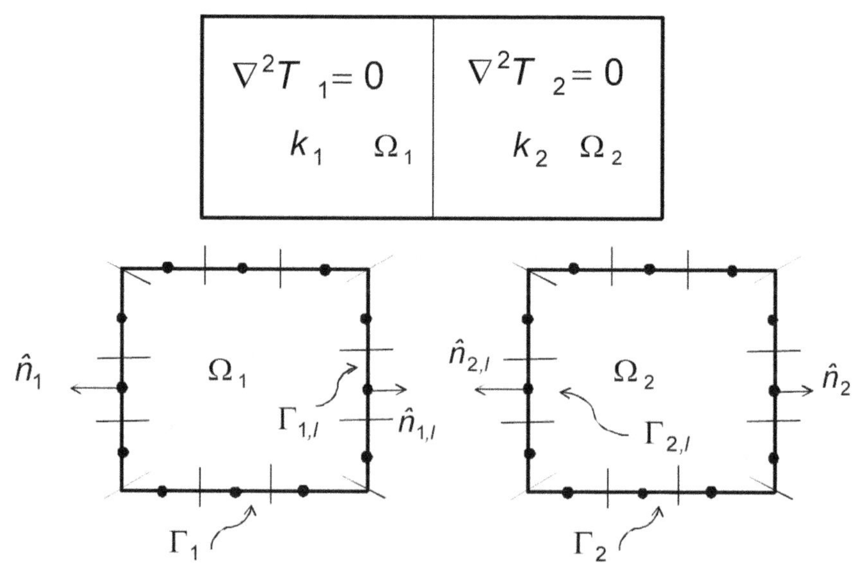

Figure 32. Piecewise non-homogeneous medium with constant element BEM discretization of each subregion.

Consequently, the BEM equations for each region can be segregated, and for region Ω_1 we have,

$$\begin{bmatrix} H_1^{\Gamma_1} & H_1^{\Gamma_{1,I}} \end{bmatrix} \begin{Bmatrix} T_1^{\Gamma_1} \\ T_1^{\Gamma_{1,I}} \end{Bmatrix} = \begin{bmatrix} G_1^{\Gamma_1} & G_1^{\Gamma_{1,I}} \end{bmatrix} \begin{Bmatrix} q_1^{\Gamma_1} \\ q_1^{\Gamma_{1,I}} \end{Bmatrix} \tag{190}$$

While for region Ω_2 we have,

$$\begin{bmatrix} H_2^{\Gamma_2} & H_2^{\Gamma_{2,I}} \end{bmatrix} \begin{Bmatrix} T_2^{\Gamma_2} \\ T_2^{\Gamma_{2,I}} \end{Bmatrix} = \begin{bmatrix} G_2^{\Gamma_2} & G_2^{\Gamma_{2,I}} \end{bmatrix} \begin{Bmatrix} q_2^{\Gamma_2} \\ q_2^{\Gamma_{2,I}} \end{Bmatrix} \tag{191}$$

Continuity of temperature and heat flux at the interface requires that

$$\begin{aligned} T_1^{\Gamma_{1,I}} &= T_2^{\Gamma_{2,I}} = T_I^{\Gamma_I} \\ k_1 q_1^{\Gamma_{1,I}} &= -k_2 q_2^{\Gamma_{2,I}} \end{aligned} \tag{192}$$

where it is recognized explicitly that physically there is only one interface, $\Gamma_{1,I} = \Gamma_{2,I} = \Gamma_I$, that the nodal interface temperatures are the same, and that the nodal heat fluxes are equal and opposite, with the convention that the normal to the interface is chosen to be the normal to $\Gamma_{1,I}$ and the interface flux $q_I^{\Gamma_I}$ is referenced to the $\Gamma_{1,I}$ side. Introducing the interface conditions, the BEM equations for the two regions are then assembled as

$$\begin{bmatrix} H_1^{\Gamma_1} & H_1^{\Gamma_{1,I}} & 0 \\ 0 & H_2^{\Gamma_{2,I}} & H_2^{\Gamma_2} \end{bmatrix} \begin{Bmatrix} T_1^{\Gamma_1} \\ T_I^{\Gamma_I} \\ T_2^{\Gamma_2} \end{Bmatrix} = \begin{bmatrix} G_1^{\Gamma_1} & G_1^{\Gamma_{1,I}} & 0 \\ 0 & -\left(\dfrac{k_1}{k_2}\right) G_2^{\Gamma_{2,i}} & G_2^{\Gamma_2} \end{bmatrix} \begin{Bmatrix} q_1^{\Gamma_1} \\ q_I^{\Gamma_I} \\ q_2^{\Gamma_2} \end{Bmatrix} \tag{193}$$

The boundary conditions are imposed on external boundaries and the above are re-arranged into the standard form [A]{x} = {b} with the external boundary nodal unknowns and the interface temperatures and fluxes gathered into the unknown vector {x}. It is worth noting that the coefficient matrix is now banded and, the more subregions we introduce, the more banded the BEM coefficient matrix becomes and thus efficient sparse matrix direct or iterative solvers can be adopted. This feature is capitalized upon in 3D codes and will be utilized in a 3D iterative method for large-scale BEM application heat transfer as described in Chapter 6.

Example 4.3: Consider a problem in a rectangular region provided to the right and a 4 constant BEM discretization for each region. The exact temperature is a function of x-only since there is nothing to cause a variation of the temperature in the y-direction. The exact solution to the problem is readily found:

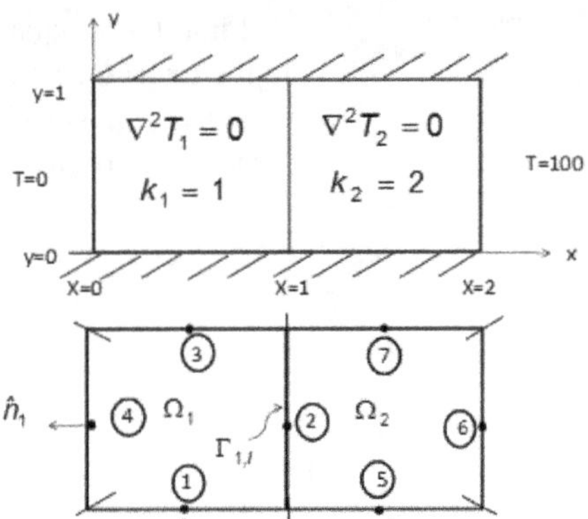

$$\begin{aligned} T_1(x) &= 100\left(\frac{2x}{3}\right) \\ T_2(x) &= 100\left(\frac{x+1}{3}\right) \end{aligned} \tag{194}$$

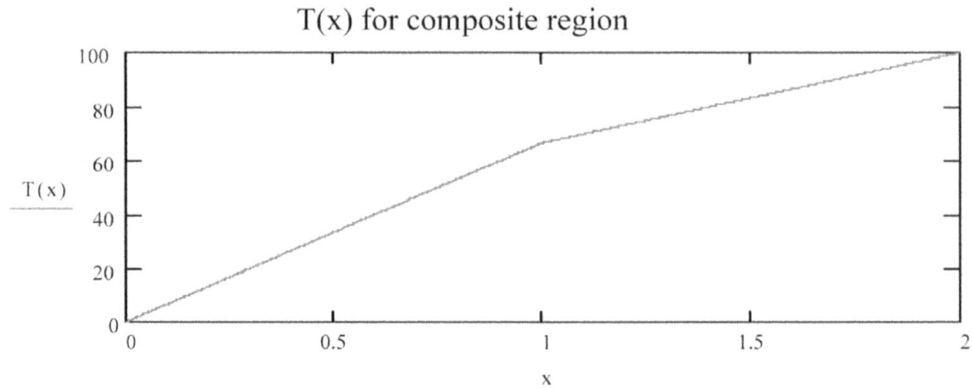

Figure 33. Temperature profile for composite medium and at the interface: temperature is continuous, temperature gradient is discontinuous, but heat flux is continuous.

Both region 1 and region 2 will generate same BEM matrix equations, that is we find that for region Ω_1 we have $[H_1]\{T_1\} = [G_1]\{q_1\}$:

$$\begin{bmatrix} 0.5 & -0.1762 & -0.1476 & -0.1762 \\ -0.1762 & 0.5 & -0.1762 & -0.1476 \\ -0.1476 & -0.1762 & 0.5 & -0.1762 \\ -0.1762 & -0.1476 & -0.1762 & 0.5 \end{bmatrix} \begin{Bmatrix} T_1 \\ T_2 \\ T_3 \\ T_4 \end{Bmatrix} = \begin{bmatrix} 0.2695 & 0.0533 & -0.0062 & 0.0533 \\ 0.0533 & 0.2695 & 0.0533 & -0.0062 \\ -0.0062 & 0.0533 & 0.2695 & 0.0533 \\ 0.0533 & -0.0062 & 0.0533 & 0.2695 \end{bmatrix} \begin{Bmatrix} q_1 \\ q_2 \\ q_3 \\ q_4 \end{Bmatrix} \tag{195}$$

and for that for region Ω_2 we have $[H_2]\{T_2\} = [G_2\{q_2\}$:

$$\begin{bmatrix} 0.5 & -0.1762 & -0.1476 & -0.1762 \\ -0.1762 & 0.5 & -0.1762 & -0.1476 \\ -0.1476 & -0.1762 & 0.5 & -0.1762 \\ -0.1762 & -0.1476 & -0.1762 & 0.5 \end{bmatrix} \begin{Bmatrix} T_5 \\ T_6 \\ T_7 \\ T_2 \end{Bmatrix} = \begin{bmatrix} 0.2695 & 0.0533 & -0.0062 & 0.0533 \\ 0.0533 & 0.2695 & 0.0533 & -0.0062 \\ -0.0062 & 0.0533 & 0.2695 & 0.0533 \\ 0.0533 & -0.0062 & 0.0533 & 0.2695 \end{bmatrix} \begin{Bmatrix} q_5 \\ q_6 \\ q_7 \\ q_2 \end{Bmatrix} \tag{196}$$

First, these equations are re-arranged for each domain segregating the external nodes and the interface nodes. To aid expose the process, we will use the following notation borrowed from MATHCAD, that the ith column of matrix [H] is $\mathbf{H}^{<i>}$.

Then for region Ω_1 we have:

$$\left[H_1^{\Gamma_1} \ \ -G_1^{\Gamma_{1,I}} \ H_1^{\Gamma_{1,I}} \right] \begin{Bmatrix} T_1^{\Gamma_1} \\ q_2^{\Gamma_{1,I}} \\ T_2^{\Gamma_1} \end{Bmatrix} = \left[G_1^{\Gamma_1} \right] \left\{ q_1^{\Gamma_1} \right\} \tag{197}$$

Expanding and indentifying components of the matrix equations, as they relate to the original $[H_1]$ and $[G_1]$ matrices

$$\underbrace{\left[H_1^{<1>} \quad H_1^{<3>} \quad -G_1^{<4>} \quad -G_1^{<2>} \quad H_1^{<2>} \right]}_{[H_1]} \begin{Bmatrix} T_1 \\ T_3 \\ q_4 \\ q_2 \\ T_2 \end{Bmatrix} = \underbrace{\left[G_1^{<1>} \quad G_1^{<3>} \quad H_1^{<4>} \right]}_{[G_1]} \begin{Bmatrix} q_1 \\ q_3 \\ T_4 \end{Bmatrix} \tag{198}$$

and filling out the entries we find

$$H_1\begin{pmatrix} 0.5 & -0.148 & -0.053 & -0.053 & -0.176 \\ -0.176 & -0.176 & 0.006 & -0.269 & 0.5 \\ -0.148 & 0.5 & -0.053 & -0.053 & -0.176 \\ -0.176 & -0.176 & -0.269 & 0.006 & -0.148 \end{pmatrix} \quad G_1 = \begin{pmatrix} 0.269 & -0.006 & 0.176 \\ 0.053 & 0.053 & 0.148 \\ -0.006 & 0.269 & 0.176 \\ 0.053 & 0.053 & -0.5 \end{pmatrix} \tag{199}$$

Moreover, the product

$$[G_1]\begin{Bmatrix} q_1 \\ q_3 \\ T_4 \end{Bmatrix} = [G_1]\begin{Bmatrix} q_1 = 0 \\ q_3 = 0 \\ T_4 = 0 \end{Bmatrix} = \begin{Bmatrix} 0 \\ 0 \\ 0 \\ 0 \end{Bmatrix} \tag{200}$$

provides the right hand-side for region. Then going through these steps for region Ω_2 we have

$$\left[-G_2^{\Gamma_{2,I}}\; H_2^{\Gamma_{2,I}}\; H_2^{\Gamma_2}\right]\begin{Bmatrix} q_2^{\Gamma_{1,I}} \\ T_2^{\Gamma_{1,I}} \\ T_2^{\Gamma_2} \end{Bmatrix} = \left[G_2^{\Gamma_2}\right]\left\{q_2^{\Gamma_2}\right\} \tag{201}$$

Expanding and indentifying components of the matrix equations, as they relate to the original $[H_2]$ and $[G_2]$ matrices

$$\underbrace{\left[\frac{k_1}{k_2}G_2^{<4>} \quad H_2^{<4>} \quad H_2^{<1>} \quad -G_2^{<2>} \quad H_1^{<3>}\right]}_{[H_2]}\begin{Bmatrix} q_2 \\ T_2 \\ T_5 \\ q_6 \\ T_7 \end{Bmatrix} = \underbrace{\left[G_2^{<1>} \quad -H_2^{<2>} \quad G_2^{<3>}\right]}_{[G_2]}\begin{Bmatrix} q_1 \\ q_3 \\ T_4 \end{Bmatrix} \tag{202}$$

and filling out the entries we find

$$H_2 = \begin{pmatrix} 0.027 & -0.176 & 0.5 & -0.053 & -0.148 \\ -0.003 & -0.148 & -0.176 & -0.269 & -0.176 \\ 0.027 & -0.176 & -0.148 & -0.053 & 0.5 \\ 0.135 & 0.5 & -0.176 & 0.006 & -0.176 \end{pmatrix} \quad G_2 = \begin{pmatrix} 0.269 & 0.176 & -0.006 \\ 0.053 & -0.5 & 0.053 \\ -0.006 & 0.176 & 0.269 \\ 0.053 & 0.148 & 0.053 \end{pmatrix} \tag{203}$$

Finally the product

$$\left[G_2^{<1>} \quad -H_2^{<2>} \quad G_2^{<3>}\right]\begin{Bmatrix} q_5 \\ T_6 \\ q_7 \end{Bmatrix} = \left[G_2^{<1>} \quad -H_2^{<2>} \quad G_2^{<3>}\right]\begin{Bmatrix} q_5 \\ T_6 = 100 \\ q_7 \end{Bmatrix} = \begin{Bmatrix} 17.621 \\ -50 \\ 17.621 \\ 14.758 \end{Bmatrix} \tag{204}$$

Allows to now put it all together into the final matrix equation: [A]{x} = {b}, where the coefficient matrix is now sparse and the vector {x} gathers all external boundary unknowns and interface unknowns into a single vector.

$$A = \begin{pmatrix} 0.5 & -0.148 & -0.053 & -0.053 & -0.176 & 0 & 0 & 0 \\ -0.176 & -0.176 & 0.006 & -0.269 & 0.5 & 0 & 0 & 0 \\ -0.148 & 0.5 & -0.053 & -0.053 & -0.176 & 0 & 0 & 0 \\ -0.176 & -0.176 & -0.269 & 0.006 & -0.148 & 0 & 0 & 0 \\ 0 & 0 & 0 & 0.027 & -0.176 & 0.5 & -0.053 & -0.148 \\ 0 & 0 & 0 & -0.003 & -0.148 & -0.176 & -0.269 & -0.176 \\ 0 & 0 & 0 & 0.027 & -0.176 & -1.148 & -0.053 & 0.5 \\ 0 & 0 & 0 & 0.135 & 0.5 & -0.176 & 0.176 & 0.176 \end{pmatrix} \quad b = \begin{pmatrix} 0 \\ 0 \\ 0 \\ 0 \\ 17.621 \\ -50 \\ 17.621 \\ 14.758 \end{pmatrix} \tag{205}$$

The solution to these equations is

$$x = \begin{pmatrix} 33.333 \\ 33.333 \\ -78.307 \\ 78.307 \\ 66.667 \\ 83.333 \\ 39.154 \\ 83.333 \end{pmatrix} \quad x_{exact} = \begin{Bmatrix} T_1 = 33.33 \\ T_3 = 33.33 \\ q_4 = -66.66 \\ q_2 = 66.66 \\ T_2 = 66.66 \\ T_5 = 83.33 \\ q_6 = 33.33 \\ T_7 = 83.33 \end{Bmatrix} \tag{206}$$

These results compare well with the exact values, with nearly exact temperatures and a maximum error of 18% in computed q's (16% for q_2 and q_4 and 18% for q_6). This has been the trend in BEM computations, namely that the temperatures are computed with much higher accuracy than the normal derivatives of the temperatures, q.

Chapter 5

HEAT CONDUCTION IN MEDIA WITH ENERGY GENERATION

The energy generation term poses a challenge to the BEM formulation as formally it leads to an area integral whose numerical evaluation leads to an FEM-like integration. However, for certain forms of the generation term, for instance for a constant generation or for generation that can be modeled as point sources/sinks, area integration is formally avoided. Moreover, should a particular solution be available, then again area integration can be avoided. There have been many approaches proposed to avoid area integration due the presence of an energy generation term with multiple reciprocity method (MRM) [68] and dual reciprocity method (DRM) [33] being the most successful. The MRM attempts to exhaust the area integral by repeated reciprocity applications and is successful when the generation term does not dominate the solution but rather acts as a perturbation. The DRM is a means of generating particular solutions based on a given choice of expansion functions. It been implemented with Fourier series and the FFT [69,70] and most successfully with radial basis function (RBF) interpolation.

The RBF-based DRM is the most general and pervasive method developed to address energy generation and any domain integral, for that matter, in BEM formulations. It is also the basis of the most effective means of solving the transient heat conduction problem. The DRM can also be utilized to extend the BEM to handle non-linear problems such as those encountered in fluid mechanics [71,72].

5.1 SPECIAL FORM OF GENERATION LEADING TO CONTOUR INTEGRALS

With generation present,

$$\nabla^2 T(x) + \frac{1}{k} u_G(x) = 0 \tag{207}$$

the BEM leads to a domain integral involving the generation term,

$$C(\xi)T(\xi) = \oint_{\Gamma(x)} \left[\frac{\partial T(x)}{\partial n} T^*(x,\xi) - T(x) \frac{\partial T^*(x,\xi)}{\partial n} \right] d\Gamma(x) + \oiint_{\Omega(x)} T^*(x,\xi) \frac{1}{k} u_G(x) d\Omega(x) \tag{208}$$

which in discretized form is

$$[H]\{T\} = [G]\{q\} + \{F\} \tag{209}$$

Here, $\{F\}$, is a known vector formally arising from the domain integral. Certainly, this term

$$I(\xi) = \oiint_{\Omega(x)} T^*(x,\xi) \frac{1}{k} u_G(x) d\Omega(x) \tag{210}$$

can be evaluated using an FEM-like discretization, however this detracts from the boundary-only feature of the BEM. However, there are certain cases where the integral can be evaluated without domain integration.

Case 1: Generation term is due to point source(s)/sink(s):
If the generation term is due Ns point source(s)/sink(s) of individual strengths Q_n located at a position $\vec{r}_n$, then the generation term is given by

$$u_G(x) = \sum_{n=1}^{Ns} Q_n \delta(\vec{r} - \vec{r}_n) \tag{211}$$

Consequently, the area integral becomes

$$I(\xi) = \oiint_{\Omega(x)} T^*(x,\xi) \frac{1}{k} u_G(x) d\Omega(x) = \frac{1}{k} \sum_{n=1}^{Ns} Q_n T^*(\vec{r}(\xi) - \vec{r}_n) \tag{212}$$

In two dimensions, for instance the above is explicitly

$$I(\xi) = \frac{-1}{2\pi k} \sum_{n=1}^{Ns} Q_n \, ln\,|\vec{r}(\xi) - \vec{r}_n| \tag{213}$$

Case 2: Generation is harmonic.

If the generation term is harmonic, that is if

$$\nabla^2 u_G(x) = 0 \tag{214}$$

Then the domain integral can be converted to boundary integrals, by setting

$$\nabla^2 V^*(x,\xi) = T^*(x,\xi) \tag{215}$$

Using Green's second identity we find that

$$I(\xi) = \oiint_{\Omega(x)} T^*(x,\xi) \frac{1}{k} u_G(x) d\Omega(x) = \frac{1}{k} \oint_{\Gamma(x)} \left(u_G(x) \frac{\partial V^*(x,\xi)}{\partial n} - V^*(x,\xi) \frac{\partial u_G(x)}{\partial n} \right) d\Gamma(x) \tag{216}$$

So that if the generation term is a constant, $u_G(x) = f_o = const.$, then the above reduces to a single contour integral,

$$I(\xi) = \oiint_{\Omega(x)} T^*(x,\xi) \frac{1}{k} u_G(x) d\Omega(x) = \frac{1}{k} f_o \oint_{\Gamma(x)} \frac{\partial V^*(x,\xi)}{\partial n} d\Gamma(x) \tag{217}$$

Determining the function $V^*(x,\xi)$ is straightforward. For instance, in 2D, solving

$$\frac{1}{r} \frac{\partial}{\partial r} \left(r \frac{\partial V^*}{\partial r} \right) = -\frac{1}{2\pi} ln\, r \tag{218}$$

results in

$$V^*(x,\xi) = \frac{r^2}{2\pi} \left[ln\left(\frac{1}{r}\right) + 1 \right] \tag{219}$$

The second required expression if the source term is harmonic is readily evaluated as

$$\frac{\partial V^*(x,\xi)}{\partial n} = \frac{\partial V^*(x,\xi)}{\partial r} \frac{\partial r}{\partial n} = \frac{r}{8\pi} \left[2\, ln\left(\frac{1}{r}\right) + 1 \right] \frac{\partial r}{\partial n} \tag{220}$$

and standard BEM methods can be adopted to evaluate these contour integrals numerically.

5.2 USE OF PARTICULAR SOLUTIONS

For certain functional dependences of the generation term, a particular solution, p, may be found and superposition can be used to recast the problem into a form that is solvable by the standard BEM. Table 7 provides some particular solutions for the Poisson equation for certain forms of the generation term.

Example 5-1: Consider the case of a heat conduction problem with constant energy generation per unit volume such that $\frac{1}{k}u_G = -4$ in a [1 × 1] square region subject to the boundary conditions listed below

G.E.: $\nabla^2 T = 4$

B.C.'s: $T(x, y = 0) = x^2$

$T(x, y = 1) = x^2 + 1$

$q(x = 0, y) = 0$

$q(x = 1, y) = 2$ (221)

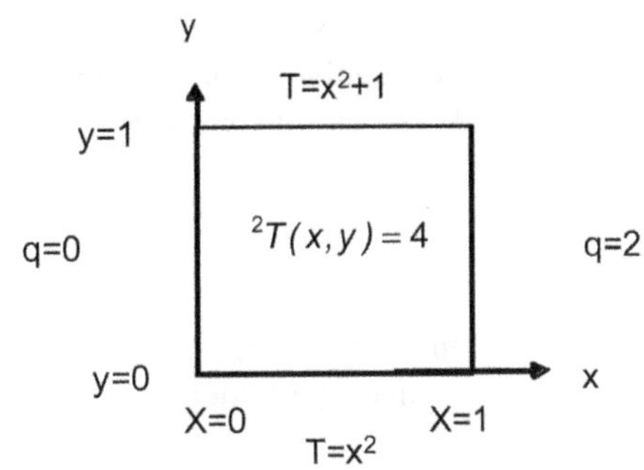

The exact temperature field for this problem is, $T(x,y) = x^2 + y^2$, and that can readily be verified by direct substitution. The particular solution for this Poisson problem can be found in Table 7 as

$$p(x) = 2x^2 \tag{222}$$

So that using superposition we can formulate the solution to the problem as

$$T(x,y) = \vartheta(x,y) + p(x) \tag{223}$$

TABLE 7. Some particular solutions of: $\nabla^2 T + \frac{1}{k}u_G = 0$

u_G	p
C	$-\frac{C}{k}\frac{x^2}{2}$
Cx	$-\frac{C}{k}\frac{x^3}{6}$
Cx^2	$-\frac{C}{k}\frac{x^4}{12}$
Cx^n	$-\frac{C}{k}\frac{x^{n+2}}{(n+1)(n+2)}$
$C\frac{x}{y^2}$	$\frac{C}{k}x\ln(y)$
$C\frac{y}{x^2}$	$\frac{C}{k}y\ln(x)$

The problem in $\vartheta(x, y)$, that is given by

$$\begin{aligned} \text{G.E.:} \quad & \nabla^2\vartheta(x,y)=0 \\ \text{B.C.'s:} \quad & \vartheta(x,y=0)=-x^2 \\ & \vartheta(x,y=1)=1-x^2 \\ & q(x=0,y)=0 \\ & q(x=1,y)=-2 \end{aligned} \tag{224}$$

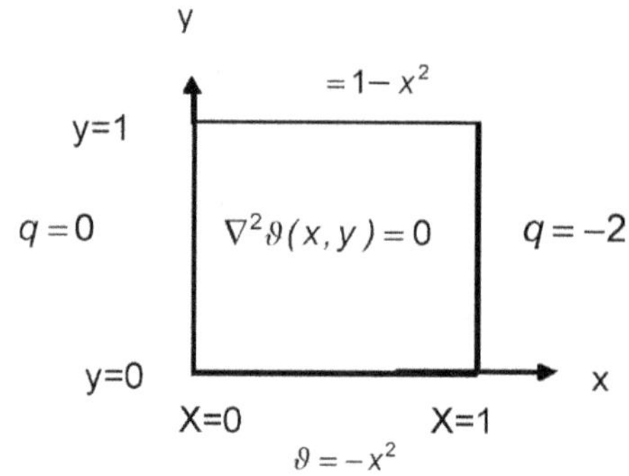

can be solved directly using standard BEM since the governing equation is homogeneous. Alternatively, the problem can be solved by using the contour integral in Eq. (217) to add a known right hand side vector to the BEM equations, namely

$$I(\xi) = -4 \oint_{\Gamma(x)} \frac{\partial V^*(x,\xi)}{\partial n} d\Gamma(x) \tag{225}$$

In general, finding a particular solution is difficult if not impossible. The Dual Reciprocity Boundary Element Method is a general approach to resolving this problem.

5.3 THE DUAL RECIPROCITY BOUNDARY ELEMENT METHOD

Without loss of generality, the basic premise of the Dual Reciprocity Method is to write the governing equation with the domain integral-generating term, $-\frac{1}{k}u_G$, on the right-hand-side as

$$\nabla^2 T = -\frac{1}{k}u_G \tag{226}$$

and to expand the forcing term in a series

$$-\frac{1}{k}u_G(x) = \sum_{m=1}^{N_e+L} \beta_m f_m(x) \tag{227}$$

where N_e is the number of BEM boundary nodes at which dual reciprocity points are collocated, L is the number of additional internal dual reciprocity (DR) collocation points, as illustrated in 2D in Fig. 34. As will be described later, the expansion coefficients can be found by collocation or by least-squares once the expansion functions are defined. The key choice to be made for the purpose of avoiding the domain integral is that expansion

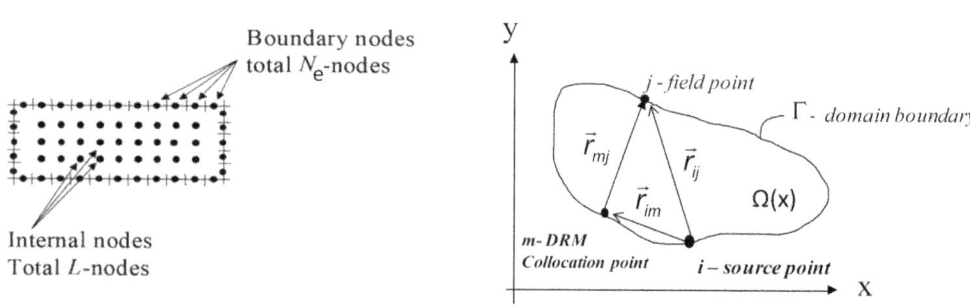

Figure 34. Dual reciprocity nodes used in the expansion of the generation term.

functions is that they are to be constructed as particular solutions of the operator used to generate the integral equation for the problem under consideration. For our case, we then require that $f_m(x)$ is a particular solution to the Laplace equation,

$$f_m(x) = \nabla^2 u_m(x) \tag{228}$$

Introducing the expansion into the governing equation, we find

$$\nabla^2 T(x) = \sum_{m=1}^{N_e+L} \beta_m \nabla^2 u_m(x) \tag{229}$$

multiplying both sides by the free space solution to the Laplace equation, $T^*(x,\xi)$, integrating over the domain, and using Green's second identity on both sides, there results the dual reciprocity boundary integral equation

$$C(\xi)T(\xi) + \oint_{\Gamma(x)} \left[T(x)\frac{\partial T^*(x,\xi)}{\partial n} - q(x)T^*(x,\xi) \right] d\Gamma(x) = \tag{230}$$

$$\sum_{m=1}^{N+L} \beta_n \left\{ C(\xi)u_m(\xi) + \oint_{\Gamma(x)} \left[u_m(x)\frac{\partial T^*(x,\xi)}{\partial n} - p_m(x)T^*(x,\xi) \right] d\Gamma(x) \right\}$$

where we called $p_m(x) = \dfrac{\partial u_m(x)}{\partial n}$. Through the judicious choice of expansion functions, $f_m(x)$, the source term has effectively been evaluated using only boundary integrals.

Using standard BEM discretization and collocating by taking the source point at the boundary and interior DR nodes leads to

$$C(\xi_i)T(\xi_i) + \sum_{j=1}^{N_e} \hat{H}_{ij}T_j - \sum_{j=1}^{N_e} G_{ij}q_j =$$

$$\sum_{m=1}^{N_e+L} \beta_m \left(C(\xi_i)u_m(\xi_i) + \sum_{j=1}^{N_e} \hat{H}_{ij}u_{jm} - \sum_{j=1}^{N_e} G_{ij}p_{jm} \right) \quad \text{for} \quad i = 1,2\ldots N_e \tag{231}$$

where $u_{jm} = u_m(x_j)$ with x_j denoting the boundary point j. In matrix form, we have for the DRBEM

$$[H]\{T\} - [G]\{q\} = \sum_{m=1}^{N_e+L} \beta_m \left([H]\{u\}_m - [G]\{p\}_m \right) \tag{232}$$

Here, $\{u\}_m$ is the vector of boundary nodal values for the $u_m(x)$ function, while $\{p\}_m$ is the vector of boundary nodal values for $p_m(x)$. Defining the matrix [U] whose columns are comprised of the vectors $\{u\}_m$ and the matrix [P] whose columns are comprised of the vectors $\{p\}_m$, so that

$$[U] = \left[\{u\}_1 \quad \{u\}_2 \quad \cdots \quad \{u\}_{N_e+L} \right] \quad \text{and} \quad [P] = \left[\{p\}_1 \quad \{p\}_2 \quad \cdots \quad \{p\}_{N_e+L} \right] \tag{233}$$

The DRBEM equations can be expressed in compact form as

$$[H]\{T\} - [G]\{q\} = ([H][U] - [G][P])\{\beta\} \tag{234}$$

Notice that the same [H] and [G] matrices developed for the standard BEM are utilized to evaluate the right-hand-side Dual Reciprocity vector for the generation. This form of the DRBEM is general and convenient to adapt to other problems where the governing differential equation can be rearranged in such a fashion that the left hand side leads to an adjoint equation for which a fundamental solution can be found and with all remaining terms appearing on the right-hand-side as a forcing term to be treated by DRBEM methodology.

At this point, it is instructive to describe how the DR expansion functions, $f_m(x)$, the derived functions, $u_m(x)$, and the expansion coefficient vector, $\{\beta\}$, are obtained. Several types of expansion functions have been used in the DRBEM, however, radial basis functions (RBF's) have proven most successful. Moreover, several RBF expansions are possible from the radially-symmetric conic RBF, to thin plate splines and Hardy multiquadrics, see Golberg et al. [73] and locally support compact RBF's, see Chen et al. [74].

We illustrate the procedure using the conic RBF that have proved successful in DRBEM applications. Here, the expansion functions are chosen as,

$$f_m(x,y) = 1 + r_m(x,y) \tag{235}$$

where, $r_m(x)$ is the radial distance from the DR collocation point to any location x, be it on the boundary or in the interior, see Fig. 34. In 2D, we have $r_m(x,y) = \sqrt{(x-x_m)^2 + (y-y_m)^2}$, and from this definition, the function, $u_m(x,y)$, is derived from the relation

$$\frac{1}{r_m}\frac{\partial}{\partial r_m}\left(r_m \frac{\partial u_m}{\partial r_m}\right) = 1 + r_m \tag{236}$$

This can be readily integrated to give

$$u_m(x,y) = \frac{r_m^2(x,y)}{4} + \frac{r_m^3(x,y)}{9} \tag{237}$$

Furthermore, the function $p_m(x,y)$ is derived from its definition as

$$\begin{aligned} p_m(x,y) &= \frac{\partial u_m(r(x,y))}{\partial r}\frac{\partial r}{\partial n} \\ &= -\left[\frac{1}{2} + \frac{r_m(x,y)}{3}\right]\left[(x-x_m)\hat{n}_x + (y-y_m)\hat{n}_y\right] \end{aligned} \tag{238}$$

where the result that

$$\frac{\partial r_m(x,y)}{\partial x} = \frac{(x-x_m)}{r_m(x,y)} \quad \text{and} \quad \frac{\partial r_m(x,y)}{\partial y} = \frac{(y-y_m)}{r_m(x,y)} \tag{239}$$

has been used. The expansion coefficient vector $\{\beta\}$ is found by collocating the RBF approximation at all $N_e + L$ DRBEM points

$$-\frac{1}{k}u_G(x_i,y_i) = \sum_{m=1}^{N_e+L} \beta_m f_m(x_i,y_i,x_m,y_m) \quad \text{for} \quad i = 1,2...N_e + L \tag{240}$$

leads to the linear system of equations,

$$[F]\{\beta\} = \left\{-\frac{1}{k}u_G\right\} \tag{241}$$

So that formally we can use

$$\{\beta\} = [F]^{-1}\left\{-\frac{1}{k}u_G\right\} \tag{242}$$

To arrive at the DRBEM equations for any forcing term as

$$[H]\{T\} - [G]\{q\} = ([H][U] - [G][P])[F]^{-1}\left\{-\frac{1}{k}u_G\right\} \tag{243}$$

The interpolating matrix $[F]$ has the following structure,

$$[F]=\begin{bmatrix} f_1(x_1,y_1,x_1,y_1) & f_2(x_1,y_1,x_2,y_2) & \cdots & f_1(x_1,y_1,x_{N_e+L},y_{N_e+L}) \\ f_1(x_2,y_2,x_1,y_1) & f_2(x_2,y_2,x_2,y_2) & \cdots & f_2(x_1,y_1,x_{N_e+L},y_{N_e+L}) \\ \vdots & \vdots & \ddots & \vdots \\ f_1(x_{N_e+L},y_{N_e+L},x_1,y_1) & f_2(x_{N_e+L},y_{N_e+L},x_2,y_2) & \cdots & f_{N_e+L}(x_{N_e+L},y_{N_e+L},x_{N_e+L},y_{N_e+L}) \end{bmatrix} \tag{244}$$

with each j-th column of the matrix comprised of the vector $\{f_j\}$ evaluated sequentially at each row i at the x-y values of the location of the i-th DRBEM point.

Example 5.2: Application of the DRBEM to transient heat conduction. In this case, the diffusion equation is recast as

$$\nabla^2 T = \frac{1}{\alpha}\frac{\partial T}{\partial t} \tag{245}$$

The right-hand-side term containing the time derivative is treated using the DRBEM methodology. Here, we expand the right hand side using the 1+r conic RBF with the expansion coefficients, $\beta_m(t)$, explicitly as function of time

$$\left.\frac{\partial T}{\partial t}\right|_i = \sum_{m=1}^{N_e+L} \beta_m(t) f_m(x_i,y_i,x_m,y_m) \quad \text{for} \quad i=1,2...N_e+L \tag{246}$$

Symbolically, we have

$$\{\dot{T}\} = [F]\{\beta\} \tag{247}$$

and inverting the expansion coefficient is

$$\{\beta\} = [F]^{-1}\{\dot{T}\} \tag{248}$$

Applying the DRBEM approach we find the matrix equations

$$[H]\{T\} - [G]\{q\} = \frac{1}{\alpha}([H][U] - [G][P])[F]^{-1}\{\dot{T}\} \tag{249}$$

These equations can be recast in a more familiar form by defining a capacitance matrix $[C]$ as

$$[C] = -\frac{1}{\alpha}\big([H][U] - [G][P]\big)[F]^{-1} \tag{250}$$

So that the DRBEM equations take on a form that is familiar from the FEM, namely,

$$[C]\{\dot{T}\} + [H]\{T\} = [G]\{q\} \tag{251}$$

Any standard time-integration scheme can be used to discretize the system. The right-hand-side vector $\{q\}$ of temperature normal derivatives renders the system of equations of a "mixed" type in contrast to the "displacement" finite element formulations. Proceeding with the Newmark-theta (θ) time integration scheme, with the superscript denoting the time step as Δt and the time level, m, as $m = m\Delta t$, we have

$$\begin{aligned} T &= (1-\theta)T^m + \theta T^{m+1} \\ q &= (1-\theta)q^m + \theta q^{m+1} \\ \dot{T} &= \frac{1}{\Delta t}\left(T^{m+1} - T^m\right) \end{aligned} \tag{252}$$

with θ taking on values between 0 and 1, most commonly, either, 0, 1/2, or 1, leading to explicit, Crank-Nicholson, and implicit methods respectively. Introducing the temporal discretizations into the DRBEM equations, there results

$$\left(\frac{1}{\Delta t}[C]+\theta[H]\right)\{T\}^{m+1}-\theta[G]\{q\}^{m+1}=\left(\frac{1}{\Delta t}[C]-(1-\theta)[H]\right)\{T\}^{m} \\ +(1-\theta)[G]\{q\}^{m} \tag{253}$$

The right side is known at time level m, since it involves values which have been specified as initial conditions or calculated previously. Upon introducing the boundary conditions at time m + 1, one can rearrange the left hand side keeping all unknowns in the solution vector and moving all known terms to add to the right hand side resulting solvable system of equations [A]{x} = {b} for each time level. Note that the elements of matrices [H], [G], and [C], depend only upon geometrical data and can thus all be computed once and stored. If the value of the time step Δt is kept constant, the system matrix can be reduced by factorization only once as well, and the time advance procedure will consist of a simple recursive scheme with only algebraic operations involved.

Example 5.3: In this industrial application, the DRBEM is used to model ablation characteristics of thrust vector control vanes (TVC), see Fig. 35, that are placed in the aft section of missiles to direct the thrust vector and thus control the direction of the missile [75]. The vanes under consideration are constructed of alloys, such as Cu-W and Al-Cu-W and are placed in a thermal environment that leads to ablation and/or erosion that, at times, can be severe. Details of the heat load computation can be found in [75]. In order to maintain sufficient thrust vector control, the size of the vane must be made larger to compensate for surface area loss and thus thermal analysis leads to the design of a TVC that is large enough to perform its intended control function but does not exhibit excessive weight.

The diffusion of heat in solids under the presence of a phase change leads to the solution of a linear parabolic partial differential equation, if the temperature dependence of the thermo-physical properties is neglected. However, since the problem is solved in a domain whose geometry is time dependent and must be resolved as part of the solution, the problem is inherently non-linear [27]. A front-tracking method is utilized to resolve the location of the receding surface. Half scale and quarter scale vanes are modeled as some recession rate data were available for these cases.

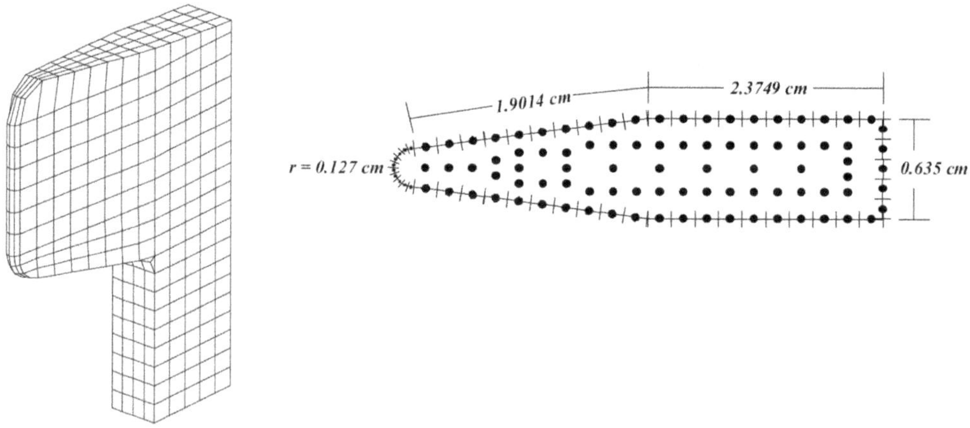

Figure 35. TVC half-scale vane profile along with cross section and BEM discretization showing internal dual reciprocity points.

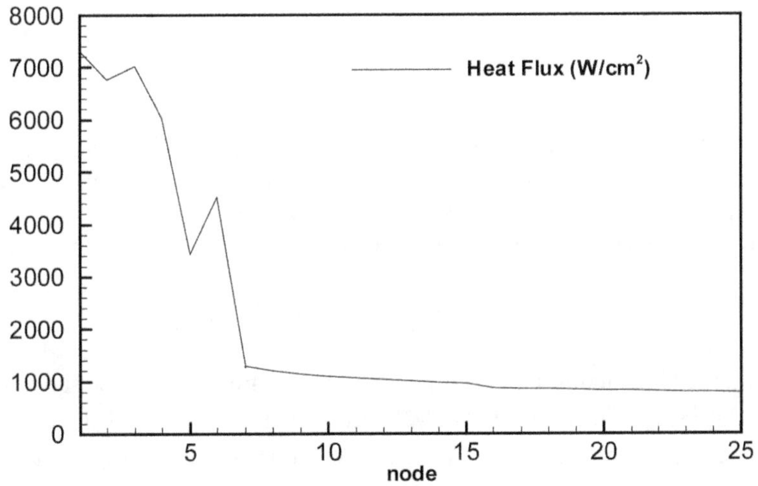

Figure 36. Surface flux distribution over quarter-scale vane [75].

Modeling of this ablation problem relies on the fully implicit DRBEM ($\theta = 1$ for both temperature and flux) to solve the diffusion equation as described above and an iterative scheme to track the ablating surface. Quadratic sub-parametric boundary elements are used and only the boundary is re-meshed as the surface ablates, with DRBEM nodes removed as needed.

The boundary condition at the ablative surface is [27]

$$\rho L_v \frac{d\vec{r}}{dt} \cdot \hat{n} = k \frac{\partial \hat{T}}{\partial n} + q_s'' \tag{254}$$

where ρ is the density, is L_v the latent heat of sublimation, $\vec{r}$ is the position of the ablating surface, $\hat{n}$ is the outward-drawn normal to the ablating surface, k is the thermal conductivity, $\hat{T}$ is the temperature at the ablating surface, and q_s'' is heat surface load including convection and radiation. Thus, the term $\frac{d\vec{r}}{dt} \cdot \hat{n}$ represents the normal velocity V_n of the moving boundary.

The moving front problem requires an iterative solution in which both the temperature field and the location of the moving front are predicted. The motion of the ablating front is found iteratively by re-casting the non-linear boundary condition at the ablating surface in two equivalent forms as

$$V_n = \frac{1}{\rho L_v}\left(k \frac{\partial \hat{T}}{\partial n} + q_s'' \right) \tag{255}$$

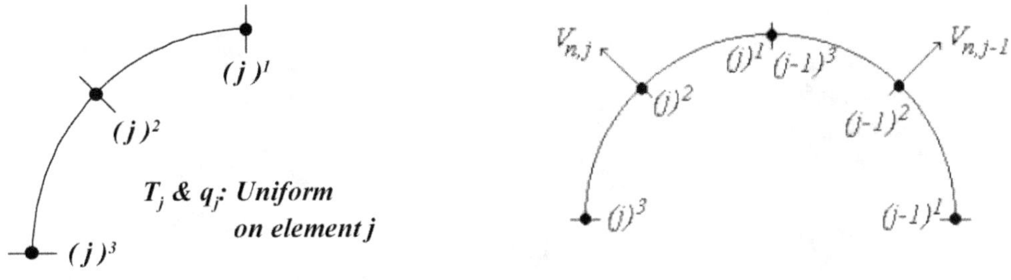

Figure 37. Sub-parametric quadratic BEM elements (quadratic geometry and constant T and q over each element) used in the discretization of the TVC and illustration of the normal velocity at the temperature and heat flux nodes.

and

$$k\frac{\partial \hat{T}}{\partial n} = \left(\rho L_v V_n - q_s''\right) \tag{256}$$

When a nodal temperature reaches above ablation temperature, the nodal temperature is reset to ablation temperature, and the normal velocity, V_n, of the ablating nodes at time step p + 1 is first estimated using the first relation provided by Eq. (255). The boundary is then moved according to the vector equation

$$x^{p+1}_{(j)^2} = x^{p+1}_{(j)^2} + \left(V_n^p\Big|_{(j)^2} \Delta t \right) \bullet \hat{n}\Big|_{(j)^2} \tag{257}$$

where $x^{p+1}_{(j)^2}$ contains the x- and y-coordinates of mid-node of the sub-parametric BEM node $(j)^2$ at the time level p + 1. The velocity of that node $V_n^p\Big|_{(j)^2}$, is found using Eq. (256). The time step taken is Δt, and $\hat{n}\Big|_{(j)^2}$ is the outward-drawn normal of mid-node node. The scheme is used to move the mid-nodes of each sub-parametric quadratic element where the normal is uniquely defined. The velocity of intermediate nodes are taken as the average of the velocities of their neighbors. The conduction problem is solved again and the boundary fluxes are computed using the new geometry. These fluxes are then compared in the second equation provided by Eq. (256). If this equation is not satisfied, then the new fluxes are introduced into Eq. (255) and the geometry is again updated. This process is continued until convergence is reached. In practice, this involves two to three iterations. Once the sub-level iteration process has converged, the solution is advanced forward in time.

Details of verification are provided in [75]. Comparison versus limited data was found to be favorable: (a) the DRBEM predicted recession of 0.102 inches after three seconds for the quarter-scale vane compares well with the experimental result of a mean recession of 0.1 in. This mean value is the recession of the half-height of the wedge portion of the vane, and (b) back-wall temperature for the half scale vane (where a thermocouple was fitted) after 3 seconds is predicted to be 2774 K, while the measured temperature was 2500 K. DRBEM computations of the temperature and receding fronts are provided in Fig. 38.

In closing, time dependent problems can be solved using the time dependent fundamental solution using a convolution scheme which is very computationally intensive, by a Laplace transform approach coupled to a numerical Laplace transform inversion scheme, which restricted to linear problems, and by the DRBEM as described in Example 5.2. The DRBEM is the most flexible approach for transient BEM and can address linear as well as non-linear and non-homogeneous problems. We now conclude the BEM portion of this book by discussing a variety of applications of BEM to heat transfer.

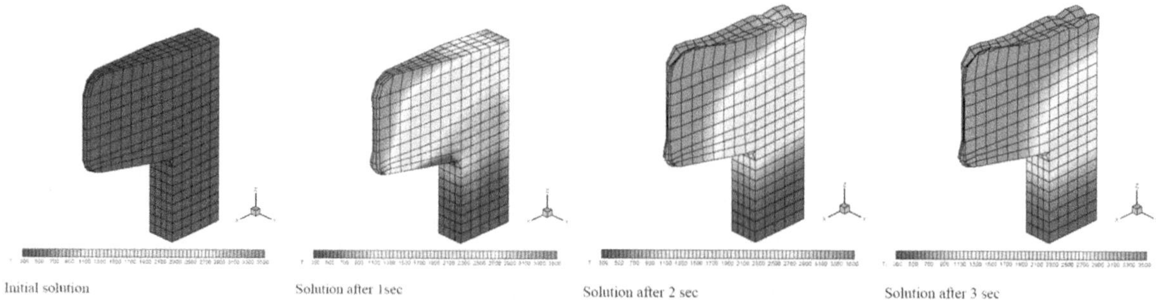

Figure 38. DRBEM predicted vane temperatures and evolution of ablated surface geometries [75].

Chapter 6

APPLICATIONS OF THE BEM TO HEAT TRANSFER AND INVERSE PROBLEMS

In this chapter we will consider specialized applications of the BEM in heat transfer. Specifically, we will consider application of heat transfer for axisymmetric problems, that lays down the groundwork for a very useful result in certain cases of heat conduction in non-homogeneous media with linearly varying thermal conductivity as well as heat transfer in thin plates and extended surfaces that utilizes a fundamental solution that can be utilized in treating linear heat diffusion problems by the Laplace transform method. This is followed by a discussion of practical applications of the BEM to large-scale three dimensional problems and detailed treatment of domain decomposition methods is provided along with some industrial applications. Finally, utilization of the BEM is presented as a suitable field solver for inverse problems problems, and consideration is given to the solution of the inverse geometric problem for the detection of subsurface flaw and cavities.

6.1 AXI-SYMMETRIC PROBLEMS

In the case of a heat conduction problem whose boundary geometry can be produce by revolving a generatix about the z-axis, and whose properties and boundary conditions are not dependent on the polar angle, then the temperature is expected to be independent of the polar angle, so that T = T(r,z). The governing equation is

$$\frac{1}{r}\frac{\partial}{\partial r}\left(r\frac{\partial T}{\partial r}\right)+\frac{\partial^2 T}{\partial z^2}=0 \tag{258}$$

In such a case, it is possible to integrate the 3D boundary integral equation in the polar direction leaving a boundary integral equation that is purely dependent on r and z [2]. The process begins by recasting the fundamental solution in cylindrical coordinates and integrating out the θ-direction, leading to an integral equation in the r-z plane

$$C(\xi)T(\xi)=\oint_{\Gamma'(x)}\left[\frac{\partial T(x)}{\partial n}T_{as}{}^{*}(x,\xi)-T(x)\frac{\partial T_{as}{}^{*}(x,\xi)}{\partial n}\right]d\Gamma'(x) \tag{259}$$

Here the superscript denotes the geometry of the generatix in the r-z plane, and the axi-symmetric fundamental solution, $T_{as}{}^{*}(x,\xi)$, corresponds to a ring source is given by

$$T_{as}^{*}(r,z,r_i,z_i)=\int_0^{2\pi}\frac{1}{4\pi}\frac{1}{|\vec{r}-\vec{r_i}|}d\theta=\frac{1}{\pi}\frac{K(m)}{\sqrt{(r+r_i)^2+(z-z_i)^2}} \tag{260}$$

Where K(m) is a complete elliptic integral of the first kind whose its argument m given by: $m^2 = 4rr_i/(r + r_i)^2 + (z - z_i)^2$. Introducing boundary elements and discretizing the integral equations leads to the standard BEM form

$$C(\xi_i)T(\xi_i) + \sum_{j=1}^{N_e} \hat{H}_{ij}^{(as)} T_j = \sum_{j=1}^{N_e} G_{ij}^{(as)} q_j \qquad (261)$$

Where $\hat{H}_{ij}^{(as)}$ and $G_{ij}^{(as)}$ are the axisymmetric influence coefficients. For the case of constant elements these are given by

$$\hat{H}_{ij}^{(as)} = \int_{\Gamma'(x)} \frac{1}{4\pi} A(r, r_i, m) r d\Gamma'(x) \qquad (262)$$

and

$$G_{ij}^{(as)} = \int_{\Gamma'(x)} \frac{1}{\pi} \frac{K(m)}{\sqrt{(r+r_i)^2 + (z-z_i)^2}} r d\Gamma'(x) \qquad (263)$$

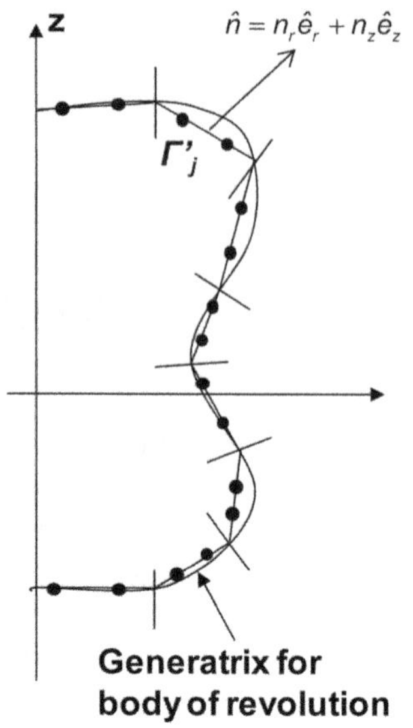

the function $A(r,r_i,m)$ can be found by taking the normal derivative of the axisymmetric fundamental solution. This leads to a lengthy expression in terms of complete elliptic integrals of the first and second kind. For small arguments of the elliptic integrals expressions in terms of Legendre functions are useful and care must be taken for elements close to and those located on the axis of symmetry. For details see [2,6,8]. It is noted that with solid bodies there are no elements on the line of symmetry as the symmetry has been accounted for in the fundamental solution Hollow axi-symmetric bodies, a pipe for instance, require complete contour discretization, as illustrated next.

Example 6.1: An axi-symmetric geometry whose generatrix is discretized with 40 constant boundary elements is displayed in Fig. 39. First kind boundary conditions

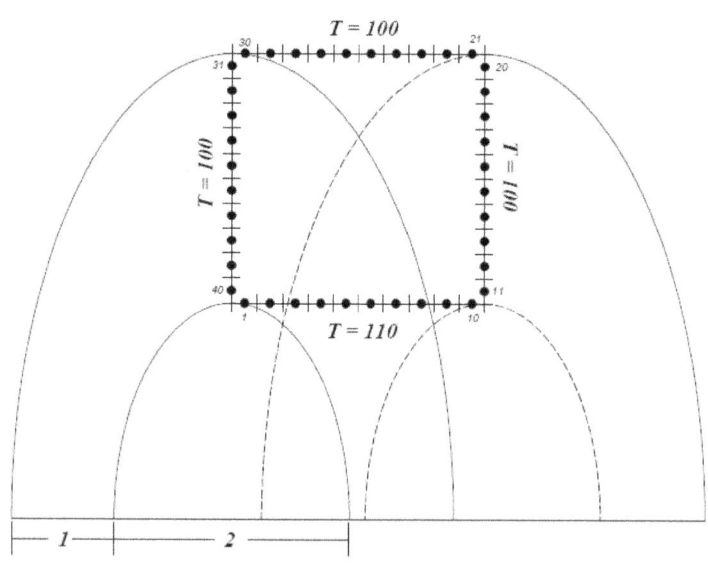

Figure 39. Axi-symmetric problem geometry, discretization and boundary conditions.

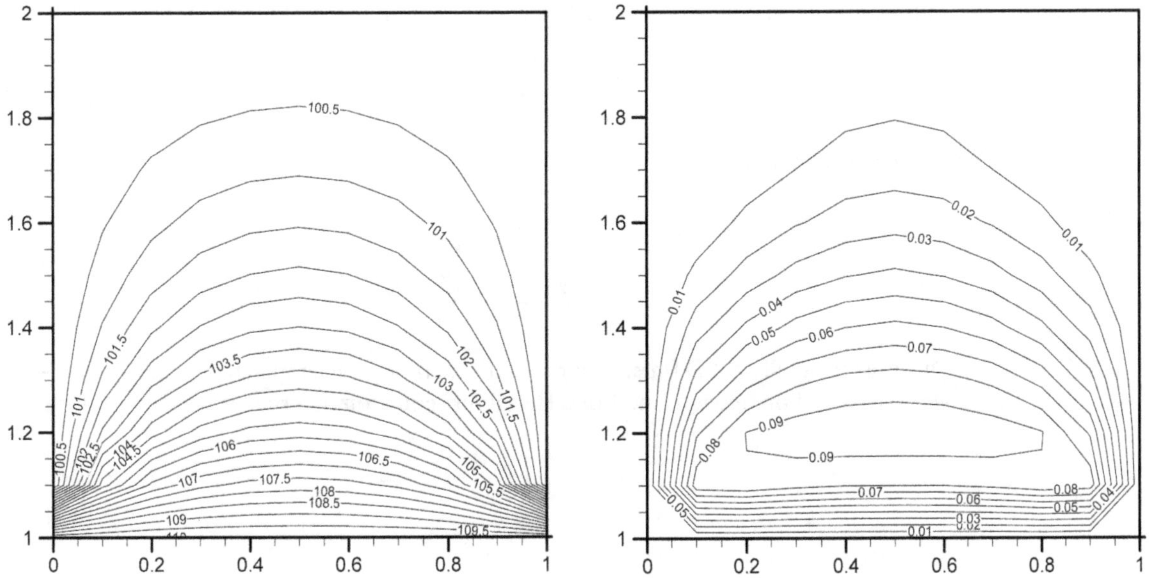

Figure 40. Plot of the BEM generated isotherms (left) and relative error (right) [6].

$$T(x,y)=100+20\sum_{n=1}^{\infty}\frac{\left[1-(-1)^{n}\right]}{n\pi}\frac{K_0(2n\pi)I_0(n\pi r)-I_0(2n\pi)K_0(n\pi r)}{K_0(2n\pi)I_0(n\pi)-I_0(2n\pi)K_0(n\pi)}\sin(n\pi z) \quad (264)$$

are imposed as shown in the figure. Taking the thermal conductivity as k = 1, using separation of variables and superposition, there results the following analytical solution, against which the computed BEM solution whose isotherms are displayed in Fig. 40 are compared revealing less than a 1% relative error.

6.2 HEAT CONDUCTION IN THIN PLATES AND EXTENDED SURFACES

Consider a thin plate of thickness δz in the z-direction. The top and bottom plate are exposed to convective heat transfer with different heat transfer coefficients and reference temperatures, see Fig. 41. One can define a mean temperature in the z-direction as

$$\bar{T}(x,y)=\frac{1}{\delta z}\int_{0}^{\delta z} T dz \quad (265)$$

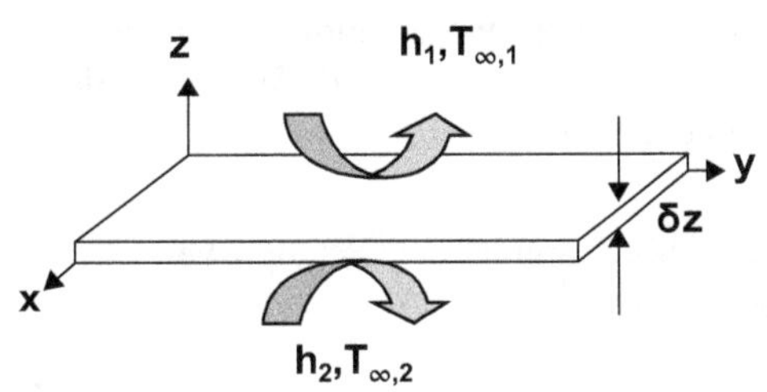

Figure 41. Heat transfer in a thin plate with convection on top and bottom surfaces.

and upon integrating the 3D heat conduction equation in the z-direction there results

$$\frac{\partial^2 \bar{T}}{\partial x^2}+\frac{\partial^2 \bar{T}}{\partial y^2}-\lambda^2\left[\bar{T}-T_{ref}\right]=0 \tag{266}$$

where we have defined

$$\begin{aligned}\lambda^2 &= \frac{h_1+h_2}{k\delta z}\\ T_{ref} &= \frac{h_1 T_{\infty 1}+h_2 T_{\infty 2}}{h_1+h_2}\end{aligned} \tag{267}$$

This is a essentially a 2-D fin equation. So that, defining the temperature excess, $\theta=\bar{T}-T_{ref}$, we arrive at a homogenous modified Helmholtz equation in terms of θ as

$$\frac{\partial^2 \theta}{\partial x^2}+\frac{\partial^2 \theta}{\partial y^2}-\lambda^2\theta=0 \tag{268}$$

This differential equation is self adjoint,

$$L=L^*=\frac{\partial^2}{\partial x^2}+\frac{\partial^2}{\partial y^2}-\lambda^2 \tag{269}$$

So that the fundamental solution satisfies

$$\frac{\partial^2 \theta^*}{\partial x^2}+\frac{\partial^2 \theta^*}{\partial y^2}-\lambda^2\theta^*=-\delta(\vec{r}-\vec{r}_i) \tag{270}$$

that is readily solved by the methods outlined for the Laplace equation to yield the radially symmetric fundamental solution

$$\theta^*(r)=\frac{1}{2\pi}K_o(\lambda r) \tag{271}$$

Consequently, the BEM methodology developed for the Laplace equation can be modified by exchanging the fundamental solution with the one above to solve the thin plate problem. Should the plate thickness Biot number be much less than one, then heat transfer is controlled by the convective resistance, the conductive resistance being negligible, resulting in negligible temperature gradients in the z-direction. In such a case, the temperature is indeed two-dimensional. Otherwise, for all other Biot numbers, the above formulation provides a two-dimensional solution in terms of mean temperature taken across the thickness of the plate.

It is also worthwhile noting, that the above is precisely the fundamental solution for the Laplace transformed diffusion equation as well as the hyperbolic heat conduction equation, so that coupled with an appropriate numerical inversion of the Laplace transform, the same code can also be used to solve linear diffusion problems [2,76] as well as hyperbolic heat conduction problems [77].

Example 6.2: In this example, the Cattaneo-Vernotte hyperbolic heat conduction equation (HHCE) is considered

$$\nabla^2 T=\frac{1}{\alpha}\frac{\partial T}{\partial t}+\frac{1}{c^2}\frac{\partial^2 T}{\partial t^2} \tag{272}$$

where α is the thermal diffusivity, and c is the speed of propagation of the thermal signal. This equation has been proposed to account for the finite speed of propagation of thermal signals and it finds applications for instance in pulsed lasers applications in eye surgery, thin film micromachining, and thermo-mechanical dynamic fracture. Taking the Laplace transform of the above equation and assuming zero initial temperature and zero initial time derivative of the temperature yields

$$\frac{\partial^2 \overline{T}}{\partial x^2}+\frac{\partial^2 \overline{T}}{\partial y^2}-\left(\frac{s}{\alpha}+\frac{s^2}{c^2}\right)\overline{T}=0 \tag{273}$$

where $\overline{T}(x,y,s)$ is the Laplace transform of the temperature and s is the Laplace transform parameter. The above equation a modified Helmholtz equation, just like the thin plate equation, and its fundamental solution is given by Eq. (271) with in $\lambda^2=\left(\frac{s}{\alpha}+\frac{s^2}{c^2}\right)$ this case. Using an appropriate numerical Laplace transform inversion method [77], the HHCE problem can be solve as illustrated in Fig. 42 for the case of a sudden thermal impulse propagated into an adiabatic chamber. Linear BEM spatial discretization was utilized in this example.

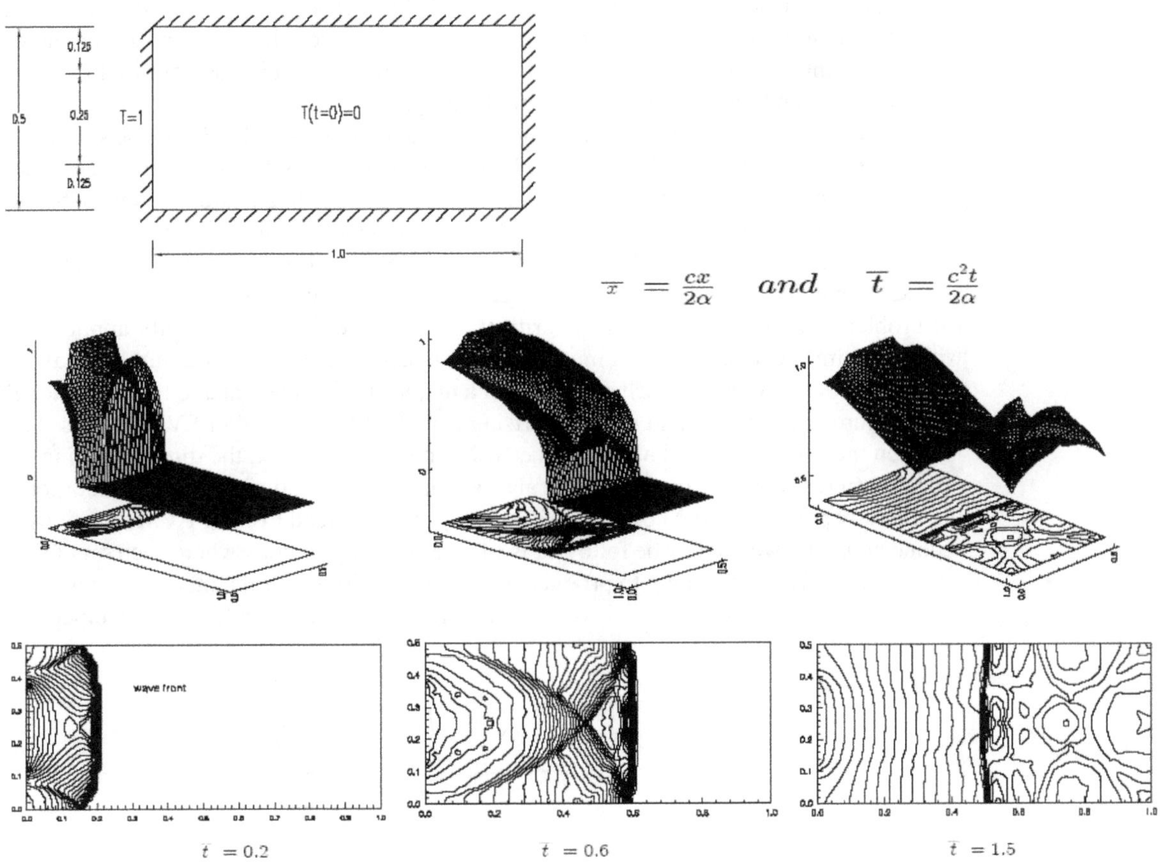

Figure 42. A two-dimensional BEM Laplace transform solution of the Cattaneo-Vernotte HHCE [43,77].

6.3 CONJUGATE HEAT TRANSFER

Engineering analysis of complex mechanical devices such as turbomachines requires ever-increasing fidelity in numerical models upon which designers rely in their efforts to attain demanding specifications placed on efficiency and durability of modern machinery. Consequently, the trend in computational mechanics is to adopt coupled-field analysis to obtain computational models which attempt to better mimic the physics under consideration, see for example Kassab and Aliabadi [78]. The coupled field problem of conjugate heat transfer (CHT) refers to the coupling of convective heat transfer external to the solid body of a thermal component coupled to conduction heat transfer within the solid body of that component. Conjugate heat transfer is of importance in thermal analysis in which multi-mode convective/conduction heat transfer is of particular importance to design. CHT examples of external and internal coupled temperature fields include, for instance, automotive engine blocks, fuel ejectors, cooled turbine blade, vanes, and end-walls, thermal protection system (TPS) for re-entry vehicles, and cooled rockets nozzles to name a few such instances. The continuity of temperature and heat flux at the solid/fluid interface must be satisfied:

$$
\begin{aligned}
T_s &= T_f \\
-k_s \frac{\partial T_s}{\partial n} &= k_f \frac{\partial T_f}{\partial n}
\end{aligned}
\tag{274}
$$

where the subscripts s and f denote the solid and fluid respectively and $\partial/\partial n$ denotes the normal derivative. In case of temperature dependent conductivity, non-linear conduction can be accounted for in the solid using the Kirchhoff transformation as both first kind and second kind boundary conditions transform linearly.

As the temperature and heat fluxes are the nodal unknowns in the BEM, the solid/fluid interface continuity is naturally provided by the BEM conduction analysis of the solid and a strategy of coupling a BEM internal heat conduction solver to an external Navier-Stokes (NS) solver can be adopted. There are two basic approaches to solving coupled field problems. In the first approach, a direct coupling is implemented in which different fields are solved simultaneously in one large set of equations. Direct coupling is mostly applicable for problems where time accuracy is critical, for instance, in aero-elasticity applications where the time scale of the fluid motion is on the same order as the structural modal frequency. However, this approach suffers from a major disadvantage due to the mismatch in the structure of the coefficient matrices arising from BEM, FEM and/or FVM solvers. That is, given the fully populated nature of the BEM coefficient matrix, the direct coupling approach would severely degrade numerical efficiency of the solution by directly incorporating the fully populated BEM equations into the sparsely banded FEM or FVM equations. A second approach which may be followed is a loose coupling strategy where each set of field equations is solved separately to produce boundary conditions for the other. The equations are solved in turn until an iterated convergence criterion, namely continuity of temperature

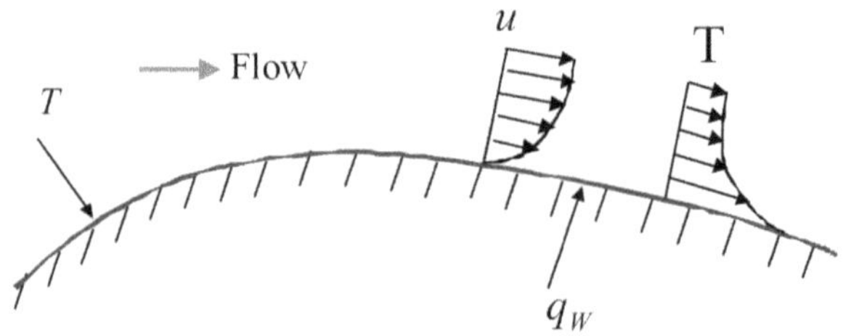

Figure 43. CHT: the coupling of the external and internal thermal fields.

and heat flux, is met at the fluid-solid interface. The loose coupling strategy is particularly attractive when coupling auxiliary field equations to computational fluid dynamics codes as the structure of neither solver interferes in the solution process [79–85].

A successful strategy for the interaction of the two fields is the temperature forward/flux back (TFFB) coupling algorithm in which the external flow and thermal fields are resolved by solving the Navier-Stokes equations using a FVM or FEM scheme and the internal thermal field is resolved by solving the heat conduction equation using the BEM in an interactive manner to reach steady-state through a time-marching algorithm, see Fig. 44 for a 2D application of such a coupling scheme for a cooled turbine blade. The surface temperature obtained from the solution of the flow solver provides the boundary condition for the BEM that in turn calculate heat flux loads on the solid surface. This heat flux is in turn used as a boundary condition for the flow solver in the next time step. This procedure is repeated until a steady-state solution is obtained. In practice, the BEM is solved every few cycles, every ten or so, of the FVM to update the boundary conditions, as intermediate solutions are not physical in this scheme. The temperature forward/flux back (TFFB) coupling algorithm as:

- FVM or FEM Navier-Stokes solver:
  1. Begins with initial adiabatic boundary condition at solid surface.
  2. Solves NS for fluid region.
  3. Provides temperature distribution to BEM conduction solver after a number of iterations.
  4. Receives flux boundary condition from the BEM as input for next set of iterations.
- BEM conduction solver:
  1. Receives temperature distribution from FVM solver.
  2. Solves steady-state conduction problem.
  3. Provides flux distribution to FVM solver.

The transfer of heat flux from the BEM to the FVM solver is accomplished after under-relaxation. The process is continued until the NS solver converges and wall temperatures and heat fluxes converge. The alternative approach of specifying the flux as a boundary condition for the BEM code leading to a flux forward temperature back (FFTB) approach

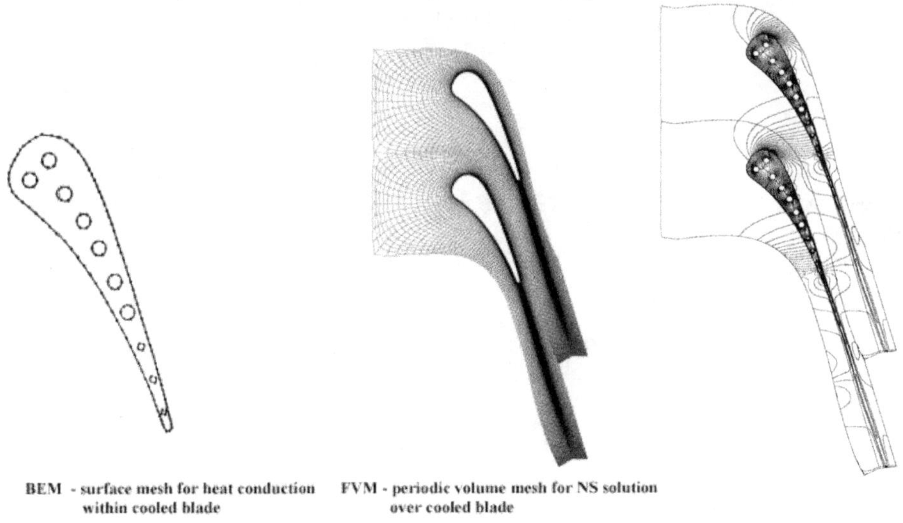

Figure 44. BEM and FVM meshes used in a CHT analysis of an internally-cooled blade [85].

would lead to specifying second kind boundary conditions completely around the surface of a domain governed by an elliptic equation, resulting in a non-unique solution.

An issue arises in information transfer between CFD and BEM as there exists a significant difference in the levels of discretization between the two meshes in a typical CHT simulation. Accurate resolution of the boundary layer requires a FVM/FEM surface grid which is much too fine to be used directly in the BEM. A much coarser surface grid is typically generated for the BEM solution of the conduction problem. The disparity between the two grids requires a general interpolation of the surface temperature and heat flux between the two solvers as it is not possible in general to isolate a single BEM node and identify a set of nearest FVM/FEM nodes. Indeed in certain regions where the CFD mesh is very fine, a BEM node can readily be surrounded by tens or more FVM nodes. A radial basis function interpolation, see Fig. 45, has been successfully used for transfer of temperature and flux values between the BEM and CFD grids in large scale turbomachinery cooled blade CHT computations [79].

An example of a CHT analysis utilizing the BEM/FVM coupled TFFB scheme of a film-cooled turbine blade is a study by Heideman et al. [86]. The authors coupled the NASA Glenn FVM multi-block turbomachinery code Glenn-HT to a 3D BEM code specially tailored through iterative domain decomposition for large-scale computation. This BEM strategy will be discussed in the next section. CHT analysis reveals that multidimensional conduction is important in the accurate determination of the thermal field of the blade and that, in general, the conjugate conduction reduces wall temperatures where heat flux is into the wall and increases them where heat flux is into the fluid. The indication is that thermal conduction is not simply one-dimensional across the thin wall of the blade, but, rather that heat also travels along the vane wall toward low temperature regions from the hotter adjacent regions. These results show that it is too simplistic to assume that increasing vane conductivity will monotonically drive all surface temperatures toward the vane mean temperature, especially for a two temperature mixing problem such as turbine film cooling.

CHT solutions can also be obtained in a time accurate manner [82]. In such a case, a transient BEM solver must be used and the DRBEM approach is suitable for this purpose. The results comparing the temperature histories from CHT computations at two thermocouple locations inside a heated blunted wedge subjected to cooling in a supersonic wind tunnel are provided in Fig. 46. The BEM mesh of an in-house DRBEM transient conduction solver and the FVM mesh utilized for two compressible NS flow solvers (GASP and NASFLO) are provided [65].

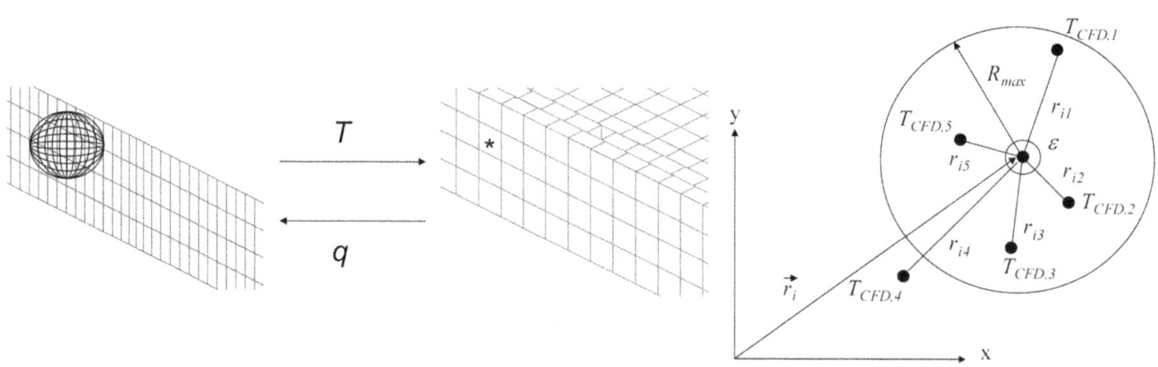

Figure 45. RBF's used to exchange information between disparately discretized solid (BEM) thermal analysis and fluid (FVM or FEM) analysis.

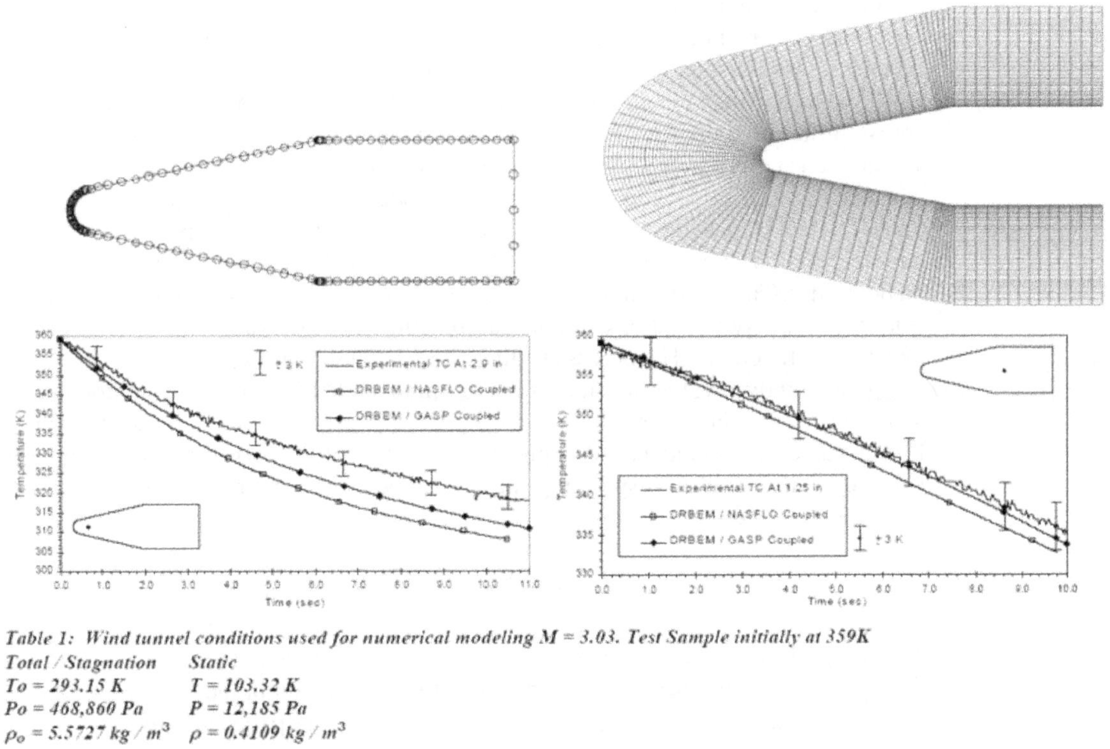

Figure 46. Time accurate CHT solution based on DRBEM/FVM coupling [82].

6.4 LARGE-SCALE HEAT TRANSFER

The BEM formally leads to a system of algebraic equations with a fully-populated coefficient matrix which is problematic for large-scale problems that occur in engineering 3-D modeling of complex structures. Addressing the serious numerical challenges due to large storage requirements and iterative solution of large sets of non-sparse equations has been approached in the BEM community by one of two means: the artificial sub-sectioning of the 3-D model into a multi-region model in conjunction with block-solvers reminiscent of the FEM frontal solvers [49,50,87] and the adoption of multipole methods in conjunction with the GMRES non-symmetric iterative solver [88,89]. The first approach of domain decomposition or sub-sectioning produces a sparse block coefficient matrix that is efficiently stored and has been successfully implemented in the context of continuous boundary elements. However, the method requires generation of complex data-structures identifying connecting regions and interfaces prior to analysis. The second approach is very efficient, however, it requires complete re-write of the BEM code to adopt multipole formulation. For a tutorial and recent review of multipole methods in BEM see [90,91]. Some authors have proposed to compress the coefficient matrix using wavelet decomposition or other such concepts without need for a major alteration of traditional BEM codes [92,93].

An effective technique to address 3D problem is to adopt a region-by-region iterative solver that has been shown has been shown to offer much savings in memory and to converge for linear and non-linear problems [76,94]. This technique does not require any complex data-structure preparation, is somewhat transparent to the user, lends itself to loosely coupled multi-physics strategies, and it is readily implemented in a parallel setting. The iterative domain decomposition artificially sub-sections the single domain under interest into K sub-domains, each of which is independently discretized and solved by standard BEM, while enforcing continuity of temperature and heat flux at the interfaces during the course of iterative updates. It is worth mentioning that the discretization of neighboring sub-domains in this method of decomposition does not have to be coincident, that is, at the connecting interface, boundary elements and nodes from the two adjoining sub-domains are not required to be structured following a sequence or particular position. The only requirement at the connecting interface is that it forms a closed boundary with the same path on both sides. The information between neighboring sub-domains separated by an interface can be effectively passed through an interpolation, for instance by compactly supported radial-basis functions. The process is illustrated in 2D in Fig. 47.

The boundary value problem is solved independently over each subdomain where an initial guessed boundary condition is iteratively updated at the interfaces boundaries. For instance, the boundary value problem of subdomain Ω_j is transformed into the algebraic analogue of corresponding influence coefficient matrices and nodal boundary values as

$$\nabla^2 T_{\Omega_j} = 0 \rightarrow \left[H_{\Omega_j}\right]\left\{T_{\Omega_j}\right\} = \left[G_{\Omega_j}\right]\left\{q_{\Omega_j}\right\} \tag{275}$$

The composition of this algebraic system requires floating point operations proportional to the square of the number boundary nodes in the subdomain, n, that is roughly equivalent to $n \approx \frac{2N}{K+1}$, where N is the number of nodes in the single BEM discretization as long as the interfaces have the same discretrization level as the external boundaries. A significant reduction is achieved in the RAM memory requirements as only the memory needs for one of the sub-domains must be allocated at a time, as the others can temporarily be stored into ROM, and when a parallel strategy is adopted the matrices for each sub-domain are stored by its assigned processor. For the illustrated case it is straightforward to compute a reduction of 16% both in memory and floating point operations to solve the problem relative to standard single discretization BEM [94].

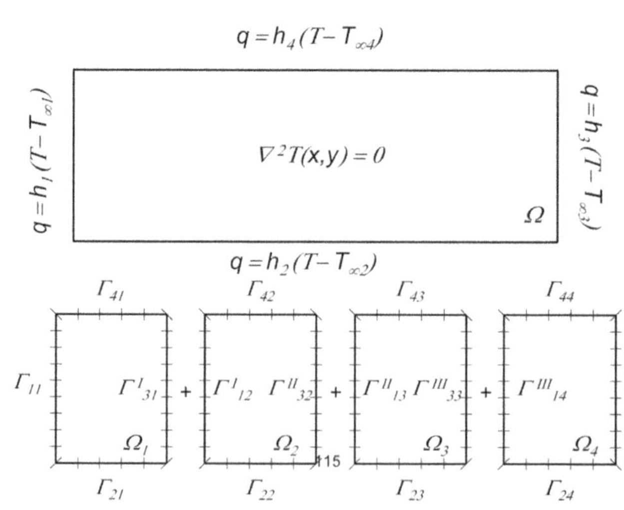

Figure 47. BEM problem domain, boundary conditions, and single region discretization with a decomposition of four (4) sub-domains.

In order to reduce the computational efforts needed with respect to the algebraic solution of the system of each subdomain, a direct approach LU factorization can be employed for all sub-domains. The LU factors of the coefficient matrices for all sub-domains are constant, as they are independent of the right-hand side vector, they can be computed only once at the first iteration step and stored on disc for later use during the iteration process. Therefore, only a forward and a backward substitution will be required for the algebraic solution at each iteration. This feature allows a significant reduction in the operational count through the iteration process, as only a number of floating point operations proportional to $O(n^2)$ as opposed to $O(n^3)$ is required at each iteration step. The access to ROM memory at each iteration step must also be added to this computation time, which is usually larger than access to RAM. Typically however, overall convergence of the problem requires few iterations, and this ROM access is not a significant addition. Additionally, iterative solvers such as GMRES may offer a more efficient alternative.

There are two critical issues in the success of the iterative approach: (a) a good initial guess at the interfaces, and (b) a physically consistent interface update. The more physical information the initial guess incorporates, the fewer the number of iterations that will be required to reduce the initial error. An efficient initial guess can be made using a physically based 1-D heat conduction argument for every node on the external surface to every node at the interface, see Fig. 48, as

$$T_i = \frac{\sum_{j=1}^{N_T} B_{ij}T_j - \sum_{j=1}^{N_q} B_{ij}R_{ij}q_j + \sum_{j=1}^{N_h} \frac{B_{ij}H_{ij}T_{\infty j}}{H_{ij}+1}}{S_j - \sum_{j=1}^{N_T} B_{ij} + \sum_{j=1}^{N_h} \frac{B_{ij}H_{ij}}{H_{ij}+1}} \tag{276}$$

where, N_T, N_q and N_h are the number of first, second, and third kind boundary conditions specified at the external (non-interfacial) surfaces and

$$B_{ij} = \frac{A_j}{|\vec{r}_{ij}|};\ R_{ij} = \frac{\vec{r}_{ij}.\hat{n}_j}{k};\ H_{ij} = \frac{h_j}{k}\vec{r}_{ij}.\hat{n}_j;\ S_i = \sum_{j=1}^{N_s} \frac{A_j}{|\vec{r}_{ij}|} \tag{277}$$

with $N_s = N_T + N_q + N_h$, the thermal conductivity of the medium is k, the film coefficient at the *j-th* convective surface is h_j, the outward-drawn normal to any surface is $\hat{n}_j$, the position vector from the interfacial node j to the external surface node i is $\vec{r}_{ij}$ while the area of interface element j denote by A_j.

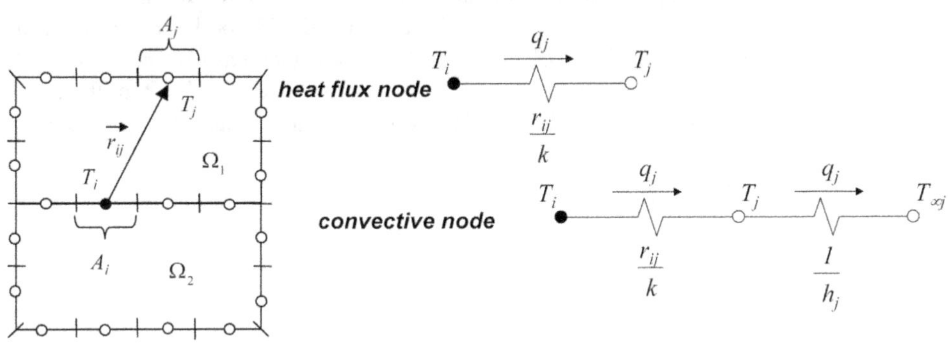

Figure 48. Initial guess at the interface node illustrated in 2-D for a 2-region subdomain decomposition. A coarse constant element surface mesh is used to provide the initial guess.

Once the initial temperatures are imposed as boundary conditions at the interfaces, a resulting set of normal heat fluxes along the interfaces will computed, and they are non-symmetically averaged. Illustrating this for a 2 domain substructure, again we have for regions 1 and region 2 interfaces

$$q_{\Omega_1}^I = q_{\Omega_1}^I - \frac{q_{\Omega_1}^I + q_{\Omega_2}^I}{2} \quad \text{and} \quad q_{\Omega_2}^I = q_{\Omega_2}^I - \frac{q_{\Omega_1}^I + q_{\Omega_2}^I}{2} \tag{278}$$

in order to maintain flux continuity after averaging during the iteration process, that is $q_{\Omega_1}^I = -q_{\Omega_2}^I$. Compactly supported radial basis interpolation can be employed for the flux average to account for unstructured grids along the interface from neighboring subdomains. Using these fluxes the BEM equations are again solved, leading to mismatched temperatures along the interfaces for neighboring subdomains. These temperatures are interpolated, if necessary, from one side of the interface to the other side using a compactly supported radial basis functions to account for the possibility of interface mismatch between the adjoining substructure grids. Once this is accomplished, the temperature is averaged out at each interface

$$T_{\Omega_1}^I = \frac{T_{\Omega_1}^I + T_{\Omega_2}^I}{2} + R'' q_{\Omega_1}^I \quad \text{and} \quad T_{\Omega_2}^I = \frac{T_{\Omega_1}^I + T_{\Omega_2}^I}{2} + R'' q_{\Omega_2}^I \tag{279}$$

Here, the possibility that a real or physical interface exists with a thermal contact resistance R'' present between the connecting subdomains is incorporated. If the interface is artificial, then $R'' = 0$. These now matched temperatures along the interfaces are used as the next set of boundary conditions, see Fig. 49.

A hierarchical strategy is useful to speed up convergence: (a) the initial guess provided by the physically-based Eq. (276) is used with a constant element BEM model using the same discretization as the bi-linear model, however, all nodes are collapsed to a single central node, and this requires 1/64th of the work required to solve the bi-linear model (b) upon convergence of the constant element model, a full bi-linear solution is subsequently computed with the constant element model providing the initial guess, and (c) the converged bilinear solution provides an initial guess for a higher order bi-quadratic solution. This iteration process is continued until a convergence criterion is satisfied. A measure of convergence may be defined as the L_2 norm of mismatched temperatures along all interfaces.

The effectiveness of this strategy is illustrated in the heat conduction analysis in a U-tube configuration whose perimeter is insulated while a heat flux of 1000 W/m^2 is imposed into one of its end caps and it removed from the other by convection with a convective coefficient of h = 100 W/m^{2}K and a reference temperature of zero degrees. A domain decomposition consisting of 44,640 elements distributed over 31 sub-domains, corresponds to 44,640 DOF for the constant elements discretization and 178,560 DOF for the bilinear discretization, as illustrated in Fig. 50. The conductivity was taken as (a) linear k = 14.9 W/mK and (b) non-linear as k = 14.9(1 + 4.7 × 10^{-4}(T-500)) W/mK. Results from the convergence and timings are reported in the Table 8 (analysis was carried out on a 12 × Pentium 4 PC cluster) [94],

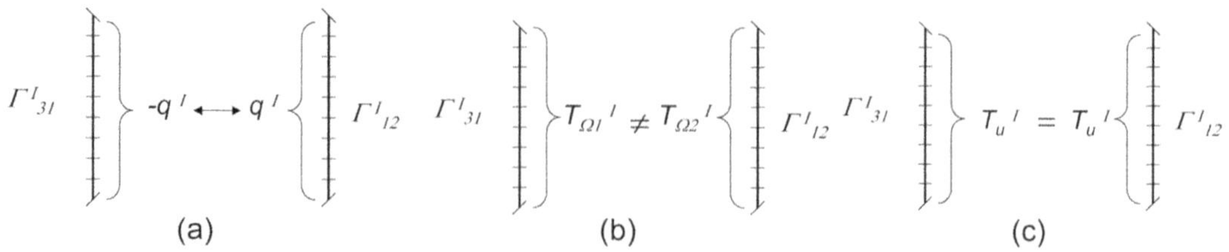

Figure 49. Iteration along the interface (a) imposed heat flux, (b) mismatched resulting temperatures, and (c) interpolated and averaged out temperatures as new boundary conditions.

TABLE 8. Number of iterations and timings for film-cooled blade problem [94]

12 × PC cluster (44,640 elements)	Case 1 (linear)	Case 2 (non-linear)
constant elements (44,640 DOF)	11 iterations	12 iterations
bilinear elements (178,560 DOF)	1 iteration	1 iteration
Total time to solution	4,307 seconds	4,353 seconds

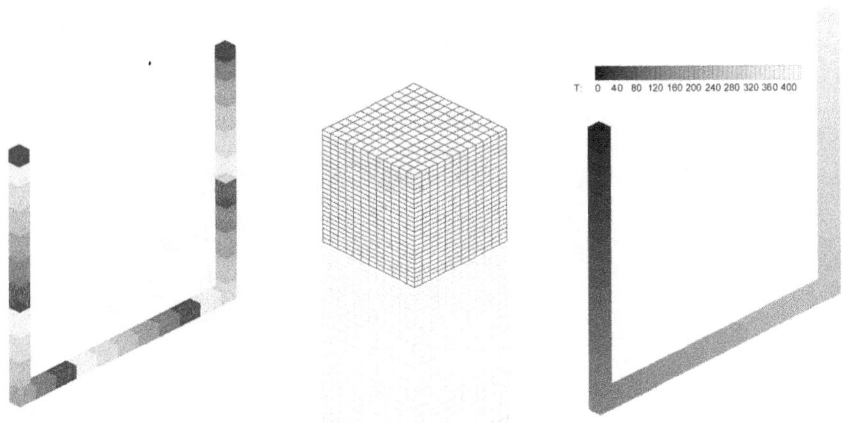

(a) overall domain decomposition. (b) close-up of mesh. (c) temperature distribution for the linear case.

Figure 50. Domain decomposition of U-tube, 44,640 element model and 31 sub-regions [94].

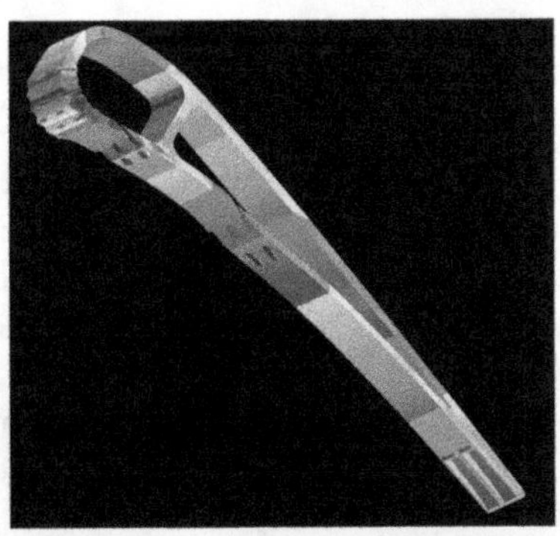

(a) overall domain decomposition of film-cooled blade.

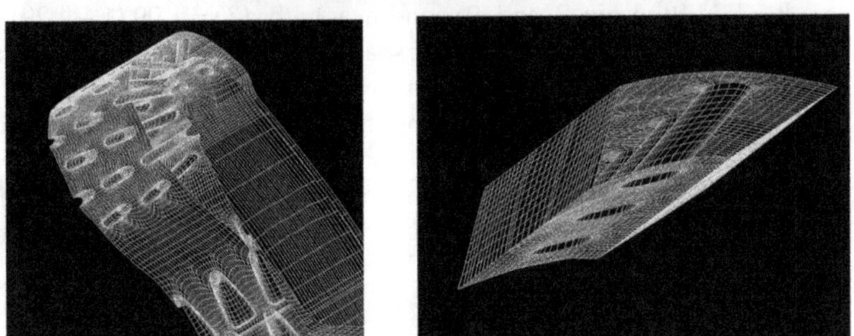

(b) close up of leading edge. (c) typical subdomain.

Figure 51. Domain decomposition of film-cooled blade, 21,306 element model and 20 sub-regions [94].

TABLE 9. Number of iterations and timings for film-cooled blade problem [94]

10 × PC cluster 21,306 elements	Case 1 (linear)	Case 2 (non-linear)
constant elements (21,306 DOF)	10 iterations	10 iterations
bilinear elements (85,224 DOF)	1 iteration	1 iteration
Total time to solution	3,222 seconds	3,230 seconds

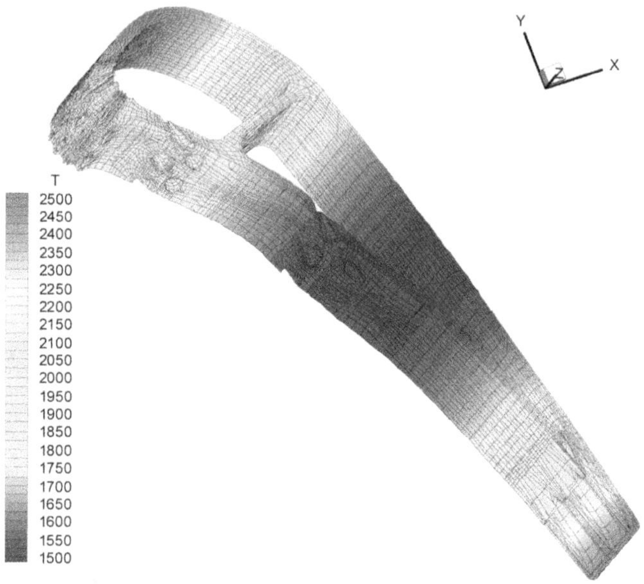

Figure 52. Converged temperature profile obtained by iterative domain decomposition [94].

see table 8. The solution in this case was not carried out beyond the bilinear model. The L_2 interface temperature iterative norm was reduced to a value of $10^{-3}\Delta T_{max}$, where ΔT_{max} is the maximum temperature difference over the field.

Another example is that of heat conduction in an intricate geometry of a film-cooled blade [94]. The domain decomposition for this blade is shown in Fig. 51. Here the discretization is comprised of 21,306 elements distributed over 20 sub-domains. This corresponds to 21,306 DOF for the constant elements discretization and 85,224 DOF for the bilinear discretization.

We consider two cases: (1) linear with a constant conductivity of Inconel, with a value of k = 1.34 Btu/hr-in-R, and (2) non-linear with $k(T) = 1.09\ (1 + 4.29) \times 10^{-4}\ (T-1620))$ Btu/hr-in-R. The endwall surfaces in the spanwise direction are taken as adiabatic, while temperature boundary conditions imposed on the blade surfaces were obtained from a conjugate analysis carried out on the blade coupling the 3D-BEM code for heat conduction to the Glenn-HT Finite Volume code for the flow analysis [86]. The temperatures varied from 1600–3100°F over the surfaces exposed to film cooling, plenum air and hot gas flowing over the external surface, see Fig. 52. The convergence and timings are reported in Table 9.

6.5 NON-HOMOGENEOUS HEAT CONDUCTION: GENERALIZED BIE

Naturally occurring materials such as sedimentary rock and wood that exhibit material heterogeneity and man-made industrial materials, for instance functionally graded materials, have tailored thermophysical heterogeneities that are obtained by careful design of their microstructure. Non-homogenous materials serve to meet ever-increasing thermo-mechanical demands placed on industrial materials. The homogenization approach which is an effective statistical macroscopic description of non-homogeneous thermal conductivity is reviewed in [95]. For heterogeneous media, the adjoint equation is a variable coefficient partial differential equation, and, the BEM formally leads to domain integrals. Some schemes, such as DRBEM or more recently, the radial integration boundary element method (RIBEM) [96], have been proposed to in an attempt to effectively integrate such domain integrals. However, in certain cases, such as layered media, the material property can be modeled as constant in certain zones of the medium. In such cases, the domain is divided into multiple zones of constant material property, and interface continuity of the field variable and its derivative are enforced to couple each zone, and the resulting system is solved using block solvers, as was discussed in 4.3.

In certain other cases, Green's free space solutions can be found, for example for potential problems in two and three-dimensional heterogeneous media whose material properties vary one-dimensionally with position [62]. The remarkable result that a linearly varying thermal conductivity problem can be converted into a radially symmetric homogeneous problem is due to Shaw [60] and comes as a natural result from the formulation of the generalized BIE [6]. Supposing that the conductivity varies linearly with position in the x-direction only, that is

$$k(x) = a + bx \tag{280}$$

then, defining new independent variables

$$x' = x + \frac{a}{b} \quad \text{and} \quad y' = y \tag{281}$$

It is straightforward to show by substitution that the non-homogenous heat conduction equation

$$\frac{\partial}{\partial x}\left[k(x,y)\frac{\partial T}{\partial x}\right] + \frac{\partial}{\partial y}\left[k(x,y)\frac{\partial T}{\partial y}\right] = 0 \tag{282}$$

is transformed into the radially-symmetric equation in (x',y')

$$\frac{1}{x'}\frac{\partial}{\partial x'}\left[x'\frac{\partial T}{\partial x'}\right] + \frac{\partial^2 T}{\partial y'^2} = 0 \tag{283}$$

which is exactly Eq. (258) and therefore can be solved by already existing axi-symmetric BEM codes. Similar types of useful variable transformations have been found for restricted classes of variations of the thermal conductivity [97,98] where the governing non-homogenous heat conduction equation is transformed into an equivalent homogenous Laplace, Helmholtz or modified Helmholtz equations thus enabling use of standard BEM codes to solve non-homogeneous problems that exhibit a requisite spatial dependence for the thermal conductivity. However, such transformations are limited by their very nature to specific spatially dependent thermal conductivities, such a linear, power law, or exponential.

A different approach is taken in the generalized BIEM where a non-symmetric fundamental solution is defined for a general variation of the thermal conductivity. Here, a fundamental solution that is defined as a locally radially symmetric response to a non-symmetric forcing functions and leads to a boundary only formulation that can be extended to transient problems using the DRM [6,65]. The variable coefficient partial differential equation for heat conduction is converted to an integral equation using the fundamental solution, $E(x)$, to the adjoint operator perturbed by a singular forcing function, $D(x,\xi)$, acting at the source point ξ, defined by

$$\nabla \bullet \left[k(x)\nabla E(x,\xi)\right] = -D(x,\xi) \tag{284}$$

that obeys the following properties:

$$\begin{aligned} &\text{(a)} \quad \int_{\Omega_{,c}} D(x,\xi) d\Omega(x) = 1 \\ &\text{(b)} \quad \int_{\Omega_{,c}} T(x)D(x,\xi) d\Omega(x) = T(\xi) + \varepsilon(\xi) \\ &\text{(c)} \quad A(\xi) = \int_{\Omega_{,c}} D(x,\xi) d\Omega(x) \end{aligned} \tag{285}$$

here, x refers to any 2-D or 3-D coordinates, and $\Omega_{,C}$ is any domain containing the source point ξ. The amplification factor, $A(\xi)$, is evaluated as a contour integral over the boundary Γ of $\Omega_{,C}$ by using the Gauss-Divergence theorem

$$A(\xi) = -\oint_{\Gamma} k(x) \frac{\partial E(x,\xi)}{\partial n} d\Gamma(x) \tag{286}$$

The amplification factor explicitly depends on the solution of the adjoint equation, $E(x)$, and the thermal conductivity. It can be shown [6] that in 2D

$$E(r,\xi_i) = -\int \frac{dr}{r \int_0^{2\pi} k(r,\theta,\xi_i) d\theta} \tag{287}$$

and in 3D

$$E(r,\xi_i) = -\int \frac{dr}{r^2 \int_0^{2\pi} \int_0^{\pi} k(r,\theta,\varphi,\xi_i) \sin\theta d\theta d\varphi} \tag{288}$$

referring to a local polar/spherical coordinate system centered at the source point. Furthermore, $D(x,\xi)$ reveals that it is actually comprised of two parts: a Dirac Delta function, $\delta(x,\xi)$, plus a non-symmetric dipole-like function, $D_d(x,\xi)$.

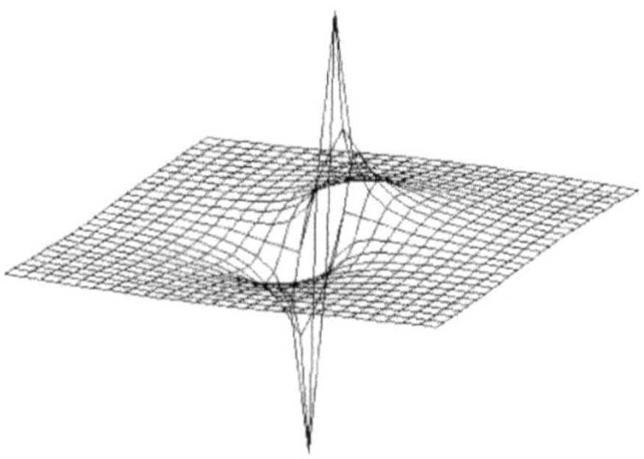

Figure 53. Plot of the dipole like, $D_d(x,\xi)$, component in proximity of a source point ξ.

The sifting term $\varepsilon(\xi)$ is given explicitly by

$$\varepsilon(\xi) = \int_{\Omega} \left[T(x) - T(\xi) \right] D_d(x,\xi) d\Omega(x) \tag{289}$$

Its value is often negligible; however, it can always be evaluated in terms of contour integrals of radial basis functions is a DRM-like manner [65]. Invoking the sampling property of the $D(x,\xi)$ function, the desired boundary integral equation for the temperature is obtained as

$$A(\xi)T(\xi) + \varepsilon(\xi) = \oint_{\Gamma} \left[E(x,\xi) k(x) \frac{\partial T(x)}{\partial n} \right] d\Gamma(x) - \oint_{\Gamma} \left[T(x) k(x) \frac{\partial E(x,\xi)}{\partial n} \right] d\Gamma(x) \tag{290}$$

This integral equation can be discretized following standard BEM procedure to lead to the following set of equations

$$A(\xi_i)T(\xi_i) + \varepsilon(\xi_i) + \sum_{j=1}^{N_e} \hat{H}_{ij} T_j = \sum_{j=1}^{N_e} G_{ij} q_j \quad \text{for} \quad i = 1,2 \ldots N_e \tag{291}$$

leading to the standard BEM form, [H]{T} = [G]{q}. The components of the influence matrices are modified to include the conductivity and the generalized fundamental solution $E(x,\xi)$. These are evaluated numerically by Gauss type quadratures as before and once the boundary conditions are introduced, the algebraic system is solved for the unknowns. Details of numerical implementation of this method can be found in [6].

The generalized BIE for heterogeneous media is applied to a 3-D thrust vector control (TVC) vane discretized with 240 constant boundary elements as shown in Fig. 54.

The spatial variation of the thermal conductivity for the medium is taken as,

$$k(x,y,x) = (2x + y + z + 20) \tag{292}$$

The generalized fundamental solution for this case is found by introducing the above thermal conductivity in the definition for $E(x,\xi)$ in Eq. (288) and evaluating the integral resulting in,

$$E(r, x_i, y_i, z_i) = \frac{1}{4\pi r(2x_i + y_i + z_i + 20)} \tag{293}$$

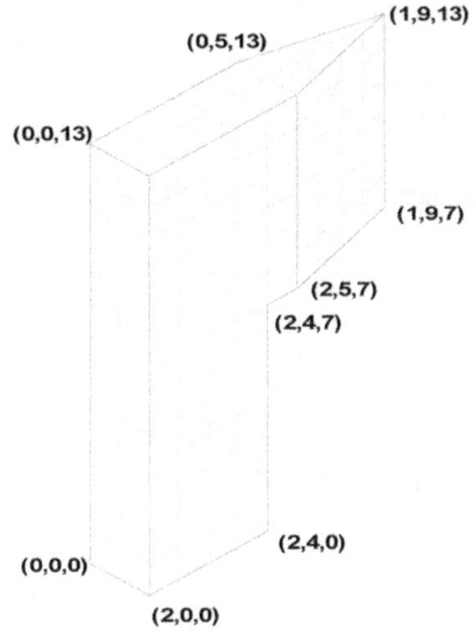

Figure 54. Thrust Vector Control vane with non-homogeneous thermal conductivity [99].

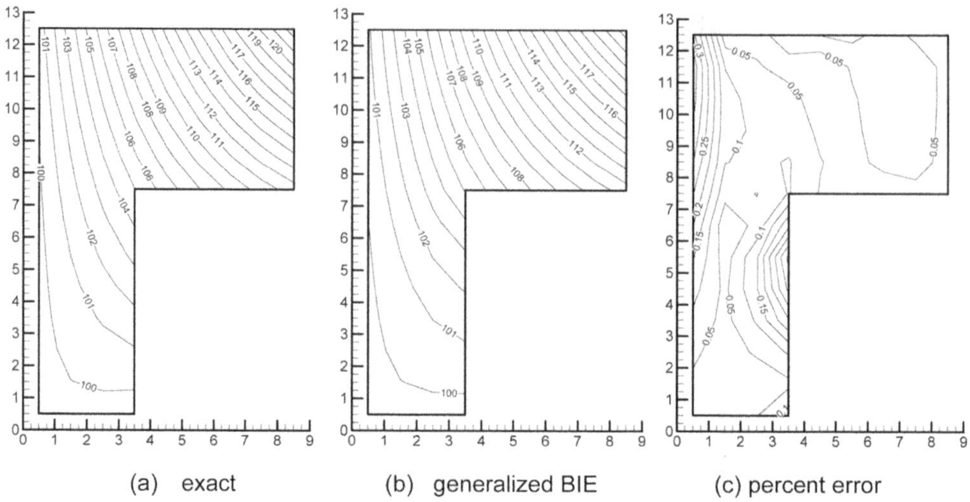

Figure 55. Results for TVC vane (a) exact (b) BEM-computed isotherm distributions and (c) relative error distribution max = 0.4%) [99].

It can verified by substitution that the following temperature

$$T(x,y,z) = 100 + \frac{(5x^2 - 5y^2 + z^2 - 9xy - 15xz + 26yz - 20x + 5y - 5z)}{100} \tag{294}$$

is an exact solution that satisfies the steady-state isotropic heterogeneous heat conduction equation. This temperature distribution is used to impose first kind boundary conditions around the surface of the TVC vane displayed in Fig. 55. Exact and BEM-computed isotherms on the mid-plane of the vane provided in along with the correspondent relative error distribution reveal excellent agreement between the exact solution and the temperature field provided by the generalized BEM with the maximum error reaching 0.4%.

6.6 INVERSE PROBLEMS APPLICATIONS OF THE BEM

The types of problems typically encountered in engineering may be broadly classified as forward or inverse with forward problems being those most commonly encountered. In a forward problem, the following are explicitly specified:

1. Governing equation for field variable.
2. Physical properties.
3. Boundary conditions.
4. Initial condition(s).
5. System geometry.

The purpose of the analysis of the forward problem is to determine the field variable(s) given these inputs. In contrast, in an inverse problem, the following are explicitly specified:

1. Part of conditions 1–5 in a forward problem.
2. Additional information provided by measurement (often an over-specified boundary condition).

The purpose of solving the inverse problem is to find the unknown in conditions 1–5 of the forward problem, using the additional measurement which may be on over-specified boundary condition, or measurement of the field variable at an accessible boundary, or at interior locations, or a combination of the two. Noise contained in the measurement is an important concern in the solution inverse problems as such problems are by their very nature ill-posed [100], that is that the results are highly sensitive to variations in the input and consequently regularization methods must be employed to obtain useful solutions. Theoretical concerns regarding inverse problems can be found in [101]. Reviews of inverse heat transfer problems can be found in [102–105] while [106] provides an introduction to inverse problems in mechanics of materials. Recent developments of inverse problems in heat transfer can be found in [107]. Applications of inverse problems in engineering analysis include acoustics, solid mechanics, heat transfer, among others. The BEM lends itself ideally to inverse problem when the underlying field forward problem can be modeled using the technique and the sought-after parameters in the inverse problem are found on the boundary, for example when a heat flux or convective coefficient is sought, or when the boundary itself or a portion of the boundary itself is to be identified.

The inverse geometric problem finds application in the nondestructive evaluation of subsurface flaws and cavities, and, it is closely related to shape optimization. The governing energy equation, the thermophysical properties, the boundary conditions, and that portion of the geometry which is exposed are all known. However, the portion of the problem geometry that is hidden from view is unknown and to be determined with the help of an over-specified (Cauchy) condition at the exposed surface, see Fig. 56. Specifically, the surface temperature and heat flux is given at the exposed surface and the geometry of the cavity(ies) that generated the measured temperature footprint is to be determined. The boundary condition at the cavity side is specified as either homogeneous or non-homogeneous first, second, or third kind of boundary condition. The solution of the inverse geometric problem by steady state thermal methods, originally proposed by Hsieh [108,109], and whose solution evolved from analytical methods that restricted the detectable geometry to BEM-based resolution of the thermal field [110–115] that allowed for detection of arbitrary geometries, illustrates the utility of the BEM in inverse problems. Ramm [116] proves mathematically that the solution of the inverse geometric problem is unique for conducting media with constant thermal conductivity.

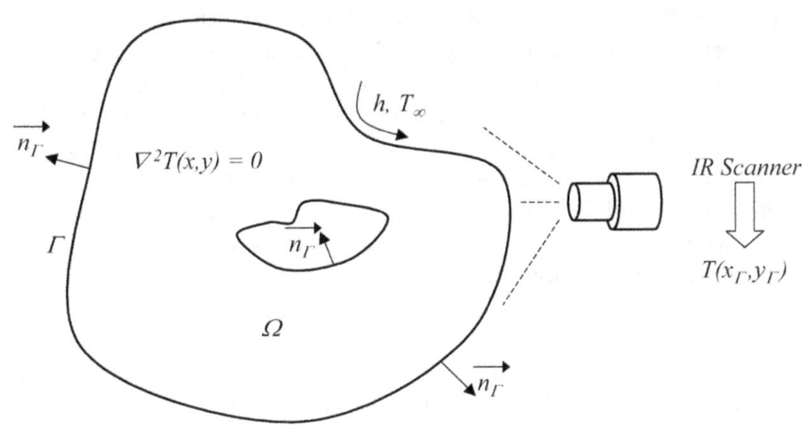

Figure 56. The inverse geometric heat conduction problem identifying a subsurface cavity utilizing knowledge of convective boundary condition and using surface temperatures measured by thermographic methods illustrated here as using an infrared scanner [115].

In the hybrid BEM/singularity superposition method [115] for the geometric inverse problem, there are three components to the solution algorithm:

(1) The forward heat steady heat conduction solver: A hybrid BEM/singularity method that resolve the thermal field under a current geometric configuration.
(2) The inverse problem solver – part 1: A minimization algorithm locates and fixes singularity(ies) thereby determining the location(s) or subsurface cavity(ies) by adjusting cluster centers and singularity strengths to match Cauchy conditions imposed at exposed boundary.
(3) The inverse problem solver – part 2: The cavity shape detection is obtained by a second mimization that locates cavity walls.

The governing equation for steady state heat conduction is the Laplace equation within domain, Ω.

***Forward problem solver*:** Borrowing from potential theory under the current geometric configuration and subject to a set of known boundary conditions, known exposed geometry, and unknown internal subsurface cavity(ies) geometries, the Poisson equation is solved assuming a number N_s of source/sink singularities distributed in cluster(s) within the exposed boundary of the domain, Ω,

$$\nabla^2 T(x) + \sum_{k=1}^{N_S} \hat{Q}_k \delta(\vec{r} - \vec{r}_k) = 0 \tag{295}$$

If they possess non-zero strengths, $\hat{Q}_n$, these singularities must be contained within the subsurface cavity(ies) otherwise the Laplace equation is violated. Utilizing the BEM, and discretizing the exposed boundary only, the above is discretized in standard form

$$\sum_{j=1}^{N_E} H_{ij} T_j = \sum_{j=1}^{N_E} G_{ij} q_j + \sum_{k=1}^{N_S} G_{ik} \hat{Q}_k \quad i = 1,2...N_E \tag{296}$$

Here, N_E, are the number of nodes on the exposed surface, and the N_S source/sinks have strengths, $\hat{Q}_k$, and location that are to be determined from part-1 solution of the inverse problem. Provided that a well-posed problem is available with a properly defined geometry and set of boundary conditions, the above equation is collocated to provide a set of linear equations of the form $A_{ij}x_j = b_i + S_i$ $i = 1,2\ldots N_E$ with and where S_i contains the effects of the added singularities $\hat{Q}_k$. The advantage of this approach is the BEM equations are never re-evaluated as the exposed surface does not change shape, and the right hand side vector {S} is readily updated with minimal computational effort. The solution to this system provides the full distribution of temperatures and fluxes around the boundary that can later be used in the same formulation to calculate these variables anywhere in the domain. Note that only an exterior discretization is required in this forward problem formulation. This is illustrated in Fig. 57.

***The inverse problem solver – part 1*:** The location, size, and shape of the cavity(ies) are unknown, the strengths of the added singularities may be adjusted to counteract the heat flow generated by the boundary conditions imposed at the exterior boundary and to consequently produce the artificial adiabatic contour enclosing them. This can be accomplished by the minimization of a functional that reduces the standard deviation of the successively BEM-computed temperatures T_i, $i = 1,2\ldots m$ at the m measuring points with respect to the $\hat{T}_i$ measured temperatures at the exposed boundaries,

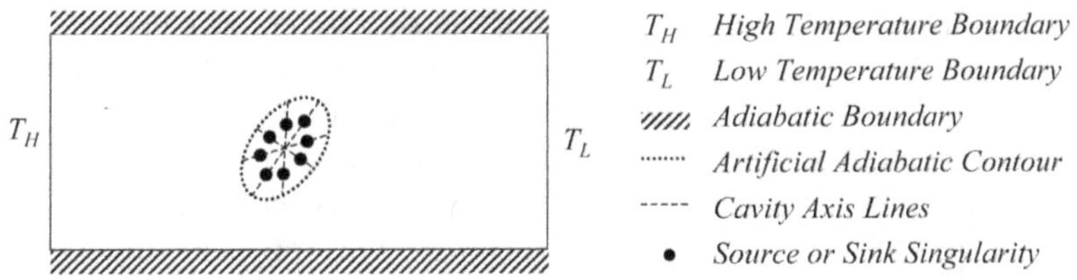

(a) Source/sink singularity cluster.

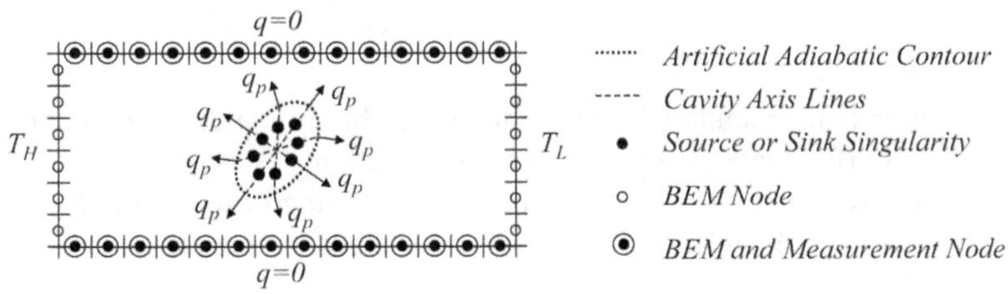

(b) BEM discretization and inverse problem configuration.

Figure 57. The singularity superposition method for the inverse geometric problem.

$$\psi(\hat{Q}_k, \rho) = \sum_{i=1}^{m} \left(T_i - \hat{T}_i\right)^2 \tag{297}$$

This objective function depends on singularity strengths, $\hat{Q}_k$, and a number of geometric parameters contained in ρ namely, the location of the centroid of the cluster that the singularities belong to as well as the axes of the ellipse (2D) or ellipsoid (3D) on which they are located as well as angles that determine the tilt relative to the reference coordinate frame. This function is readily minimized using non-gradient based methods such as the Nelder-Mead non-linear simplex or genetic algorithms. As single set of measurements may be used or multiple sets of measurements may be used with an augmented objective function to bring in more information to bear in the solution of the inverse problem as illustrated in Fig. 58 [117]. This part of the inverse problem effectively determines the existence or non-existance of subsurface cavities and holes as well as their number and their general locations and size (at least larger than any one singularity cluster).

The inverse problem solver – part 2: Locating the cavity is now the problem of searching the continuous domain enclosed by the exterior boundary for the location of the boundary condition at the surface the cavities and these may be assumes to be adiabatic. A geometry for the cavity(ies) must be assumed parametrically in order to control the dimensionality of the inverse problem to render its solution tractable. For instance one may use ellipses

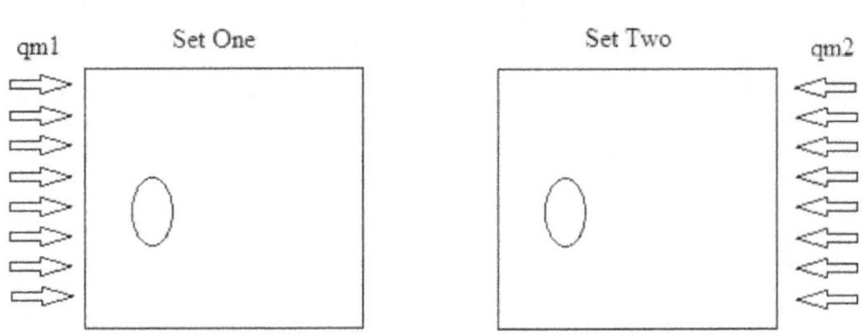

Figure 58. Multiple measurements for the inverse problem for the detection of a subsurface cavity.

(2D) or ellipsoids (3D) that enclose the singularity clusters [115], or may parametrically represent the geometry by means of periodic cubic splines [112–114,117]. The objective of the second part of the inverse problem is then to utilize these geometric control parameters for the geometry (ellipse centroids, major/minor axes, rotation angles in case of ellipse and ellipsoids, or spline knots in case of cubic splines) are collected in a vector β as independent parameters to be adjusted in a constrained optimization to identify the cavity geometries by minimizing the following second objective function,

$$\Phi(\beta) = \sum_{j=1}^{N_q} q_j^2 \tag{298}$$

To this end, the boundary integral equation is differentiated in order to provide the heat flux required to evaluate the second objective function.

An example for the location of a cavity contained within a doubly connected domain is shown in Fig. 59 and Fig. 60 where genetic algorithms were used to for minimization and

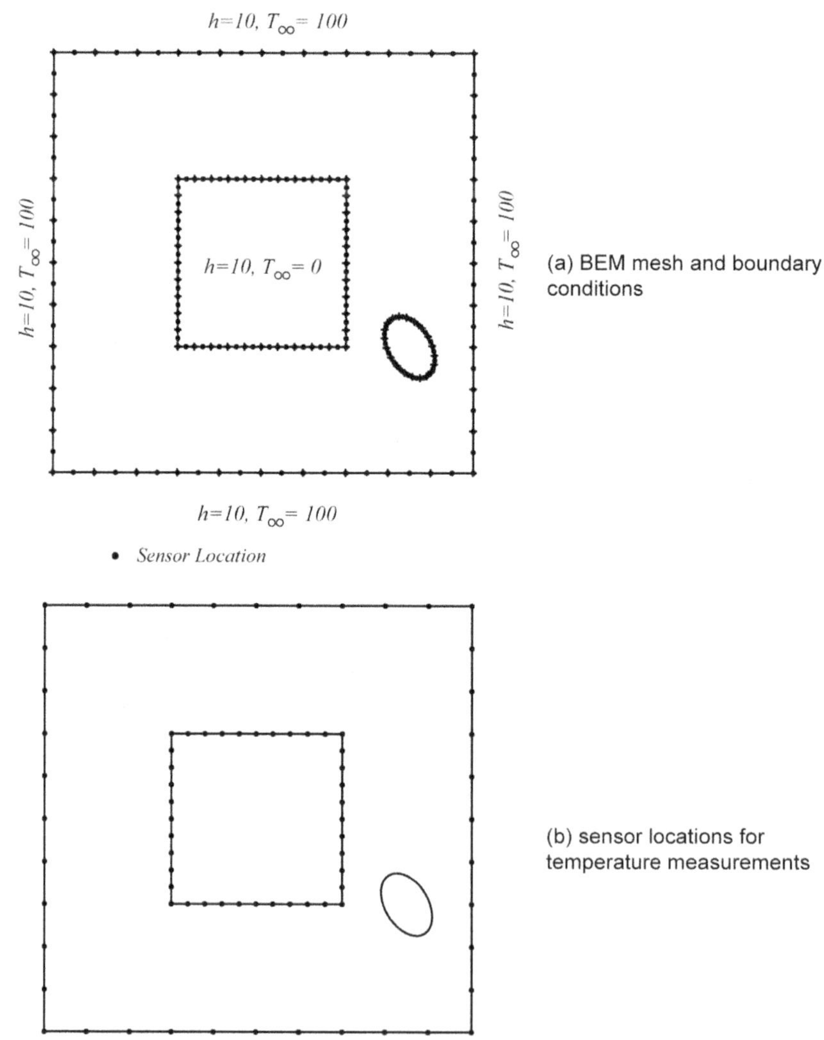

Figure 59. BEM mesh and boundary conditions used for the problem of locating a tilted elliptical cavity enclosed in a doubly-connected square plate [115].

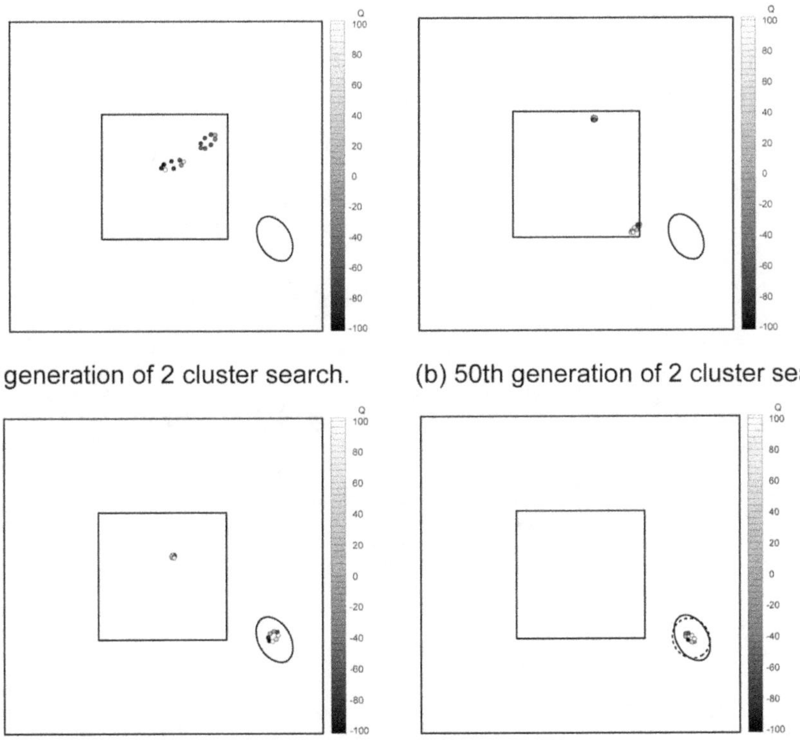

(a) 1st generation of 2 cluster search. (b) 50th generation of 2 cluster search.

(c) 1000th generation of 2 cluster search. (d) final location of the cavity.

Figure 60. Cluster identification with +/–0.25 simulated input random error in temperatures for a single tilted elliptical cavity within a doubly-connected square plate [115].

the sought after cavity was estimated using an ellipse. The measuring points at the exposed surface where the Cauchy conditions are imposed as well as the BEM discretization are displayed. Here, quadratic discontinuous elements were used in all calculations. Two clusters are initially used to search for the single cavity. One of the cluster locates itself inside the cavity, while the other exits the doubly-connected domain and shrinks to a very small area by the 1000th generation of the genetic algorithm and the strengths of the singularities is nearly zero. Consequently, such, that cluster is eliminated, as by all indications there is only cavity in this problem. The search is continued with the single surviving cluster, and, finally, the cavity is successfully located. Similar results are found in seeking multiple cavities in multiply connected domains and in locating cavities in 3D, indicating the robustness of this method.

This singularity superposition method has also been successfully applied to NDE using surface displacements in an elasticity-based BEM approach to solving the inverse geometric problem [118] and to identification of convective film coefficients [119]. Other applications of BEM to inverse problems in heat transfer, such as identification of surface heat fluxes, convective film coefficients, variable thermal conductivities, and thermal contact resistance can be found in [120–124].

Chapter 7

CONCLUSION

The Boundary Element Method (BEM) has been developed from the perspective of the Method of Weighted Residuals following the approach taken in the Finite Element Method. The BEM is a singular integral equation based method and discretization methods utilized to solve these equations have common features to those used in the FEM expect that singularities have to be addressed from the computational point of view. The distinguishing feature of the technique is the reduction of the problem dimension by one for those problems for which a fundamental solution can be found.

Certain problems of interest, for instance non-linear heat conduction and heat conduction in non-homogenous media, formally lead to domain integrals, and, in an effort to avoid evaluating such integrals, use of particular solutions developed by radial-basis function (RBF) interpolation leads to the Dual Reciprocity BEM. The latter development provides a powerful a technique that greatly expands the range of applications of the BEM and indeed the DRBEM has found wide of usage in a variety of engineering problems from transient conjugate problems, fluid mechanics, to thermo-elasticity. In the next section of this book, the integration step in the BEM is avoided altogether in a strong-form formulation of the meshless method that utilizes an underlying RBF interpolation to solve a variety of field problems encountered in thermo-fluids applications.

REFERENCES

[1] Brebbia, C.A. and Walker, S., *Boundary Element Techniques in Engineering*, Newnes-Butterworth, London, 1980.

[2] Brebbia, C.A., Telles, J.C.F., and Wrobel, L., *Boundary Element Technique in Engineering: Theory & Application in Engineering*, Springer-Verlag, New York, 1984.

[3] Gipson, G.S., *Boundary Element Fundamentals – Basic Concepts & Recent Developments in the Poisson Equation*, Computational Mechanics, Boston, 1987.

[4] Liggett, J.A. and Liu, P.L.F., *The Boundary Integral Equation Method for Porous Media Flow*, Allen & Unwin, Boston, 1983.

[5] Banerjee, P.K., *Boundary Element Method*, McGraw Hill Book Co., New York, 1992.

[6] Divo, E. and Kassab, A.J., *Boundary Element Method for Heat Conduction with Applications in Non-Homogeneous Media*, Wessex Institute of Technology (WIT) Press, Southampton, UK, and Boston, USA, 2003.

[7] Brebbia, C.A. and Dominguez, J.J., *Boundary Elements & Introductory Course*, Computational Mechanics & McGraw Hill Book Co., Boston & New York, 1992.

[8] Wrobel, L.C. and Aliabadi, M.A., *The Boundary Element Method*, Vols. 1 & 2, Wiley, New York, 2002.

[9] Kassab, A.J. and Wrobel, L.C., "Boundary Element Methods in Heat Conduction," *Chapter 5 in Recent Advances in Numerical Heat Transfer*, Mincowycz, W.J. and Sparrow, E.M. (eds.), Vol. 2, Taylor and Francis, New York, pp. 143–188, 2000.

[10] Kassab, A.J., Wrobel, L.C., Bialecki, R., and Divo, E., "Boundary Elements in Heat Transfer," *Chapter 4 in Handbook of Numerical Heat Transfer*, Vol. 1, 2nd Edition, Minkowycz, W., Sparrow, E.M., and Murthy, J.Y. (eds.), John Wiley and Sons, pp. 125–166, 2006.

[11] Hess, J.L. and Smith, A.M.O., "Calculations of Nonlifting Potential Flow about Arbitrary Three-Dimensional Bodies," *J. Ship. Res.*, Vol. 8, pp. 22–44, 1964.

[12] Jaswon, M.A., "Integral Equation Methods in Potential Theory I," *Proc. Royal Society, A*, Vol. 275, pp. 23–32, 1963.

[13] Jaswon, M.A. and Symm, G.T., *Integral Equation Methods in Potential Theory and Elastostatics*, Academic Press, London, 1977.

[14] Cruse, T.A., "A Direct Formulation and Numerical Solution of the General Transient Elastodynamic Problem—II," *J. Math. Anal. Appl.*, Vol. 22, pp. 341–355, 1968.

[15] Cruse, T.A., "Numerical Solutions in Three Dimensional Elastostatics," *International Journal of Solids Structures*, Vol. 5, pp. 1259–1274, 1969.

[16] Cruse, T.A., "Lateral Constraint in a Cracked, Three-Dimensional Elastic Body," *International J. Fract. Mech.*, Vol. 6, pp. 326–328, 1970.

[17] Cruse, T.A. and VanBuren, W., "Three-Dimensional Elastic Stress Analysis of a Fracture Specimen with an Edge Crack," *Int. J. Fract. Mech.*, Vol. 7, pp. 1–15, 1971.

[18] Rizzo, F.J., "An Integral Equation Approach to Boundary Value Problems of Classical Elastostatics," *Q. Appl. Math.*, Vol. 25, pp. 83–95, 1967.

[19] Rizzo, F.J. and Shippy, D.J., "Formulation and Solution Procedure for the General Non-Homogeneous Elastic Inclusion Problem," *Int. J. Solids Struct.*, Vol. 4, pp. 1161–1179, 1968.

[20] Rizzo, F.J. and Shippy, D.J., "A Method for Stress Determination in Plane Anisotropic Bodies," *J. Compos. Mater.*, Vol. 4, pp. 36–61, 1970.

[21] Rizzo, F.J. and Shippy, D.J., "A Method of Solution for Certain Problems of Transient Heat Conduction," *AIAA Journal*, Vol. 8, pp. 2004–2009, 1970.

[22] Shaw, R.P., "Diffraction of Acoustic Pulses by Obstacles of Arbitrary Shape with a Robin Boundary Condition," *J. Acoust. Soc. Am.*, Vol. 41, pp. 855–859, 1967.

[23] Shaw, R.P., "Boundary Integral Equation Methods Applied to Water Waves," In: Cruse, T.A. and Rizzo, F.J., editors. *Boundary-Integral Equation Method: Computational Applications in Applied Mechanics*, ASME AMD, Vol. 11, pp. 7–14, 1975.

[24] Brebbia, C.A., *The Boundary Element Method for Engineers*, Pentech Press/Halstead Press, London/New York, 1978.

[25] Cheng, A.H.D. and Cheng, D.T., "Heritage and Early History of the Boundary Element Method," *Engineering Analysis with Boundary Elements*, Vol. 1, pp. 1–35, 2005.

[26] Kellogg, O.D., *Foundations of Potential Theory*, Dover, New York, 1953.

[27] Özisik, N.M., *Heat Conduction*, John Wiley and Sons, New York, 1980.

[28] Greenberg, M.D., *Application of Green's Functions in Science and Engineering,* Prentice Hall, Englewood Cliffs, New Jersey, 1971.

[29] Beck, J.V., Cole, K.D., Haji-Sheik, A., and Litkouthi, B., *Heat Conduction Using Green's Function*, Hemisphere Publishing Co., Washington, 1992.

[30] Zauderer, E., *Partial Differential Equations of Applied Mathematics,* John Wiley and Sons, New York, 1989.

[31] Hoskins, R.F., *Generalized Functions*, John Wiley and Sons, New York, 1979.

[32] Gel'fand, I.M. and Shilov, G.E., *Generalized Forcing Functions*, Academic Press, New York, 1964.

[33] Partridge, P.W., Brebbia, C.A., and Wrobel, L.C., *The Dual Reciprocity Boundary Element Method*, Computational Mechanics, Boston, 1992.

[34] Butler, B., Kassab, A., Chopra, M., and Chaitanya, V., "Boundary Element Method for Electrochemical Dissolution with Geometric Non-Linearities," *Engineering Analysis* (in press).

[35] Cabral, J.J.S.P., Wrobel, L.C., and Brebbia, C.A., "A BEM Formulation Using B-Splines: I-Uniform Blending Functions," *Engineering Analysis*, Vol. 7, No. 4, pp. 136–144, 1990.

[36] Cabral, J.J.S.P., Wrobel, L.C., and Brebbia, C.A., "A BEM Formulation Using B-Splines: II-Multiple Knots and Non-Uniform Blending Functions," *Engineering Analysis*, Vol. 8, No. 1, pp. 51–55, 1991.

[37] Watson, J.O., "Hermetian Cubic Boundary Elements for Plane Problems of Fracture Mechanics," *Res. Mechanica*, Vol. 4, pp. 23–42, 1982.

[38] Watson, J.O., "Hermetian Cubic and Singular Elements for Plane Strain," *Developments in Boundary Element Methods-4*, eds. Banerjee, P.K. and Watson, J.O., Elsevier Applied Science Publishers, London, pp. 1–28, 1986.

[39] Beer, G. and Watson, J.O., *Introduction to Finite and Boundary Element Methods for Engineers*, John Wiley and Sons, New York, 1992.

[40] Durodola, F.F. and Fenner, R.T., "Hermetian Cubic Boundary Elements for Two-Dimensional Potential Problems," *International Journal for Numerical Methods in Engineering*, Vol. 30, pp. 1051–1062, 1990.
[41] Silva and Wrobel, L.C., "A New Family of Continuous/Discontinuous Three-Dimensional Boundary Elements with Application to Acoustic Wave Propagation," *International Journal for Numerical Methods in Engineering*, Vol. 36, pp. 1661–1679, 1993.
[42] Kane, J.H., Kashava-Kumar, B.L., and Saigal, S., "An Arbitrary Condensing, Non-Condensing Strategy for Large Scale, Multi-Zone Boundary Element Analysis," *Computer Methods Applied Mechanics and Engineering*, Vol. 79, pp. 219–244, 1990.
[43] Kassab, A.J. and Nordlund, R.S., "Addressing the Corner Problem in BEM Solution of Heat Conduction Problems," *Communications in Numerical Methods in Engineering*, Vol. 10, pp. 385–392, 1995.
[44] Mitra, A. and Ingber, M., "A Multiple Node Method to Resolve the Difficulties in the Boundary Integral Equation Method Caused by Corners and Discontinuous Boundary Conditions," *International Journal of Numerical Methods in Engineering*, Vol. 36, pp. 1735–1746, 1993.
[45] Ganguly, S., Layton, J.B., Balakrishna, C., and Kane, J.H., "A Fully Symmetric Multi-Zone Galerkin Boundary Element Method," *Int. J. Num. Methods in Engineering*, Vol. 44, pp. 991–1009.
[46] Mullen, R.L. and Rencis, J.J., "Iterative Methods for Solving Boundary Element Equations," *Computers & Structures*, Vol. 25, No. 5, pp. 713–723, 1997.
[47] Barra, L.P.S., Coutinho, A.L.G.A., Mansur, W.J., and Telles, J.C.F., "Iterative Solution of BEM Equations by GMRES Algorithm," *Computers and Structures*, Vol. 44, No. 6, pp. 1249–1253, 1992.
[48] Prasad, K.G., Kane, J.H., Keyes, D.E., and Balakrishna, C., "Preconditioned Krylov Solvers for BEA," *Int. J. Num. Methods in Engineering*, Vol. 37, pp. 1651–1672, 1994.
[49] Kane, J.H., Kumar, B.L., and Saigal, S., "An Arbitrary Condensing, Non-Condensing Solution Strategy for Large Scale, Multi-Zone Boundary Element Analysis," *Comput. Meth. Appl. Mech. Engng*, Vol. 79, pp. 219–244, 1990.
[50] Merkel, M., Bulgakov, V., Bialecki, R., and Kuhn, G., "Iterative Solution of Large-Scale 3-D BEM Industrial Problems," *Engineering Analysis*, Vol. 22, pp. 183–197, 1998.
[51] Divo, E., Kassab, A.J., and Rodriguez, F., "Parallel Domain Decomposition Approach for Large-Scale 3D Boundary Element Models in Linear and Non-Linear Heat Conduction," *Numerical Heat Transfer, Part B, Fundamentals*, Vol. 44, No. 5, pp. 417–437, 2003.
[52] Piessens, R., De-Dönker-Kapenga, E., Überhuber, C.W., and Kahaner, D.K., *QUADPACK - A Subroutine Package for Automatic Integration*, Springer Verlag, New York, 1983.
[53] Zienkiewycz, O.C. and Taylor, R.L., *The Finite Element Method*, McGraw Hill Book Co., New York, 1987.
[54] Bialecki, R.A., Dallner, R., and Kuhn, G., "New Application of Hypersingular Equations in the Boundary Element Method," *Computer Methods in Applied Mechanics and Engineering*, Vol. 103, pp. 399–416, 1993.
[55] Saad, Y. and Schultz, M.H., "GMRES: A Generalized Minimal Residual Algorithm for Solving Nonsymmetric Linear Systems," *SIAM J. Sci. Stat. Comput.*, Vol. 7, pp. 856–869, 1986.
[56] Temple, G., *Proc. Of Royal Society of London*, A169, p. 476, 1938.
[57] Bialecki, R.A. and Nowak, A.J., "Boundary Value Problems in Heat Conduction with Nonlinear Material and Nonlinear Boundary Conditions," *Applied Mathematical Modelling*, Vol. 5, pp. 417–421, 1981.
[58] Azevedo, J.P.S. and Wrobel, L.C., "Non-Linear Heat Conduction in Composite Bodies: A Boundary Element Formulation," *International Journal for Numerical Methods in Engineering*, Vol. 26, pp. 19–38, 1988.
[59] Bialecki, R.A. and Nhalik, R., "Solving Nonlinear Steady State Potential Problems in Inhomogeneous Bodies Using the Boundary Element Method," *Numerical Heat Transfer, Part B*, Vol. 15, pp. 79–96 1989.
[60] Shaw, R.P., "Green's Functions for Heterogeneous Media Potential Problems," *Engineering Analysis*, Vol. 13, pp. 219–221, 1994.
[61] Shaw, R.P. and Gipson, G.S., "A BIE Formulation of Linearly Layered Potential Problems," *Engineering Analysis*, Vol. 16, pp. 1–3, 1996.

[62] Shaw, R.P., Manolis, G., and Gipson, G.S., "The 2-D Free-Space Green's Function and BIE for a Poisson Equation with Linearly Varying Conductivity in Two Directions," (in) Ertekin, E.G., Brebbia, C.A., Tanaka, M., and Shaw, R.P., (eds.), *Boundary Element Technology XI: Proc. 10th Int. Conf. on Boundary Element Technology*, Computational Mechanics, Boston, pp. 327–334, 1996.

[63] Kassab, A.J. and Divo, E.A., "A Generalized Boundary Integral Equation for Isotropic Heat Conduction with Spatially Varying Thermal Conductivity," *Engineering Analysis with Boundary Elements*, Vol. 18, pp. 273–286, 1996.

[64] Divo, E.A. and Kassab, A.J., "A Boundary Integral Equation for Steady Heat Conduction in Anisotropic and Heterogeneous Media," *Numerical Heat Transfer, Part B*, Vol. 32, pp. 37–61, 1997.

[65] Divo, E.A. and Kassab, A.J., "A Generalized BIE for Transient Heat Conduction in Heterogeneous Media," *AIAA Journal of Thermophysics Heat Transfer*, Vol. 12, No. 3, pp. 364–373, 1998.

[66] Rizzo, F. and Shippy, D.J., "A Formulation and Solution Procedure for the General Non-Homogeneous Elastic Inclusion Problem," *Int. J. Solids Struct.*, Vol. 4, pp. 1161–1179, 1968.

[67] Korn, G. and Korn, T., *Mathematical Handbook for Scientists and Engineers*, Dover Publication, Inc., Mineola, New York, p. 877, (21.9-11), 1961.

[68] Nowak, A.J. and Brebbia, C.A., "The Multiple Reciprocity Method: A New Approach for Transforming BEM Domain Integrals into Boundary Integrals," *Engineering Analysis with Boundary Elements*, Vol. 6, pp. 164–167, 1989.

[69] Kassab, A.J. and Nordlund, R.S., "Efficient Implementation of the Fourier Dual Reciprocity Boundary Element Method Using Two-Dimensional Fast Fourier Transforms," *Engineering Analysis with Boundary Elements*, Vol. 12, No. 2, pp. 93–102, 1993.

[70] Kassab, A.J. and Nordlund, R.S., "Application of the FDRBEM to Nonlinear Heat Conduction with Generation Using a Two-Dimensional FFT," *Boundary Element Communications*, Vol. 4, No. 5, pp. 195–198, 1993.

[71] Rahaim, C.P. and Kassab, A.J., "Pressure Correction DRBEM Solution for Heat Transfer and Fluid Flow in Incompressible Viscous Fluids," *Engineering Analysis with Boundary Elements*, Vol. 18, No. 4, pp. 265–272, 1996.

[72] Rahaim, C.P. and Kassab, A.J., "A DRBEM Solution for Incompressible Viscous Flows and Heat Transfer," *AIAA Paper 95-2113*.

[73] Golberg, M.A., Chen, C.S., and Power, H., "Some Comments on the Use of Radial Basis Functions in the Dual Reciprocity Method," *Computational Mechanics*, Vol. 22, pp. 61–69, 1998.

[74] Chen, C.S., Brebbia, C.A., and Power, H., "Dual Reciprocity Method Using Compactly Supported Radial Basis Functions," *Communications in Numerical Methods in Engineering*, Vol. 15, pp. 137–150, 1999.

[75] Divo, E., Kassab, A.J., and Cavalleri, R.J., "Application of the DRBEM to Model Ablation Characteristics of a Thrust Vector Control Vane," *Engineering Analysis*, Vol. 23, No. 8, pp. 693–702, 1999.

[76] Erhart, K., Divo, E., and Kassab, A.J., "Parallelization for Large-Scale Heat Conduction Problems," *Chapter in Domain Decomposition Techniques for Boundary Elements-Application to Fluid Flow*, (eds.), Popov, V., Power, H., and Skerget, L., WIT Press, Southampton, U.K., pp. 147–182, 2007.

[77] Nordlund, R.S. and Kassab, A.J., "Non-Fourier Heat Conduction: A boundary Element Solution of the Hyperbolic Heat Conduction Equation," *Proceedings of the 17th International Conference on Boundary Elements, BEM 17, July 17–19, 1995, Madison, Wisconsin*, pp. 279–286.

[78] Kassab, A.J. and Aliabadi, M.H. (eds.), *Advances in Boundary Elements: Coupled Field Problems*, WIT Press, Southampton and Boston, 2001.

[79] Kassab, A., Divo, E., Heidmann, J., Steinthorsson, E., and Rodriguez, F., "BEM/FVM Conjugate Heat Transfer Analysis of a Three-Dimensional Film Cooled Turbine Blade," *International Journal for Numerical Methods in Heat and Fluid Flow*, Vol. 13, No. 5, pp. 581–610, 2003.

[80] Divo, E., Steinthorsson, E., Kassab, A., and Bialecki, R., "An Iterative BEM/FVM Protocol for Steady-State Multi-Dimensional Conjugate Heat Transfer in Compressible Flows," *Engineering Analysis*, Vol. 26, No. 5, pp. 447–454, 2002.

[81] Bialecki, R., Ostrowski, Z., Kassab, A.J., Yin, Q., and Scubbia, E., "Coupling FEM, BEM, and Analytical Solutions in Steady-State Potential Problems," *Engineering Analysis*, Vol. 26, No. 7, pp. 597–612, 2002.

[82] Rahaim, C.P., Kassab, A.J., and Cavalleri, R., "A Coupled Dual Reciprocity Boundary Element/Finite Volume Method for Transient Conjugate Heat Transfer," *AIAA Journal of Thermophysics and Heat Transfer*, Vol. 14, No. 1, pp. 27–38, 2000.

[83] He, M., Bishop, P., Kassab, A.J., and Minardi, A., "A Coupled FDM/BEM Solution for the Conjugate Heat Transfer Problem," *Numerical Heat Transfer, Part B: Fundamentals*, Vol. 28, No. 2, pp. 139–154, 1995.

[84] Li, H. and Kassab, A.J., "Numerical Prediction of Fluid Flow and Heat Transfer in Turbine Blades with Internal Cooling," *AIAA/ASME 94-2933*.

[85] Li, H. and Kassab, A.J., "A Coupled FVM/BEM Solution to Conjugate Heat Transfer in Turbine Blades," *AIAA Paper 94-1981*.

[86] Heidman, J., Kassab, A.J., Divo, E., Rodriguez, F., and Steinthorssen, E., "Conjugate Heat Transfer Effects on a Realistic Film-Cooled Turbine Vane," *IGTI Paper GT2003-38553*.

[87] Bialecki, R.A., Merkel, M., Mews, H., and Kuhn, G., "In-and Out-Of-Core BEM Equation Solver with Parallel and Nonlinear Options," *International Journal for Numerical Methods in Engineering*, Vol. 39, pp. 4215–4242, 1996.

[88] Greengard, L. and Strain, J., "A Fast Algorithm for the Evaluation of Heat Potentials," *Communications in Pure and Applied Mathematics*, Vol. 43, pp. 949–963, 1990.

[89] Hackbush, W. and Nowak, Z.P., "On the Fast Multiplication in the Boundary Element Method by Panel Clustering," *Numerische Mathematik*, Vol. 54, pp. 463–491, 1989.

[90] Liu, Y.J. and Nishimura, N., "The Fast Multipole Boundary Element Method: A Tutorial," *Engineering Analysis*, Vol. 30, pp. 371–381, 2006.

[91] Liu, Y.J., Mukherjee, S., Nishimura, N., Schanz, M., Ye, W., Sutradhar, A., Pan, E., Dumont, N.A., Frangi, A., and Saez, A., "Recent Advances and Emerging Applications of the Boundary Element Method," *Applied Mechanics*, Vol. 64, pp. 1–38, 2011.

[92] Bucher, H. and Wrobel, L.C., "A Novel Approach to Applying Wavelet Transforms in Boundary Element Method," *Advances in Boundary Element Techniques, II*, Denda, M., Aliabadi, M.H., and Charafi, A., (eds.), Hogaar Press, Switzerland, pp. 3–11, 2000.

[93] Xiao, J., Tausch, J., and Hu, Y., "A-Posteriori Compression of Wavelet-BEM Matrices," *Computational Mechanics*, Vol. 44, Issue 5, pp. 705–715.

[94] Divo, E., Rodriguez, F., and Kassab, A.J., "Parallel Domain Decomposition Approach to Large-Scale Three Dimensional Boundary Element Models in Linear and Non-Linear Heat Conduction," *Numerical Heat Transfer, Part B, Fundamentals*, Vol. 44, No. 5, pp. 417–437, 2003.

[95] Furmański, P., Heat Conduction in Composites: Homogenization & Macroscopic Behavior," *ASME Applied Mechanics Reviews*, Vol. 50, No., 1997.

[96] Peng, H.F., Yang, K., and Gao, X.W., "Element Nodal Computation-Based Radial Integration BEM for Non-Homogeneous Problems," *Acta Mechanica Sinica*, Vol. 29, No. 33, pp. 429–436, 2013.

[97] Cheng, A.H.-D., "Darcy's Flow with Variable Permeability: A Boundary Integral Solution," *Water Resources Research*, Vol. 20, pp. 980–984, 1984.

[98] Sutradhar, A. and Paulino, G.H., "A Simple Boundary Element Method for Problems of Potential in Non-Homogeneous Media," *International Journal for Numerical Methods in Engineering*, Vol. 60, pp. 2203–2230, 2004.

[99] Divo, E., A New Boundary Integral Method for Heat Conduction in Anisotropic Media with Space Dependent Thermal Conductivity, Ph.D. Dissertation, University of Central Florida, 1998.

[100] Hammard, J., *Lectures on Cauchy's Problem in Linear Partial Differential Equations*, Dover Publications, New York, 1952.

[101] Thikonov, A.N. and Arsenin, *Solution of Ill-Posed Problems*, John Wiley and Sons, New York, 1994.

[102] Beck, J.V., Blackwell, B., and St. Clair, C.R., *Inverse Heat Conduction: Ill-Posed Problems*, John Wiley and Sons, New York, 1985.

[103] Alifanov, O.M., *Inverse Heat Transfer Problems*, Springer Verlag, New York, 1994.

[104] Kurpisz, K. and Nowak, A. J., (eds.), *Inverse Thermal Problems*, Wessex Institute of Technology Press, Boston, 1995.

[105] Woodbury, K., ed., *Inverse Engineering Handbook*, CRC Press, Boca Raton, 2003.

[106] Bui, H.D., *Inverse Problems in Mechanics of Materials: An Introduction*, CRC Press, Boca Raton, 1994.

[107] Kassab, A.J. and Divo, E.A., (eds.), *Proceedings of ICIPE2011, 7th International Conference on Inverse Problems*, Centecorp Press, Orlando, 2011.

[108] Hsieh, C.K. and Su, K.C., "A Methodology of Predicting Cavity Geometry Based on the Scanned Surface Temperature Data-Prescribed Surface Temperature at the Cavity Side," *Journal of Heat Transfer*, Vol. 102, pp. 324–329, 1980.

[109] Hsieh, C.K. and Kassab, A.J., "A General Method for the Solution of Inverse Heat Conduction Problems with Partially Specified System Geometries," *International Journal Heat and Mass Transfer*, Vol. 29, pp. 47–58, 1986.

[110] Hsieh, C.K., Choi, C.Y., and Liu, K.M., "A Domain Extension Method for Quantitative Detection of Cavities by Infrared Scanning," *Journal of Nondestructive Evaluation*, Vol. 8, pp. 195–211, 1989.

[111] Kassab, A.J., Hsieh, C.K., and Pollard, J., "Solution of the Inverse Geometric Problem for the Detection of Subsurface Cavities by the IR CAT Method," Chapter 2, Boundary Integral Formulations in Inverse Analysis, L.C. Wrobel and D.B. Ingham, (eds.), Computational Mechanics, Boston, pp. 33–65, 1997.

[112] Kassab, A.J. and Pollard, J., "Automated Algorithm for the Nondestructive Detection of Subsurface Cavities by the IR-CAT Method," *Journal of Nondestructive Evaluation*, Vol. 12, No. 3, pp. 175–187, 1993.

[113] Kassab, A.J. and Pollard, J., "Automated Cubic Spline Anchored Grid Pattern Algorithm for the High Resolution Detection of Subsurface Cavities by the IR-CAT Method," *Numerical Heat Transfer, Part B: Fundamentals*, Vol. 26, No. 1, pp. 63–78, 1994.

[114] Kassab, A.J. and Pollard, J., "An Inverse Heat Conduction Algorithm for the Thermal Detection of Arbitrarily Shaped Cavities," *Inverse Problems in Engineering*, Vol. 1, No. 3, pp. 231–245, 1995.

[115] Divo, E., Kassab, A., and Rodriguez, F., "A Hybrid Singularity Superposition/BEM Method for the Solution of the Inverse Geometric Problem," *Numerical Heat Transfer, Part B: Fundamentals*, Vol. 46, No. 1, pp. 1–30, 2004.

[116] Ramm, A.G., "A Geometrical Inverse Problem," *Inverse Problems*, Vol. 2, pp. L19–L21, 1986.

[117] Ni, M., Automated Hybrid Singularity Superposition and Anchored Grid Pattern BEM Algorithm for the Solution of the Inverse Geometric Problem, MSME Thesis, University of Central Florida, 2013.

[118] Ojeda, D., Divo, E., Kassab, A., and Cerrolaza, M., "Cavity Detection in Biomechanics by an Inverse Evolutionary Point Load BEM Technique," *Inverse Problems in Science and Engineering*, Vol. 16, No. 8, pp. 981–993, 2008.

[119] Silieti, M., Kassab, A.J., and Divo, E.A., "Singular Superposition/Boundary Element Method for Reconstruction of Multi-Dimensional Heat Flux Distributions with Application to Film Cooling Holes," *Computers, Materials and Continua, CMC*, Vol. 12, No. 2, pp. 121–143, 2009.

[120] Bialecki, R., Divo, E., and Kassab, A., "Reconstruction of Time Dependent Boundary Heat Flux by a BEM-Based Inverse Algorithm," *Engineering Analysis*, Vol. 30, No. 9, pp. 767–773, 2006.

[121] Kassab, A.J., Divo, E., Kapat, J.S., and Chyu, M.K., "Retrieval of Multi-Dimensional Heat Transfer Coefficient Distributions Using an Inverse-BEM-Based Regularized Algorithm: Numerical and Experimental Examples," *Engineering Analysis*, Vol. 29, No. 2, pp. 150–160, 2005.

[122] Silieti, M., Divo, E., and Kassab, A.J., "An Inverse Boundary Element Method/Genetic Algorithm Based Approach for Retrieval of Multi-Dimensional Heat Transfer Coefficients within Film Cooling Holes/Slots," *Inverse Problems in Engineering and Science*, Vol. 13, No. 1, pp. 79–98, 2005.

[123] Divo, E., Kassab, A.J., and Rodriguez, F., "Characterization of Space Dependent Thermal Conductivity with a BEM-Based Genetic Algorithm," *Numerical Heat Transfer, Part A: Applications*, Vol. 37, No., pp. 845–877, 2000.

[124] Gill, J., Divo, E., and Kassab, A.J., "Estimating Thermal Contact Resistance Using Sensitivity Analysis and Regularization," *Engineering Analysis*, Vol. 33, No. 1, pp. 54–62, 2009.

PART III
THE MESHLESS METHOD

Chapter 1

INTRODUCTION AND BACKGROUND

In recent history, the area of physics-based engineering simulation has seen rapid increases in both computer workstation performance as well as common engineering model complexity, both driven largely in part by advances in memory density and availability of clusters and multi-core processors. While the increase in computation time due to model complexity has been largely offset by the increased performance of modern workstations, the increase in model setup time due to model complexity has continued to rise. As such, the major time requirement for solving an engineering model has transitioned from computation time to problem setup time. This is due to the fact that developing the required mesh for complex geometries can be an extremely complicated and time consuming task, and consequently, new solution techniques that can reduce the required amount of human interaction are desirable.

While the finite element method (FEM) and the boundary element method (BEM) have been developed to a mature stage such that they are now utilized routinely to model complex multi-physics problems, they require significant effort in formulation, mesh generation, and data management. Meshless methods are a relative newcomer to the field of computational methods. The term "Meshless Methods" refers to the class of numerical techniques that rely on either global or localized interpolation on non-ordered spatial point distributions. As such, there has been much interest in the development of these techniques as they have the hope of reducing the effort devoted to model preparation [1–7]. The approach finds its origin in classical spectral or pseudo-spectral methods [8–12] that are based on global orthogonal functions such as Legendre or Chebyshev polynomials requiring a regular nodal point distribution. In contrast, Meshless methods use a nodal or point distribution that is not required to be uniform or regular due to the fact that most such techniques rely on global radial-basis functions (RBF) [13–23]. RBF have proved quite successful in their application to an earlier mesh-reduction method, namely the dual reciprocity boundary element method (DRBEM) [24,25]. However, global RBF-based Meshless methods have some drawbacks including poor conditioning of the ensuing algebraic set of equations which can be addressed to some extent by domain decomposition and appropriate pre-conditioning [20]. Moreover, care must be taken in the evaluation of derivatives in global RBF-based Meshless methods. Although, very promising, these techniques can also be computationally intensive. Recently, localized collocation Meshless methods [26–28] have been proposed to address many of the issues posed by global RBF Meshless methods.

For more than a decade the co-authors have been involved in numerical modeling of heat transfer, fluid flow, and solid mechanics problems using the Boundary Element Methods (BEM), its derivative, the Dual Reciprocity BEM, and other mesh reduction techniques, see [24,25,29–32]. In a series of recent publications [33–51], the co-authors have developed a Localized Collocation Meshless Method (LCMM) based on Radial-Basis Function (RBF) interpolation for modeling of coupled viscous fluid flow, heat transfer, and fluid-structure interaction problems. The LCMM features Hardy Multiquadrics RBF augmented by polynomial expansions over a local topology of points for the sought-after unknowns with an efficient formulation for computing the interpolations in terms of vector products. This approach is applicable to explicit or implicit time marching schemes as well as steady-state iterative methods. The LCMM technique lends itself very well to parallel computations. In recent publications [41–43,45–51], the LCMM is shown to be computationally more efficient than a comparative Finite Volume Method (FVM) code whilst affording the distinct advantage of solving the partial differential conservation field equations of fluid flow and heat transfer on a non-ordered set of points. The method has been extensively verified against benchmarks and validated finite volume codes for several cases.

Chapter 2

RADIAL-BASIS FUNCTION (RBF) INTERPOLATION

Assume a general field variable, $\phi(x)$, may be interpolated in space over a finite number, N, of collocation points in terms of the same number of predefined expansion functions, $\chi_j(x)$, as:

$$\phi(x) = \sum_{j=1}^{N} \alpha_j \chi_j(x) \tag{1}$$

Here, the coefficients, α_j, are found through a standard collocation process of the field variable, $\phi(x)$, at the N collocation points, $x_i, \because i = 1 \cdots N$, leading to:

$$\phi(x_i) = \sum_{j=1}^{N} \alpha_j \chi_j(x_i), \because i = 1 \ldots N \tag{2}$$

or, in matrix-vector form:

$$\{\phi\} = [C]\{\alpha\} \tag{3}$$

where the elements, C_{ij}, of the matrix $[C]$, are found through the evaluation of the expansion functions, $\chi_j(x)$, at the collocation points, x_i, such that:

$$C_{ij} = \chi_j(x_i), \because i, j = 1 \ldots N \tag{4}$$

or,

$$[C] = \begin{bmatrix} \chi_1(x_1) & \cdots & \chi_N(x_1) \\ \vdots & \ddots & \vdots \\ \chi_1(x_N) & \cdots & \chi_N(x_N) \end{bmatrix} \tag{5}$$

The coefficients, α_j, are found by solving the linear system in Eq. (3) as:

$$\{\alpha\} = [C]^{-1}\{\phi\} \tag{6}$$

The expansion functions, $\chi_j(x)$, may be selected from the family of Radial-Basis functions (RBF). Such functions consist of algebraic expressions uniquely defined in terms of the Euclidean distance, $r_j(x)$, from a general field point, x, to an expansion point, x_j. Examples of these functions are:

(i) Polyharmonics RBF:

$$\begin{aligned} \chi_j(x) &= r_j^{2n}(x) \ln\left[r_j(x)\right], \quad \text{in } 2D \\ \chi_j(x) &= r_j^{2n-1}(x), \quad \text{in } 3D \end{aligned} \tag{7}$$

(ii) Multiquadrics RBF:

$$\chi_j(x) = \left[r_j^2(x) + c^2\right]^{n-\frac{3}{2}} \tag{8}$$

(iii) Gaussian RBF:

$$\chi_j(x) = \exp\left[-\frac{r_j^2(x)}{c^2}\right] \tag{9}$$

where n is a positive exponent parameter and c is a shape parameter. The effects of these parameters on the interpolation will be discussed later. In all cases, the radial distance, $r_j(x)$, is defined in the Cartesian coordinate system as:

$$\begin{aligned} r_j(\mathbf{x}) &= \sqrt{(x-x_j)^2+(y-y_j)^2}, \quad \text{in } 2D \\ r_j(\mathbf{x}) &= \sqrt{(x-x_j)^2+(y-y_j)^2+(z-z_j)^2}, \quad \text{in } 3D \end{aligned} \tag{10}$$

An example of an RBF collocation is now presented over a 1 × 1 square representing the domain of interest. A total of $N = 140$ points are distributed in the domain with $NI = 100$ equally distributed points in the interior and $NB = 40$ equally distributed points on the boundary. The domain and point distribution are shown in Fig. 1.

The following test function will be expanded in the domain using the collocation points shown in Fig. 1:

$$f(\mathbf{x}) = \frac{\sin(\pi y)\cosh\left(\frac{1-x}{2}\right)}{\cosh\left(\frac{1}{2}\right)} \tag{11}$$

The Multiquadrics RBF will be employed to perform the expansion using an exponent parameter $n = 1$ and a shape parameter $c = 0.5$. In the case of the choice of $n = 1$, the well established inverse Multiquadrics is arrived at as:

$$\chi_j(x) = \frac{1}{\sqrt{r_j^2(x)+c^2}} \tag{12}$$

The behavior of this function has been largely studied in the literature [17–23]. However, the shape parameter, c, single-handedly dictates the behavior of the expansion and more importantly, its derivatives. For a specific expansion over a specific set of points, the larger the shape parameter, c, the smoother the derivative field. However, the magnitude of the shape parameter, c, cannot be increased boundless as the expansion functions become flatter and, hence, the coefficient matrix, $[C]$, resulting from the collocation becomes

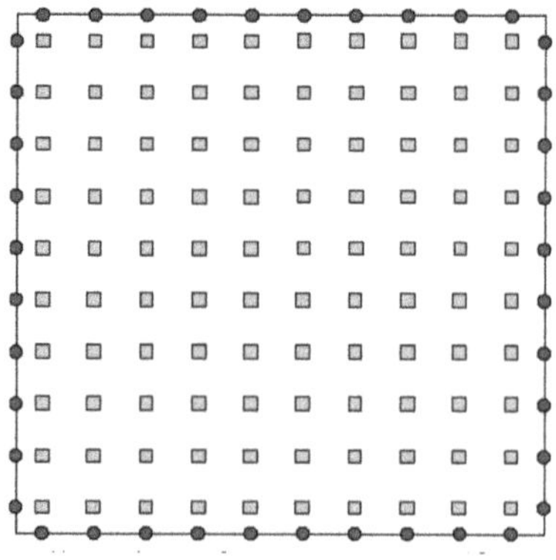

Figure 1. Geometry and point collocation of RBF test case (from [33]).

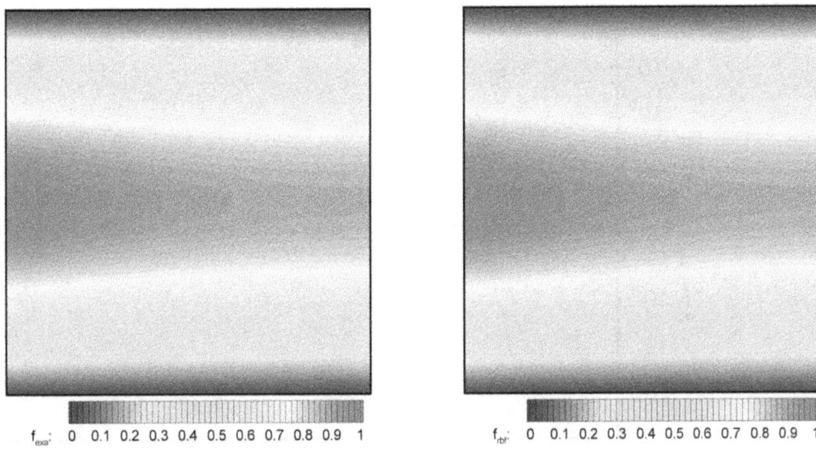

Figure 2. Contour plots of the exact and RBF-expanded test function (from [33]).

ill-conditioned. A simple optimization process to determine the value of this shape parameter, c, will be discuss in the next section.

The test function in Eq. (11) and its corresponding RBF expansion are evaluated at 441(21 × 21) equally distributed points in the 1 × 1 square domain to check the quality of the interpolation. Fig. 2 shows the contour plots of the exact and RBF-expanded test function at these points revealing a very accurate representation evident in the virtually indistinguishable qualitative plots.

However, when the x-derivative of the test function is approximated using the same expansion with the already determined expansion coefficients, α_j, such that:

$$\frac{\partial f}{\partial x}(x) = \sum_{j=1}^{N} \alpha_j \frac{\partial \chi_j}{\partial x}(x) \tag{13}$$

The results reveal far less accurate behavior as evidenced in Fig. 3 where a very noticeable deviation is seen between the exact and RBF-expanded x-derivative fields.

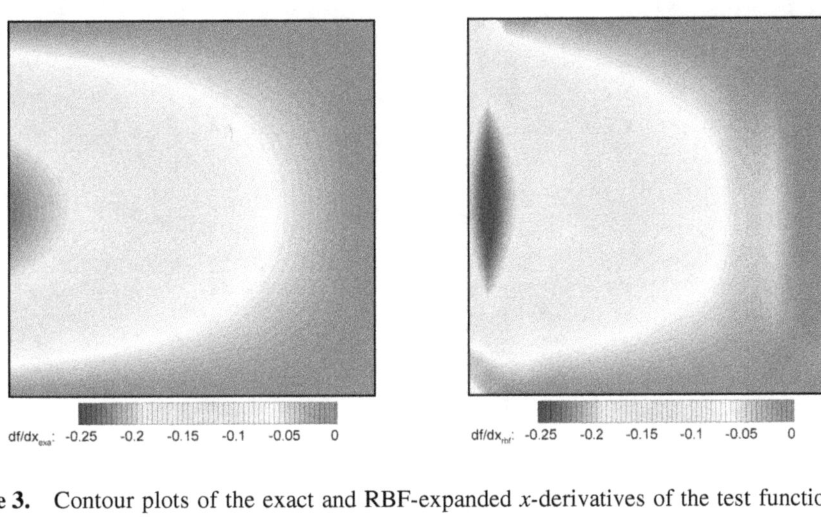

Figure 3. Contour plots of the exact and RBF-expanded x-derivatives of the test function (from [33]).

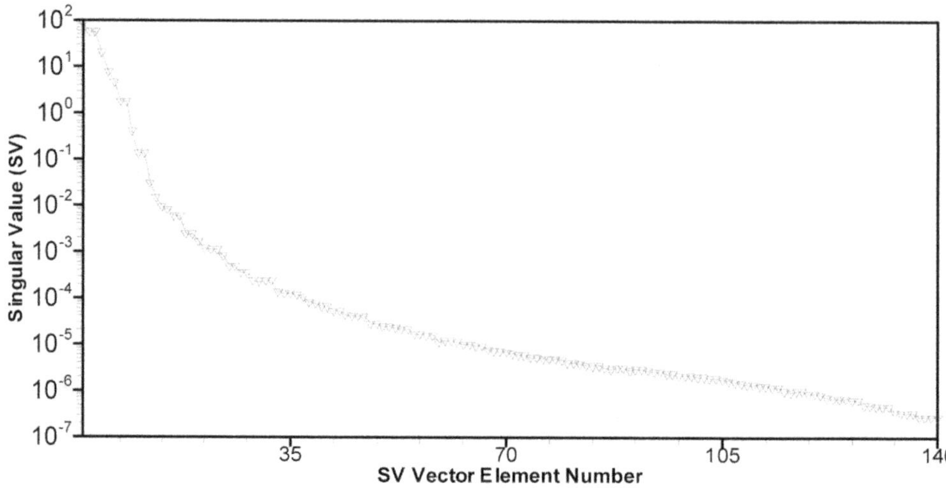

Figure 4. Singular values of the coefficient matrix, [*C*] (from [33]).

To analyze the behavior of the RBF expansion a Singular-Value decomposition of the coefficient matrix, [*C*], is performed revealing a maximum singular value in the order of 10^2 and a minimum singular value in the order of 10^{-7} as seen in Fig. 4. These singular values produce a ratio (condition number of [*C*]) in the order of 10^9.

The condition number of the coefficient matrix, [*C*], is found to be proportional to the shape parameter, *c*, used in the Inverse Multiquadrics RBF. Notice that the larger the shape parameter, c, the 'Flatter' the Inverse Multiquadrics RBF become. Flatter interpolation functions produce better derivative representations but highly ill-conditioned coefficient matrices. These results suggest that for an accurate representation of a field variable and its derivatives using a global RBF expansion, a highly ill-conditioned and fully-populated coefficient matrix must be produced. Therefore, for numerical reasons, it becomes imperative to mitigate this issue by:

- Efficient preconditioning
- Domain decomposition
- Localized expansion

Chapter 3

THE LOCALIZED COLLOCATION MESHLESS METHOD (LCMM) FRAMEWORK

The Meshless formulation begins by defining a set of data centers, *NC*, comprised of points on the boundary, *NB*, and points on the interior, *NI*. These data centers will serve as collocation points for the localized expansion of the different field variables in the domain, Ω, and on the boundary, Γ, see Fig. 5. The essential difference between boundary points and internal points is simply that boundary conditions will be applied at the first while governing equations will be applied at the last.

To illustrate the Meshless formulation the diffusion equation for a general field variable, ϕ, in a generalized coordinate system, x, time, t, and a general diffusion coefficient, κ, will be taken into consideration as the governing equation valid in the domain, Ω, as:

$$\frac{\partial \phi}{\partial t}(x,t) = \kappa \nabla^2 \phi(x,t) \tag{14}$$

In addition, a set of generalized boundary conditions for the variable, ϕ, on the boundary, Γ, are given by:

$$\hat{\beta}_1 \frac{\partial \phi}{\partial n} + \hat{\beta}_2 \phi = \hat{\beta}_3 \tag{15}$$

where $\hat{\beta}_1, \hat{\beta}_2$, and $\hat{\beta}_3$ are imposed coefficients of (x,t) that dictate the boundary condition type and constrain values.

A linear localized expansion over a group or topology of influence points, *NF*, around each data center is sought such that:

$$\phi(x) = \sum_{j=1}^{NF} \alpha_j \chi_j(x) + \sum_{j=1}^{NP} \alpha_{j+NF} P_j(x) \tag{16}$$

The terms α_j represent the unknown expansion coefficients while the terms $\chi_j(x)$ are expansion functions defined a-priori. Here *NP* is a number of additional polynomial functions, $P_j(x)$, added to the expansion to guarantee that constant and linear fields can be retrieved by

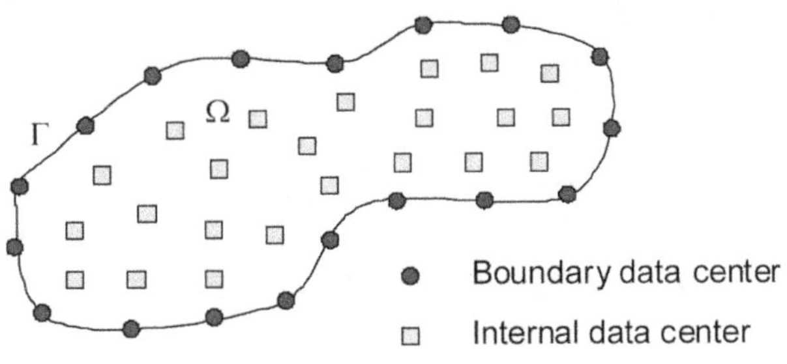

Figure 5. Scattered point distribution in a generalized domain (from [48]).

the expansion exactly. Notice that the time dependency has been dropped as a different expansion will be performed for each time level and, therefore, the expansion coefficients, α_j, will vary as time progresses.

The selection of an influence region or localized topology of expansion around each data center is easily accomplished by a circular (spherical in 3D) search around each data center. The search is automated to guarantee that a minimum number of points will be included and additional criteria, such as including all directions around internal data centers, are met. In addition, this search must guarantee that topologies around boundary data centers do not include opposing boundaries or points around a re-entry corner. Fig. 6 shows typical collocation topologies for internal and boundary data centers including re-entry corners and opposing boundaries. Fig. 7 shows an example of the circular search to build the topology around an internal data center of a typical non-uniform point distribution.

The collocation of the known field variable, ϕ, (from previous time level or iteration step) at the points within the localized topology, leads to the following in matrix-vector form:

$$\{\phi\} = [C]\{\alpha\} \tag{17}$$

And, therefore, the expansion coefficients can be determined as:

$$\{\alpha\} = [C]^{-1}\{\phi\} \tag{18}$$

where the resulting collocation matrix is given by:

$$[C] = \begin{bmatrix} \chi_1(x_1) & \cdots & \chi_{NF}(x_1) & P_1(x_1) & \cdots & P_{NP}(x_1) \\ \vdots & \ddots & \vdots & \vdots & \ddots & \vdots \\ \chi_1(x_{NF}) & \cdots & \chi_{NF}(x_{NF}) & P_1(x_{NF}) & \cdots & P_{NP}(x_{NF}) \\ P_1(x_1) & \cdots & P_1(x_{NF}) & 0 & \cdots & 0 \\ \cdots & \ddots & \cdots & \vdots & \ddots & \vdots \\ P_{NP}(x_1) & \cdots & P_{NP}(x_{NF}) & 0 & \cdots & 0 \end{bmatrix}_{NF+NP,\, NF+NP} \tag{19}$$

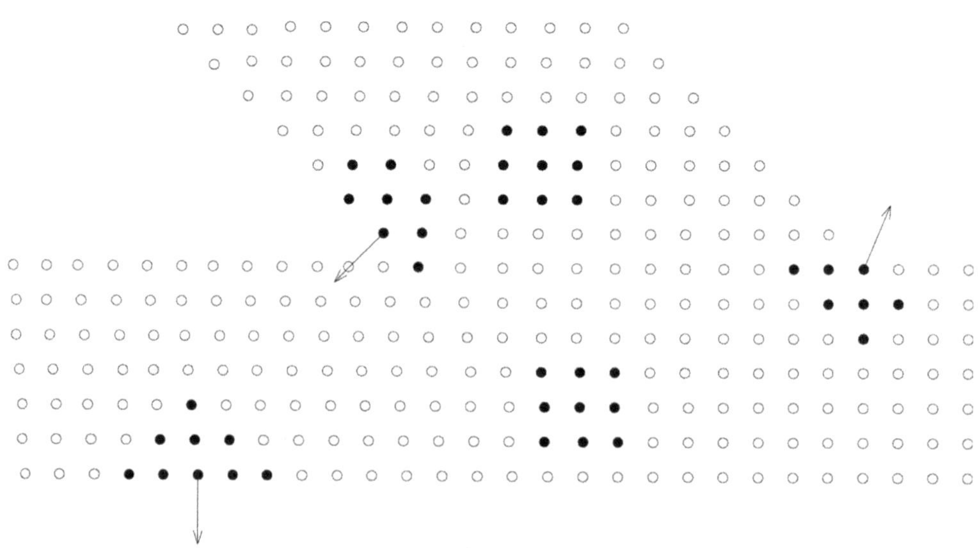

Figure 6. Collocation topology for internal, boundary, and corner data centers (from [43]).

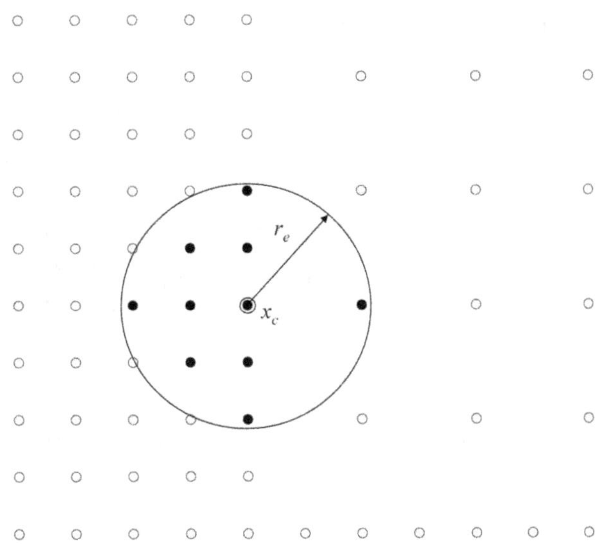

Figure 7. Collocation topology selection on a non-uniform point distribution (from [43]).

and the right-hand side known vector is augmented as:

$$\{\phi\} = \begin{Bmatrix} \phi(x_1) \\ \vdots \\ \phi(x_{NF}) \\ 0 \\ \vdots \\ 0 \end{Bmatrix}_{NF+NP,1} \tag{20}$$

The NF expansion functions, $\chi_j(x)$, may be written in terms of the inverse Multiquadrics RBF defined in Eq. (12). A simple optimization search is employed to determine the value of the shape parameter, c, used in every expansion over the different local topologies that cover the entire field. An initial guess for c is based on the ratio of the average distance between data centers in a topology to the number of points in the topology. A line search is performed to slightly modify the value of the shape parameter, c, until the resulting collocation matrix, $[C]$, yields a condition number in the range between 10^{11} and 10^{12} (in double-precision). This range of condition number for the collocation matrix, $[C]$, has been documented to produce interpolations that render smooth derivative fields for a wide range of test functions. It is important to mention that the resulting collocation matrix, $[C]$, depends only on the geometrical distribution of the points within each localized topology and therefore, the optimization of the shape parameter, c, is performed at a setup stage before the solution process begins. However, there may be instances when running-time optimization of the shape parameter, c, may be necessary as, for example, when adaptive refinement is performed or when sharp discontinuities in the solution field are found.

The real advantage of the localized collocation approach is capitalized in the way the derivatives of the field variable are calculated at the data center, x_c, of each topology. For instance, any linear differential operator, L, can be applied over the localized expansion equation as:

$$L\phi(x_c) = \sum_{j=1}^{NF} \alpha_j L\,\chi_j(x_c) + \sum_{j=1}^{NP} \alpha_{j+NF} LP_j(x_c) \tag{21}$$

Or, in matrix-vector form:

$$L\phi_c = \{L_c\}^T\{\alpha\} \tag{22}$$

where

$$\{L_c\} = \begin{Bmatrix} L\chi_1(x_c) \\ \vdots \\ L\chi_{NF}(x_c) \\ LP_1(x_c) \\ \vdots \\ LP_{NP}(x_c) \end{Bmatrix}_{NF+NP,1} \tag{23}$$

Substitution of the expansion coefficients, $\{\alpha\}$, leads to:

$$L\phi_c = \{L_c\}^T[C]^{-1}\{\phi\} \tag{24}$$

Finally,

$$L\phi_c = \{L\}^T\{\phi\} \tag{25}$$

where:

$$\{L\}^T = \{L_c\}^T[C]^{-1} \tag{26}$$

The coefficients of the vector $\{L\}$ of size $(NF,1)$ directly interpolate the derivative of the field variable distribution, $\{\phi\}$, at the data center of the topology, x_c. Therefore, evaluation of the field variable derivatives at everyone of the data centers, x_c, is provided by a simple inner product of two small vectors: $\{L\}$ which can be pre-built and stored and $\{\phi\}$ which is the updated field variable distribution in the topology of the data center.

To illustrate this approach, the generalized diffusion equation, Eq. (14), will be marched explicitly in time at each data center, x_c, using a first-order finite-differencing approximation as:

$$\phi_c^{k+1} = \phi_c^k + \kappa\Delta t\left(\nabla^2\phi_c^k\right) \tag{27}$$

where the superscript, k, denotes the time level and Δt denotes the size of the time step. The Laplace operator on the right-hand side can be replaced by the localized RBF interpolation as:

$$\phi_c^{k+1} = \phi_c^k + \kappa\Delta t\{L\}^T\{\phi\}^k \tag{28}$$

Therefore, knowledge of the solution field, $\{\phi\}$, at every point at the previous time level, k, can very efficiently yield the updated field variable value at each data center, x_c, through a simple inner product of small vectors.

Furthermore, imposition of the generalized boundary conditions in Eq. (15), at the boundary data centers, x_c, can be accomplished in a similar fashion as:

$$\hat{\beta}_1 \left.\frac{\partial\phi}{\partial n}\right|_{x_c} + \hat{\beta}_2\phi_c = \hat{\beta}_3 \tag{29}$$

Replacing the normal derivative, $\partial/\partial n$, at the data center, x_c, with a localized expansion, yields:

$$\hat{\beta}_1\{\partial n\}^T\{\phi\} + \hat{\beta}_2\phi_c = \hat{\beta}_3 \tag{30}$$

where the interpolation vector, $\{\partial n\}$, yields the normal derivative of the field variable, $\{\phi\}$, at the boundary topology data center, x_c. Ultimately, through some algebraic manipulation of the expression in Eq. (30), a simple relation can be arrived at to determine the boundary field variable at the current time level, $k + 1$, as:

$$\phi_c^{k+1} = \{\Gamma\}^T \{\phi\}^k \tag{31}$$

where the boundary interpolation vector, $\{\Gamma\}$, is composed by a combination of the normal derivative interpolation vector, $\{\partial n\}$, and the boundary condition coefficients, $\hat{\beta}_1, \hat{\beta}_2$, and $\hat{\beta}_3$, and, therefore, can be pre-built at a setup stage for every boundary data center, x_c.

The normal derivative interpolation vector, $\{\partial n\}$, may be pre-built in one of two ways. A simple approach is to express this vector as a combination of derivative vectors in all directions times their corresponding unit normal vector components. For instance, in 3D Cartesian coordinates, this is:

$$\{\partial n\} = \{\partial x\}n_x + \{\partial y\}n_y + \{\partial z\}n_z \tag{32}$$

A slightly more involved approach but one that yields more stable results is to generate additional internal points that "shadow" each boundary point in the direction of the normal vector into the domain, Ω, and use these shadow points to directly approximate the normal derivatives at each boundary data center. This approach is illustrated in Fig. 8.

The shadow point approach will mitigate the inherent inaccuracies of the directional derivative interpolation vectors of the truncated topologies of boundary data centers especially of those around corners and highly curved boundaries. Following this approach, the normal derivative interpolation vector, $\{\partial n\}$, can be computed simply as:

$$\{\partial n\} = \begin{Bmatrix} 1/r_s \\ 0 \\ \vdots \\ 0 \\ -1/r_s \\ 0 \\ \vdots \\ 0 \end{Bmatrix} \tag{33}$$

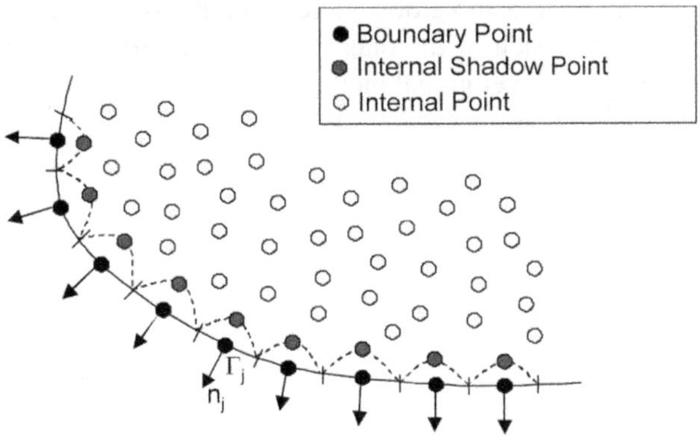

Figure 8. Distribution of internal shadow points to compute normal derivatives (from [43]).

where r_s is the distance from the boundary data center to its corresponding internal shadow point. Notice that all of the elements of the interpolation vector except two vanish. Additionally, higher order differentiation can be accomplished by simply inserting multiple layers of internal shadow points in the normal direction of each boundary data center.

In addition, this approach can be implemented to estimate not only the normal derivatives but the tangential derivatives as well. That is, an interpolation in the tangential direction can be formulated such that the tangential derivative at the boundary data center, x_c, is estimated as:

$$\frac{\partial \phi_c}{\partial s} = \{\partial s\}^T \{\phi\} \tag{34}$$

where the tangential derivative interpolation vector, $\{\partial s\}$, is found through direct finite-differencing in the tangential direction, similarly to the way it is done in the normal direction. This approach is very useful as it can be employed to estimate other directional derivatives by simple rotations. For instance, in 2D Cartesian coordinates, the x and y derivatives can be computed from the normal and tangential derivatives as:

$$\begin{Bmatrix} \dfrac{\partial \phi_c}{\partial x} \\ \dfrac{\partial \phi_c}{\partial y} \end{Bmatrix} = \begin{bmatrix} n_x & -n_y \\ n_y & n_x \end{bmatrix} \begin{Bmatrix} \dfrac{\partial \phi_c}{\partial n} \\ \dfrac{\partial \phi_c}{\partial s} \end{Bmatrix} \tag{35}$$

This localized expansion approach [41–43] reduces the burden of the more common global interpolation methods [18–20] by expanding the field variable locally around each data center to obtain its derivatives that are then used in time-marching or iterative schemes. This approach yields the generation of multiple but small interpolation matrices rather than the large and fully-populated global interpolation matrix of the standard global interpolation methods. However, since the approach relies on expanding known values of the field variables, it is applicable as long as an explicit time-marching or iterative scheme is formulated and inapplicable directly to steady problems. This is not a drawback, as time marching can always be considered as a relaxation scheme for the iterative solution of steady-state problems.

The particular feature of the LCMM that allows estimation of the field variables and their derivatives by simple inner products of vectors that can be pre-built and stored is precisely what makes this approach attractive. Notice that the Multiquadrics RBF only need to be evaluated at a setup stage when these vectors are being built, reducing the CPU burden of having to evaluate fractional powers and complicated functions at every step of an iteration or time-marching scheme. In addition, the memory demands of this approach are minimal, as no global collocation matrix is allocated, and, only very small vectors are stored for each of the data centers. This offers tremendous advantages in terms of data preparation over global methods.

Chapter 4

THE MOVING LEAST-SQUARES (MLS) SMOOTHING SCHEME

Despite the efforts in optimizing the shape parameter, c, of the RBF expansion for every topology, it is found that derivative fields, in particular odd one-sided derivatives, such as those found in convective terms of transport equations as well as in divergence operators, tend to oscillate specially towards the areas of large gradients such as re-circulating zones, corners, impingement planes, etc. For this reason, additional care must be taken when formulating the direct derivative expansion vectors within each topology.

An effective method consists in the application of moving least-squares smoothing over the data center topology to approximate the derivative value at the data center. This method can be easily adapted to the localized Meshless technique and, in addition, extended to be formulated in the same form where the derivative value at the data center of the topology is retrieved by a simple inner product of a vector that can be pre-built and stored, and the vector of field variable within the topology. This particular scheme retains the attractive efficiency feature of the localized Meshless method.

For instance consider the topology of NF influence points around the data center, x_c, as seen in Fig. 9.

Then, a least-squares expansion of the field variable, $\phi(x)$, using NP polynomials, $P_j(x)$, (where $NP < NF$) may be formulated as:

$$\phi(x) = \sum_{j=1}^{NP} \alpha_j P_j(x) \tag{36}$$

with the expansion coefficients, α_j, found through a least-squares minimization process over all the NF influence points, leading to:

$$\sum_{j=1}^{NP} \alpha_j \left[\sum_{k=1}^{NF} P_i(x_k) P_j(x_k) \right] = \sum_{k=1}^{NF} P_i(x_k) \phi(x_k) \tag{37}$$

or, in matrix-vector form:

$$[C]_{NP,NP} \{\alpha\}_{NP,1} = [P]_{NP,NF} \{\phi\}_{NF,1} \tag{38}$$

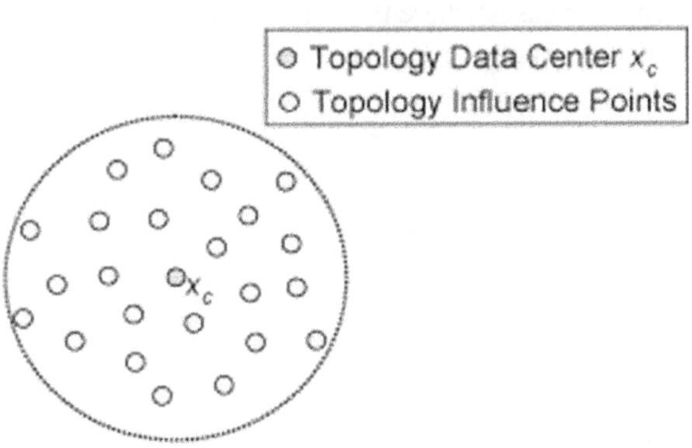

Figure 9. Illustration of a topology around a data center, x_c (from [43]).

where the coefficients $C_{i,j}$ of the least-squares matrix $[C]$ of size (NP, NP) are given by:

$$C_{i,j} = \sum_{k=1}^{NF} P_i(x_k)P_j(x_k) \tag{39}$$

Therefore, the expansion coefficients, α_j, can be determined as:

$$\{\alpha\}_{NP,1} = [C]^{-1}_{NP,NP}[P]_{NP,NF}\{\phi\}_{NF,1} \tag{40}$$

The application of a linear differential operator, L, over the field variable, $\phi(x)$, at the data center, x_c, is given by:

$$L\phi(x_c) = \sum_{j=1}^{NP} \alpha_j LP_j(x_c) \tag{41}$$

Or, in matrix-vector form:

$$L\phi_c = \{LP_c\}^T_{1,NP}\{\alpha\}_{NP,1} \tag{42}$$

Substitution of the expansion coefficients, α_j, leads to:

$$L\phi_c = \{LP_c\}^T_{1,NP}[C]^{-1}_{NP,NP}[P]_{NP,NF}\{\phi\}_{NF,1} \tag{43}$$

or:

$$L\phi_c = \{L_{ls}\}^T_{1,NF}\{\phi\}_{NF,1} \tag{44}$$

where the least-squares operator vector, $\{L_{ls}\}$, is explicitly built as:

$$\{L_{ls}\}^T_{1,NF} = \{LP_c\}^T_{1,NP}[C]^{-1}_{NP,NP}[P]_{NP,NF} \tag{45}$$

Notice that this smoothing scheme is performed over the same topology as the localized RBF collocation which renders the methodology general in its application. The least-squares operator vectors, $\{L_{ls}\}$, can be pre-built at the same pre-processing stage as the topology generation, shape parameter optimization, and RBF collocation. In addition, the least-squares expansion polynomials, $P_j(x)$, are the same as those employed in the RBF collocation augmented formulation.

The moving least-squares smoothing scheme is implemented to approximate one-sided derivatives present in transport equations to add stability to the iteration process. It should be noted that this approach is not used in place of upwinding schemes that must be implemented in cases where the convective thresholds are surpassed as it is commonly found in forced convection problems.

Chapter 5

THE FINITE-DIFFERENCING ENHANCED LCMM

Even after optimization of the shape parameter, c, and careful reduction of oscillative behavior through smoothing and conditioning, the LCMM still has some difficulties when dealing with steep gradients and highly convective fields. This is in part due to the fact that there is little control over the way the information is passed to the data center from the scattered points in the topology. A practical and general way around this problem is to take full advantage of the highly accurate field variable interpolation capabilities of the RBF but do not use them directly to determine the derivative fields. Instead, the field variable can be RBF-interpolated to locations where it can be used to pass information to the data center in a controlled fashion, like, for instance, to a set of locations where finite-differencing approximation of derivatives can be performed. This approach is general in the sense that it can be implemented in the same localized topologies defined for the LCMM and formulated, as it will be seen shortly, to yield derivative interpolation vectors in the same way the LCMM formulation was rendered. In addition, it will provide the necessary degree of control over the topology to pass the desired information to the data center as it will be shown in the following section when upwinding schemes are presented. This is the approach, in combination with smoothing and refinement schemes, that will be followed from here on to implement the LCMM method in all the applications presented after the framework is fully formulated.

To begin, a set of "Virtual" Points are distributed in the Topology at the locations required by finite-differencing evaluation of the derivatives. i.e. n, s, e, w,... These are not actual internal data centers but rather Virtual locations at which the field will be interpolated. The virtual spacing employed is consistent with the average spacing of the actual points within the localized topology. This is illustrated in Fig. 10.

Evaluation of the derivatives by finite-differencing is performed through the RBF interpolation at the Virtual locations. For example, the evaluation of the Laplace operator at the data center, x_c, assuming uniform virtual spacing, h, in all directions, is:

$$\nabla^2\phi(x_c) = \frac{\phi(x_e)+\phi(x_n)+\phi(x_w)+\phi(x_s)-4\phi(x_c)}{h^2} \tag{46}$$

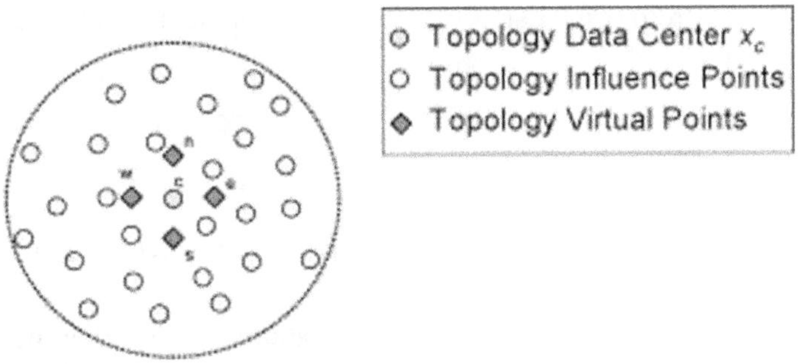

Figure 10. Illustration of a topology with Virtual points around the data center, x_c (from [48]).

The assumption of uniform virtual spacing, h, is of course not necessary and an equivalent form of the operator can be arrived at should this not be the case.

The evaluation of the field variable, $\phi(x)$, at any of the Virtual locations, for instance, x_e, using the polynomial-augmented RBF interpolation described in the previous sections, is reduced to:

$$\phi(x_e)=\sum_{j=1}^{NF}\alpha_j\chi_j(x_e)+\sum_{j=1}^{NP}\alpha_{j+NF}P_j(x_e) \tag{47}$$

or, in matrix-vector form:

$$\phi_e=\{\Psi_e\}^T\{\alpha\} \tag{48}$$

where:

$$\{\Psi_e\}=\begin{Bmatrix}\chi_1(x_e)\\ \vdots\\ \chi_{NF}(x_e)\\ P_1(x_e)\\ \vdots\\ P_{NP}(x_e)\end{Bmatrix} \tag{49}$$

Substitution of the expansion coefficients, α_j, found in Eq. (18), leads to:

$$\phi_e=\{\Psi_e\}^T[C]^{-1}\{\phi\} \tag{50}$$

and, therefore:

$$\phi_e=\{\mathrm{I}_e\}^T\{\phi\} \tag{51}$$

where the east interpolation vector, $\{\mathrm{I}_e\}$, is given by:

$$\{\mathrm{I}_e\}^T=\{\Psi_e\}^T[C]^{-1} \tag{52}$$

The expression in Eq. (51) can be easily extended to the other Virtual points as:

$$\begin{aligned}\phi_w&=\{\mathrm{I}_w\}^T\{\phi\}\\ \phi_n&=\{\mathrm{I}_n\}^T\{\phi\}\\ \phi_s&=\{\mathrm{I}_s\}^T\{\phi\}\end{aligned} \tag{53}$$

Therefore, the Laplace operator over the field variable, $\phi(x)$, at the data center, x_c, reduces to:

$$\nabla^2\phi_c=\{L\}^T\{\phi\} \tag{54}$$

where:

$$\{L\}^T=\frac{1}{h^2}\left(\{\mathrm{I}_e\}^T+\{\mathrm{I}_n\}^T+\{\mathrm{I}_w\}^T+\{\mathrm{I}_s\}^T-4\{\mathrm{I}_c\}^T\right) \tag{55}$$

which retains the feature and capability of rendering the derivative of the field variable, $\phi(x)$, at the data center, x_c, through a simple inner product of two small vectors. The derivative interpolation vector, $\{L\}$, can be pre-built and stored at a setup stage of the problem and implemented over the exact same topology employed for the Localized RBF interpolation and Moving Least-Squares smoothing. It is important to mention that any derivative operator can be expressed in this same form through a simple finite-differencing approximation over a set of Virtual points in the topology.

It is worth mentioning also that in the event that a Virtual point coincides with an actual data point in the topology, the interpolation vector, {I}, at that Virtual point, reduces to all zeroes except for unity at the vector entry of the coincident point. For instance, in the case of the interpolation vector, $\{I_c\}$, where the Virtual point always coincides with the actual point, x_c, the interpolation vector is simply:

$$\{I_c\} = \begin{Bmatrix} 1 \\ 0 \\ \vdots \\ 0 \end{Bmatrix} \tag{56}$$

Similarly for the cases where any other Virtual point $(x_e, x_w, x_n, x_s, \cdots)$ coincides with an actual point in the topology. This dramatically reduces the setup time especially around areas where the topologies are composed of uniformly-distributed points, as no shape parameter, c, optimization or matrix inversion is necessary to generate the derivative interpolation vectors.

The main advantage of the Finite-Differencing Enhanced LCMM approach will be evident in the following section when Upwinding schemes for highly convective flows are discussed. This is true because the Finite-Differencing Enhanced LCMM approach affords the possibility of having control over the location of the Virtual points within the topology which will allow for higher-order and offset finite-differencing approximations of the derivatives.

Chapter 6

UPWINDING SCHEMES

In the cases when the field variable is not only diffused within the domain but convected by a flow field, $\vec{V}$, the governing equation for the field variable becomes a transport equation as:

$$\frac{\partial \phi}{\partial t} + (\vec{V} \cdot \nabla)\phi = \kappa \nabla^2 \phi \tag{57}$$

Special attention should be given to the convective derivative term, $(\vec{V} \cdot \nabla)\phi$, in the event that a threshold is surpassed where the convective forces dominate the diffusion forces. A grid-based Reynolds number is employed for this purpose as:

$$\mathrm{Re}_h = \frac{|V|h}{\kappa} \tag{58}$$

where h can be assumed to be the average point spacing within the topology. The usual threshold that requires special treatment of the convective derivative is when $\mathrm{Re}_h > 1$, in which case an upwinded or offset derivative in the opposite direction of the flow field, $\vec{V}$, is necessary to achieve stability of the numerical solution scheme.

For example, in the case of 2D Cartesian coordinate system, the convective derivative has the following form:

$$(\vec{V} \cdot \nabla)\phi = V_x \frac{\partial \phi}{\partial x} + V_y \frac{\partial \phi}{\partial y} \tag{59}$$

where V_x and V_y are the x and y components of the flow field, $\vec{V}$. Concentrating on the x-direction and placing a set of Virtual points in the topology as illustrated in Fig. 11, three types of Finite-Differencing schemes for offset derivatives in the x-direction may be formulated:

First-order:

$$\frac{\partial \phi}{\partial x}^{1^{st}} = \frac{\phi_c - \phi_{w1}}{h} \tag{60}$$

Second-order:

$$\frac{\partial \phi}{\partial x}^{2^{nd}} = \frac{3\phi_c - 4\phi_{w1} + \phi_{w2}}{2h} \tag{61}$$

Third-order:

$$\frac{\partial \phi}{\partial x}^{3^{rd}} = \frac{2\phi_e + 3\phi_c - 6\phi_{w1} + \phi_{w2}}{6h} \tag{62}$$

As it is well known, high-order finite-differencing approximation schemes are more accurate but tend to yield unnatural oscillations that can render the solution process unstable, while first-order finite-differencing approximation schemes are very stable at the expense of artificially diffusing the solution. Therefore, a blend between low and high-order approximation schemes is necessary to compromise between accuracy and stability. In any case, all these operators can be pre-built and stored at the problem setup stage in the same

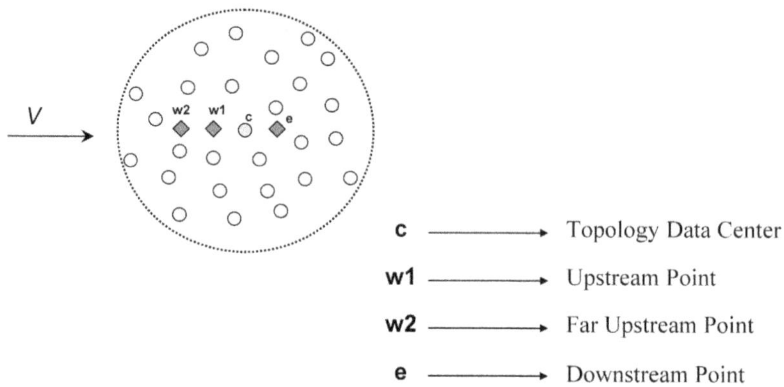

Figure 11. Localized topology with virtual point distribution for Upwinding scheme.

way it is done for the standard differential operators discussed in previous sections, that is, for example:

$$\left.\frac{\partial \phi_c}{\partial x}\right|_{w}^{1^{st}} = \left\{\partial x_w^1\right\}^T \{\phi\}$$
$$\left.\frac{\partial \phi_c}{\partial x}\right|_{w}^{2^{nd}} = \left\{\partial x_w^2\right\}^T \{\phi\} \tag{63}$$
$$\left.\frac{\partial \phi_c}{\partial x}\right|_{w}^{3^{rd}} = \left\{\partial x_w^3\right\}^T \{\phi\}$$

where $\left\{\partial x_w^1\right\}^T$, $\left\{\partial x_w^2\right\}^T$, and $\left\{\partial x_w^3\right\}^T$ are the first, second, and third-order x-derivative interpolation vectors upwinded from the west to approximate the derivative at the topology data center, x_c.

A limiter is introduced to the current upwinding technique to suppress potential oscillations. Compared to the cell-based Finite Volume Method (FVM), the LCMM is a point-based method where no cell face values are reconstructed. Hence, a direct application of the Flux Corrected Transport (FCT) [52,53], the Total Variation Diminishing (TVD) [54–56], or Normalized Variable Diagrams (NVD) [57,58] limiter schemes is not possible. However, the FCT, TVD and NVD concepts of monotonicity could be used as basis to form a limiter for the LCMM. To this end, the field variable is first evaluated at the Virtual points (downstream point, the data center, the upstream point, and one far upstream point) to detect if a minimum or maximum occur at the data center or upstream point, meaning that the field is not monotonic on that particular direction. If neither a minimum nor a maximum is detected at these locations no limiting action is needed and major upwinding (third-order approximation of the derivative) is performed. If a minimum or a maximum is detected at any of these locations a limiting action is introduced and minor upwinding (first-order approximation of the derivative) is performed. This minor upwinding leads to stable solutions, however, these are highly dissipative. Similar upwinding techniques were reported in [59,60].

6.1 ONE-DIMENSIONAL LCMM UPWINDING TEST

Studying the validity of any upwinding/limiter technique as applied to a linear convection equation is a necessary step before applying this technique to more complex transport equations. The present validation is based on studying the time decay of a signal that is being linearly convected by a constant convection speed, a. The governing equation for this problem is then:

$$\frac{\partial \phi}{\partial t} + a\frac{\partial \phi}{\partial x} = 0 \tag{64}$$

An explicit time-marching scheme will be employed such that:

$$\phi^{k+1} = \phi^k - a\Delta t\left(\frac{\partial \phi}{\partial x}\right)^k \tag{65}$$

This can be expressed in terms of the localized expansion at the data center, x_c, as:

$$\phi_c^{k+1} = \phi_c^k - a\Delta t\{\partial x\}^T\{\phi^k\} \tag{66}$$

with $\{\partial x\}$ being the x-derivative interpolation vector upwinded in the opposite direction of the flow field, a, using the different schemes and limiting factors discussed in this section. The 1D space is discretized using a Meshless point distribution equally spaced by a distance $h = 0.01$ m. The chosen signal consists of a unit-step profile and a unit-amplitude squared-sine profile extending over a width of 20 h. The signal is shown in Fig. 12 at its initial condition. The robustness of the LCMM upwinding/limiter scheme is tested with two different Courant numbers: $Co = 0.03$ and $Co = 0.1$. Recall that the Courant number is expressed as $Co = a\Delta t/h$.

Fig. 13 shows a comparison between the exact solution of the convected signal and the numerical solutions obtained by the major (third-order) and minor (first-order) upwinding techniques with $Co = 0.03$. The comparison is made at two different times: $t = 0.1$ s and $t = 0.4$ s. Another comparison between the exact solution of the convected signal and the LCMM upwinding solutions for $Co = 0.1$ is shown in Fig. 14 at times: $t = 0.025$ s and $t = 0.1$ s. It is remarkable that the major upwinding is not exhibiting any oscillation thanks to the incorporation of the limiter strategy by which some dissipation is borrowed from the minor upwinding. The stable yet over-damped solution profile of the minor upwinding is prominent which justifies why the upwinding process should start with a major upwinding and then revert to the minor upwinding should the monotonicity criteria is not satisfied.

As pointed out in the discussion of the limiter strategy, the minor upwinding should have an interventional role such that it provides dissipation to the solution just enough

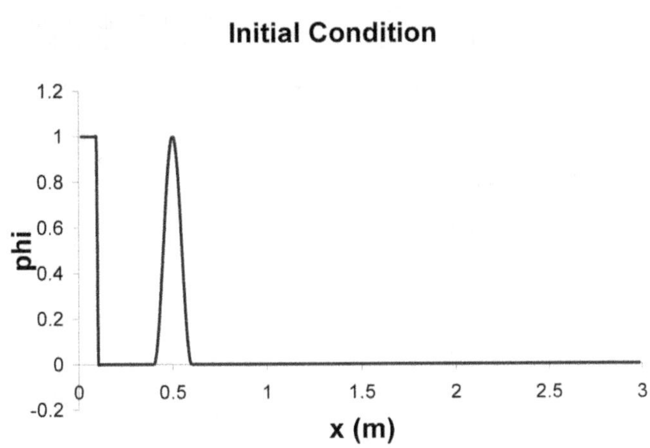

Figure 12. The initial condition of the linear wave profile.

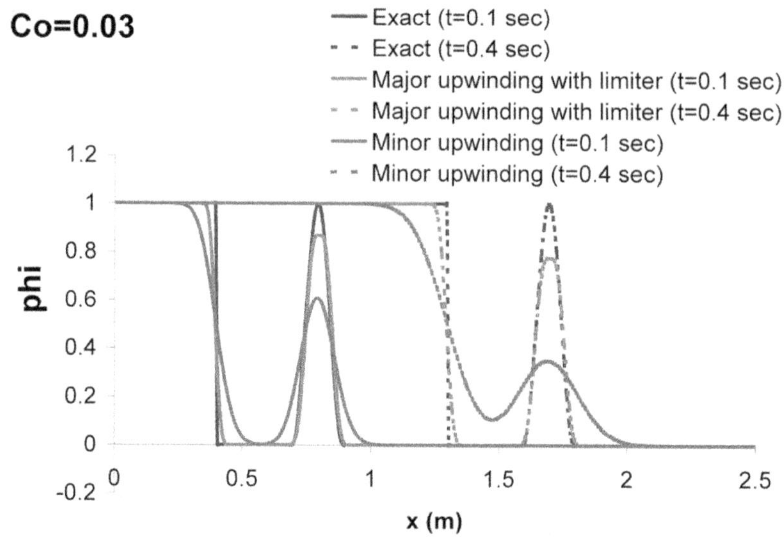

Figure 13. The performances of the minor and major LCMM upwinding techniques as benchmarked with the exact solution at $Co = 0.03$ (from [45]).

to suppress the oscillations. A comparison among the exact solution of the convected signal, the numerical solution obtained by the major LCMM upwinding technique, and the numerical solution obtained by the FVM Muscl TVD scheme are shown in Fig. 15 and Fig. 16. According to the results from the LCMM limiter-enhanced major upwinding technique seems to be superior to the Muscl TVD scheme. At both test Courant numbers, the step profile resolution is better preserved by the LCMM upwinding. Besides, the LCMM upwinding technique presents less diffusion as illustrated by the squared-sine wave amplitude decay.

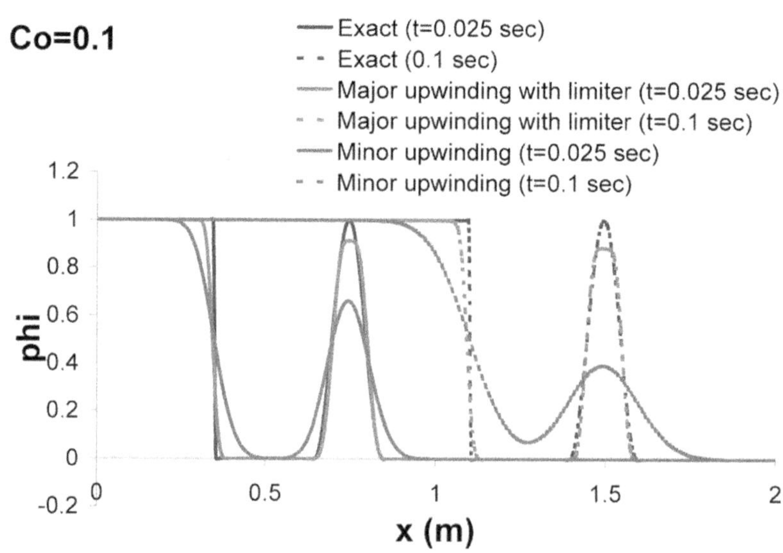

Figure 14. The performances of the minor and major LCMM upwinding techniques as benchmarked with the exact solution at $Co = 0.1$ (from [45]).

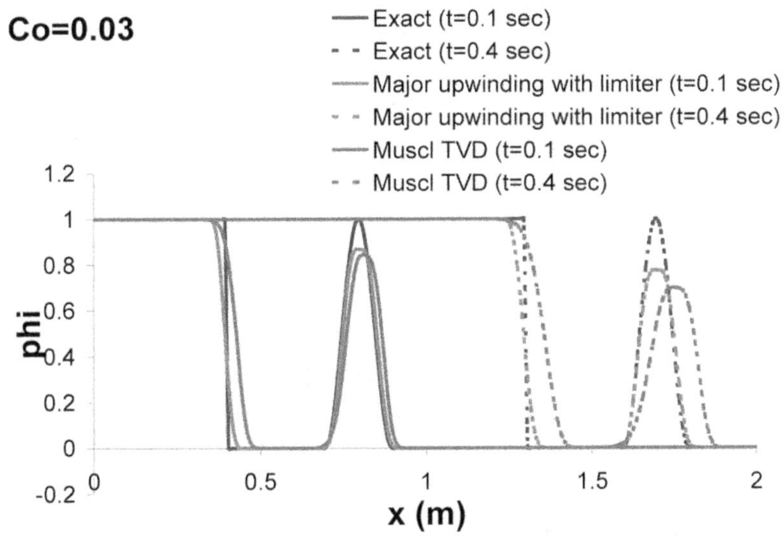

Figure 15. The performances of the major LCMM upwinding technique and the FVM Muscl TVD scheme as benchmarked with the exact solution at $Co = 0.03$ (from [45]).

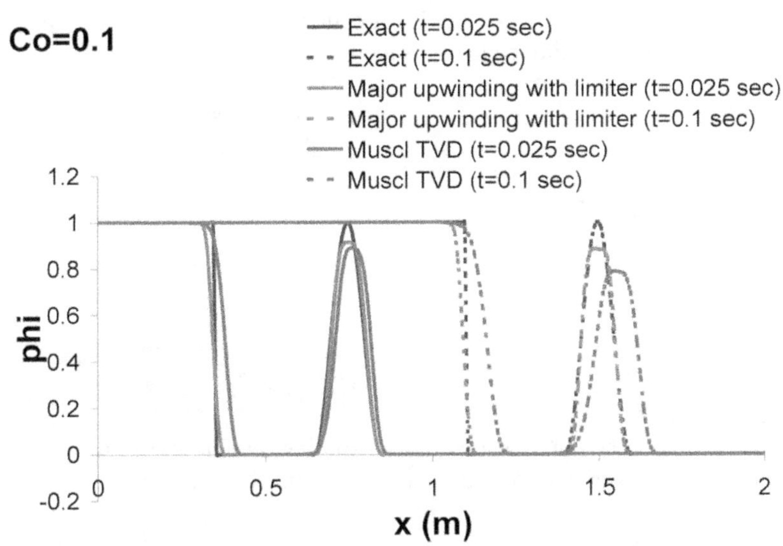

Figure 16. The performances of the major LCMM upwinding technique and the FVM Muscl TVD scheme as benchmarked with the exact solution at $Co = 0.1$ (from [45]).

6.2 TWO-DIMENSIONAL LCMM UPWINDING TEST FOR AN INCLINED WAVE

The next step in the validation of the LCMM upwinding technique is to examine the time evolution of wave profiles propagating on 2D planes. Fig. 17 shows the geometry of a $1\ m \times 1\ m$ square domain where a box-shaped profile of a field variable $\phi(x,y,t)$ is purely

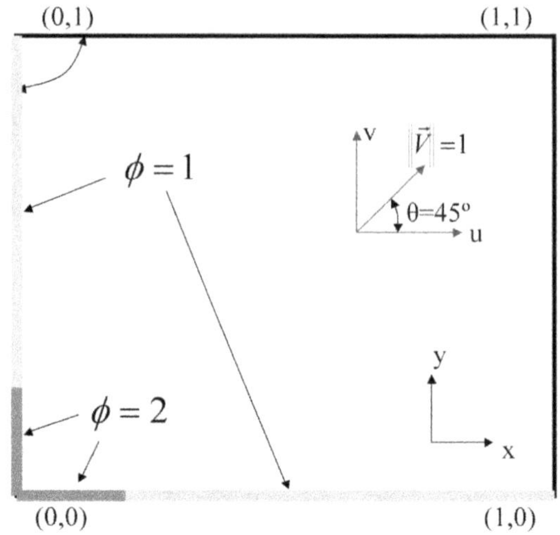

Figure 17. The geometry of box profile case in the square (from [45]).

convected. The flow enters the square domain from the bottom and left sides and exits through the top and right sides. The field variable ϕ takes an inlet value of 1 on the boundary green lines and a value of 2 on the boundary red lines. The constant convection speed magnitude is $V = 1\ m/s$ and its direction is at a 45° angle with respect to the horizontal x-axis. The governing equation is therefore:

$$\frac{\partial \phi}{\partial t} + V_x \frac{\partial \phi}{\partial x} + V_y \frac{\partial \phi}{\partial y} = 0 \tag{67}$$

where:

$$V_x = \frac{\sqrt{2}}{2} \frac{m}{s}$$
$$V_y = \frac{\sqrt{2}}{2} \frac{m}{s} \tag{68}$$

The same explicit time-marching scheme shown in Eq. (66) is employed to evolve the field variable, ϕ, in time and generalized in 2D as:

$$\phi_c^{k+1} = \phi_c^k - \Delta t \left(V_x \{\partial x\}^T \{\phi^k\} + V_y \{\partial y\}^T \{\phi^k\} \right) \tag{69}$$

Again with $\{\partial x\}$ and $\{\partial y\}$ being the x-derivative and y-derivative interpolation vectors upwinded in the opposite direction of the flow field using the different schemes and limiting factors discussed in this section. The point distribution is uniform in both directions with $h = 0.025\ m$ except for the shadow points where $h = 0.0125\ m$. The contour of ϕ for the box-shaped profile convection problem is shown in Fig. 18. This contour is obtained by the LCMM upwinding technique involving a limiter. Fig. 19 shows the distribution of ϕ on a vertical line traced at $x = 0.5\ m$ for the exact solution, the minor LCMM upwinding technique, and the major LCMM upwinding technique with and without a limiter. Notice the oscillatory behavior of the major upwinding when the limiter is not employed.

Figure 18. The field variable ϕ contour for the inlet box-shaped profile pure convection (from [45]).

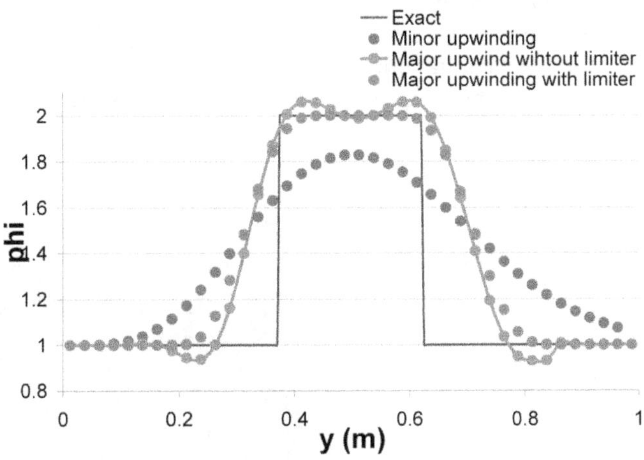

Figure 19. The field variable ϕ distribution at $x = 0.5\ m$ (from [45]).

6.3 TWO-DIMENSIONAL LCMM UPWINDING TEST FOR A TURNING WAVE

Another example for the 2D validation of the LCMM upwinding technique is the rotational convection of an inlet box-shaped profile of the field variable ϕ. Fig. 20 shows the geometry of a $2\ m \times 1\ m$ rectangular domain where the pure advection of ϕ occurs. The convection is carried out by a rotational (Smith-Hutton) flow field with an analytical solution expressed for the horizontal and vertical components as:

$$V_x(x,y) = 2y\left[1-(x-1)^2\right]$$
$$V_y(x,y) = -2(x-1)(1-y^2) \tag{70}$$

The flow enters the rectangular domain from the bottom and left sides and after performing a 180-degree rotation exits through the bottom and right sides. The field variable ϕ takes an inlet value of 0 on the boundary green lines and a value of 2 on the boundary red line. Again, the explicit time-marching and upwinding scheme discussed in the previous examples

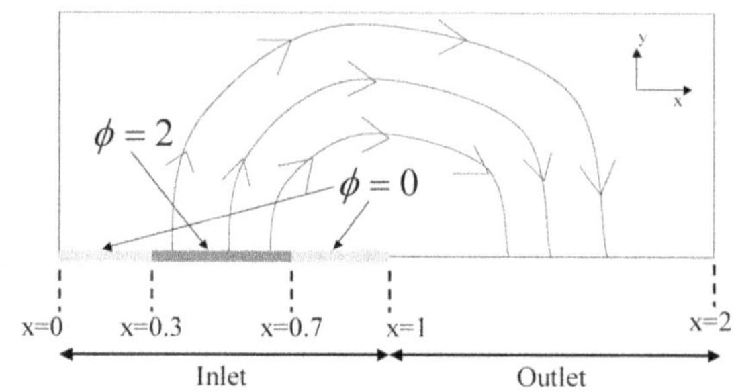

Figure 20. The geometry for the rotational pure convection problem (from [45]).

are employed and the point distribution is uniform in both directions with h = 0.025 m except for the shadow points where h = 0.0125 m. The contour of ϕ for this rotational convection problem is shown in Fig. 21. This contour solution is obtained by the major LCMM upwinding technique involving a limiter. Fig. 22 reveals the distribution of ϕ at the outlet of the rectangular domain as computed through the exact solution, the minor LCMM upwinding technique, and the major LCMM upwinding technique with and without a limiter.

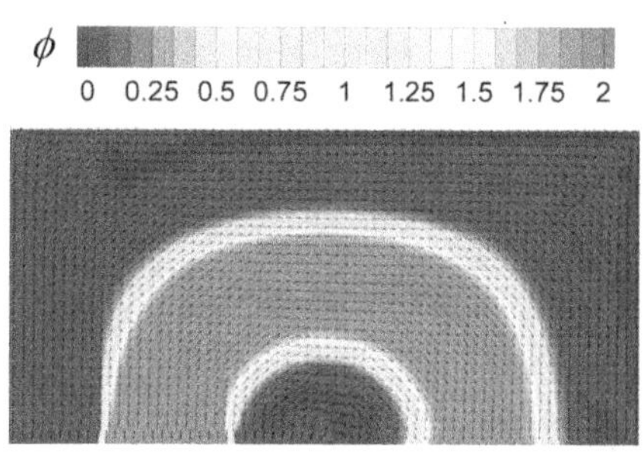

Figure 21. The field variable ϕ contour for the rotational convection problem (from [45]).

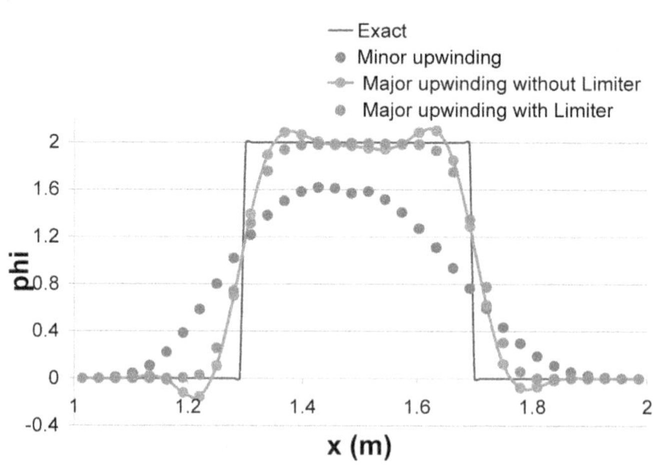

Figure 22. The field variable ϕ distribution at the outlet of the rectangular geometry involving the rotational flow field (from [45]).

Chapter 7

AUTOMATIC POINT DISTRIBUTION

A fully autonomous procedure to generate the Meshless point distribution in regular and irregular geometries may be devised with the use of Quadtree (Octree in 3D). A Quadtree is a tree data structure whose construction is based on the recursive decomposition of the Cartesian plane [44,61]. Specifically, a region-based Quadtree is defined by the recursive partitioning of the Cartesian plane into four equally-sized quadrants which lie parallel to the coordinate axis. This recursion can be repeated until a model resolution criterion is achieved.

The Quadtree method has the ability to model irregular geometries and refinement of areas of interest and it lends itself the ability for complete automation. Complete automation leads to a reduced amount of effort and time by the user. The use of a Quadtree method has the capability to trace the boundary by clustering Quadtrees to gain resolution for good model geometry approximation. Fig. 23 shows an example of the procedure that is undergone to bring an initial Quadtree discretization to the final Quadtree discretization followed by the interior point insertion. The corners of the Quadtree squares that lie within the enclosed boundary become an interior point. The boundary points are directly defined by the imported model (through CAD) or by standard boundary (surface in 3D) primitive curve distributions defined by the user. Any vertex of the Quadtree that lies within half a Quadtree square distance (round points) from the boundary is discarded. Instead, the

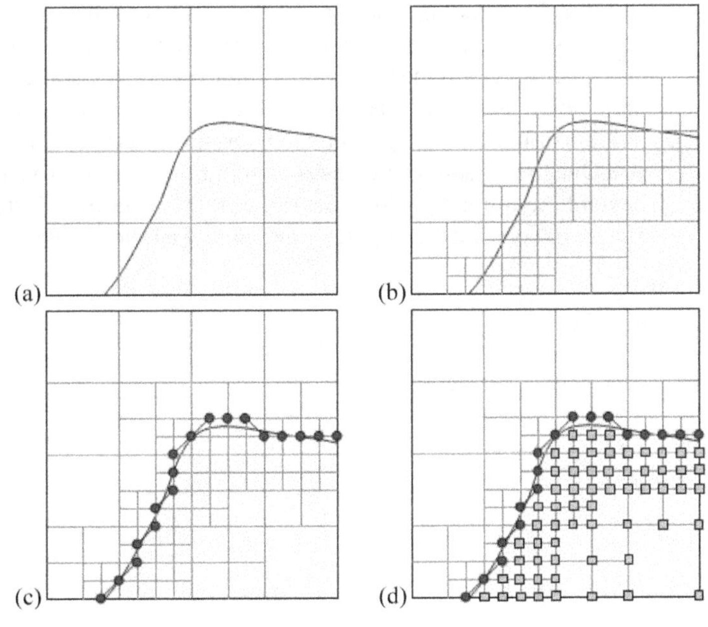

Figure 23. Application of Quadtree to an arbitrary boundary; (a) initial discretization, (b) first level, (c) second level, (d) third level with internal point distribution (from [44]).

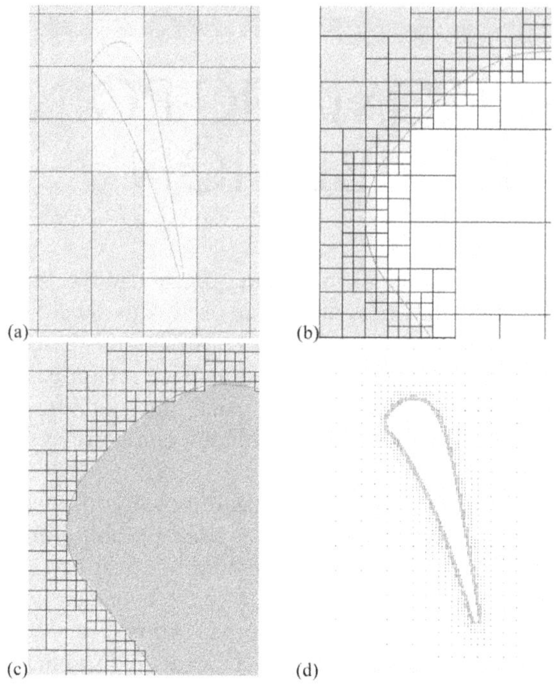

Figure 24. Point generation over a cavity; (a) initial Quadtree distribution, (b) third level Quadtree distribution, (c) final Quadtree distribution, (d) final Meshless point distribution (from [44]).

original boundary points from the curve definitions are employed along with the set of 'shadow' points defined in previous sections.

The Quadtree approach can be employed for static (at problem setup) refinement of the point distribution based on geometry irregularities as well as for adaptive (during running-time) refinement of the point distribution based on significant changes of the solution field variables and areas of steep gradients. In addition, the approach is suitable for generating point distributions in external and internal geometries. For example, a cavity is subjected to a Quadtree discretization as shown in Fig. 24. After five levels of automatic refinement beyond the initial Quadtree distribution the resulting Meshless point distribution is well suited to represent the geometry and model a field problem.

Chapter 8

PARALLELIZATION

An effective way to minimize the computational time necessary to solve a given problem is to implement a parallel approach. Parallelizing computational solvers have made efficient use of multi-processor clusters as well as multi-core processors for traditional mesh-based methods as well as for Meshless methods. The burden of parallelizing a domain lies with the model preprocessing stage which requires domain decomposition techniques. Traditionally, decomposed Meshless models required the use of artificial interior boundaries, or interfaces, in order to iteratively solve independent well-posed problems [33,34]. With the introduction of the Localized Collocation Meshless Method (LCMM), fully bounded sub-domains are not a necessity due the nature of explicit solution schemes. This allows the solution of a partial domain for each individual iteration given that the information from the previous iteration is made available.

An automated parallel segmentation is developed for this purpose through the use of Voronoï cells, or Dirichlet tessellation to break up the domain into segments for distribution to multiple processors. Dirichlet [62] first proposed a method whereby given a set of points, P_i, in arbitrary space, this could be systematically decomposed into a set of convex regions, R_i, such that the region R_i is the space closer to point P_i than any other point. This geometrical construction, known as Dirichlet tessellation, results in a set of non-overlapping convex regions called Voronoï regions as shown in Fig. 25. Each Voronoï region and all points within it are then assigned to a processor. With the nodes grouped into regions only the nodes that require information from nodes in other regions need to be communicated, thereby creating a need to minimize the amount of nodes that need their data passed between processors. Also, the number of nodes in each Voronoï region needs to be optimized to fit the processor in a given cluster. The geometric placement of the processor points, P_i,

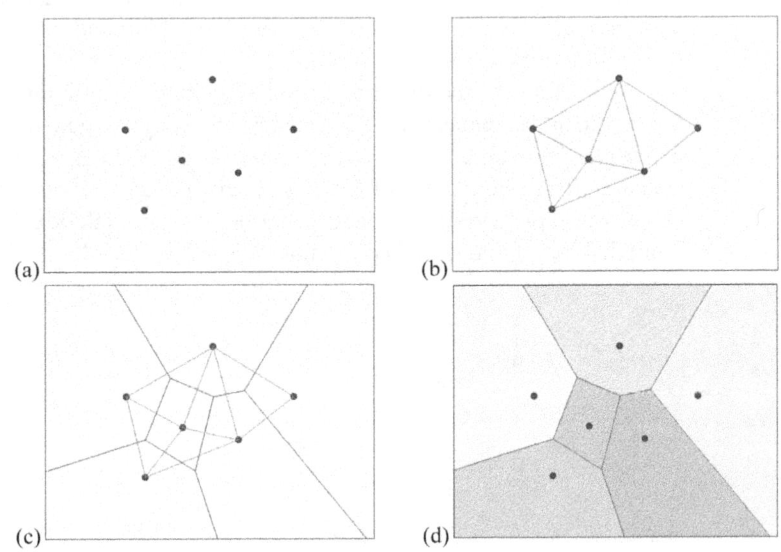

Figure 25. Auto-segmentation through Voronoï cells; (a) Point distribution, (b) Delaney triangulation, (c) triangulation and Voronoï diagram, (d) Voronoï cells (from [44]).

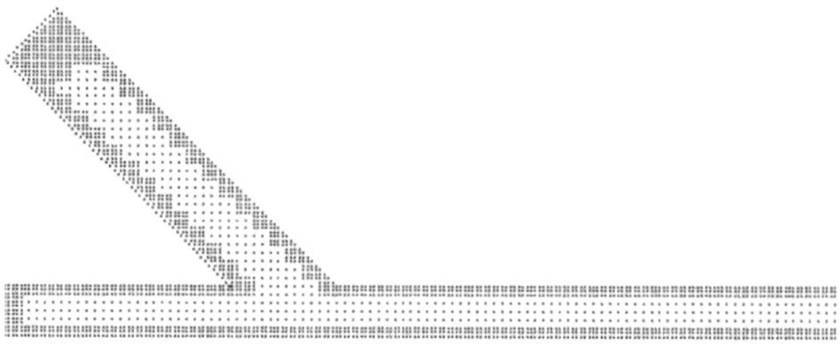
(a)

(b)

Figure 26. Auto-segmentation applied to a Meshless model; (a) one region, (b) five regions (from [44]).

governs both the size of each region and the communication effort between processors. Therefore, the geometric placement of the processor nodes can be optimized by employing a genetic algorithm with an objective function defined as:

$$S = \sqrt{\frac{1}{N}\sum\left[NP_{i,optimal} - NP_{i,actual}\right]^2} + \beta \cdot NC_{actuals} \tag{71}$$

Here, N is the number of Voronoï processor regions, NP_i is the number of nodes within region R_i, NC is the number of nodes that are required to be communicated between processors and β is a tuning parameter dependent upon the performance and communication abilities of a given cluster. The goal is to minimize the objective function, S, such that the nodes are distributed proportionally based on individual machine performance and communication of nodal data is minimized.

An example of the resulting Quadtree point distribution with automated segmentation for an irregular geometry as generated by the adaptive genetic algorithm described herein is presented in Fig. 26. Each region to be assigned to a specific processor is color (shade) coded, to illustrate the use of five (5) equally capable processors in a parallel cluster. The objectives of proportional node distribution based on individual processor performance and minimization of nodal data communication are achieved.

Chapter 9

APPLICATIONS

A series of multi-disciplinary field problem applications are now presented to demonstrate the robustness and range of the LCMM. The formulation of the field problem solution approach as well as the implementation on a variety of relevant problems is presented. Field problems ranging from Incompressible Fluid Flows, Conjugate Heat Transfer, Natural Convection, Turbulent Flows, Compressible Flows, Supersonic Flows, Solid Mechanics, Fluid-Structure Interaction, Two-Phase Flow, and Poro-Elasticity are among those discussed herein and applied to situations involving realistic problems in turbomachinery and bioengineering cases.

9.1 INCOMPRESSIBLE FLUID FLOW AND CONJUGATE HEAT TRANSFER

Incompressible laminar fluid flows and conjugate heat transfer is governed by the standard set of Continuity, Momentum, and Energy equations as:

$$
\begin{aligned}
&\nabla \cdot \vec{V} = 0 \\
&\rho \frac{\partial \vec{V}}{\partial t} + \rho\left(\vec{V} \cdot \nabla\right)\vec{V} = \mu \nabla^2 \vec{V} - \nabla p + \rho \vec{f} \\
&\rho c \frac{\partial T}{\partial t} + \rho c\left(\vec{V} \cdot \nabla\right)T = k \nabla^2 T + \dot{u}_g''' + \Phi
\end{aligned}
\tag{72}
$$

where ρ is the density, μ is the dynamic viscosity, $\vec{f}$ is a specific body force (neglected from here onward), c is the specific heat, k is the thermal conductivity, $\dot{u}_g'''$ is the energy generation (neglected from here onward), and $\mathbf{\Phi}$ is the viscous thermal dissipation (neglected from here onward). The field variables in question are the velocity field, $\vec{V}$, the pressure, p, and the temperature, T.

The set of partial differential equations in Eq. (72) are coupled and nonlinear and therefore require a specialized approach to guarantee their satisfaction along the time-evolution process. An explicit time-marching pressure-velocity correction scheme [63,64] is chosen to decouple these equations. This scheme departs from an initial velocity field, $\vec{V}^0$, required to satisfy the continuity equation, such that:

$$\nabla \cdot \vec{V}^0 = 0 \tag{73}$$

This can be accomplished by simply imposing a constant initial velocity field or one derived from an ideal fluid flow solution which automatically satisfies this constraint. Following this, the momentum equation is explicitly advanced as follows:

$$\vec{V}^{k+1} = \vec{V}^k + \Delta t\left[\frac{\mu}{\rho}\nabla^2 \vec{V}^k - \left(\vec{V}^k \cdot \nabla\right)\vec{V}^k - \frac{1}{\rho}\nabla p^{k+1}\right] \tag{74}$$

The time integration of the velocity field can be accomplished by explicit (as shown), implicit, and/or fractional stepping schemes [65–66]. Notice that the pressure field is not taken to be that of the previous time level, k, but instead is taken to be the unknown pressure field of the new time level, $k + 1$. In order to advance Eq. (74) explicitly, the unknown pressure field is expressed in terms of the known pressure field from the previous time level as:

$$p^{k+1} = p^k + \phi \tag{75}$$

where ϕ is a pressure correction field variable to be determined shortly. Introducing this definition into Eq. (74) yields:

$$\vec{V}^{k+1} + \frac{\Delta t}{\rho}\nabla\phi = \vec{V}^k + \Delta t\left[\frac{\mu}{\rho}\nabla^2\vec{V}^k - \left(\vec{V}^k\cdot\nabla\right)\vec{V}^k - \frac{1}{\rho}\nabla p^k\right] \tag{76}$$

Furthermore, a new intermediate velocity field, $\vec{V}^*$, can be defined as:

$$\vec{V}^* = \vec{V}^{k+1} + \frac{\Delta t}{\rho}\nabla\phi \tag{77}$$

Therefore, Eq. (76) reduces to:

$$\vec{V}^* = \vec{V}^k + \Delta t\left[\frac{\mu}{\rho}\nabla^2\vec{V}^k - \left(\vec{V}^k\cdot\nabla\right)\vec{V}^k - \frac{1}{\rho}\nabla p^k\right] \tag{78}$$

The expression in Eq. (78) can now be evaluated explicitly because all of the terms on the right-hand side are known from the previous time level, k. These terms on the right-hand side are evaluated using the LCMM interpolation schemes presented in the previous sections including the upwinding techniques for the convective derivative.

The next step is to guarantee that the velocity field at the new time level, $\vec{V}^{k+1}$, satisfies the continuity equation which is not done by the intermediate velocity field, $\vec{V}^*$, obtained by explicitly evaluating the momentum equation. This is accomplished by requiring:

$$\nabla\cdot\vec{V}^{k+1} = 0 \tag{79}$$

which is introduced in the definition of the intermediate velocity field of Eq. (77) as:

$$\nabla\cdot\vec{V}^* = \nabla\cdot\vec{V}^{k+1} + \nabla\cdot\left(\frac{\Delta t}{\rho}\nabla\phi\right) \tag{80}$$

Leading to:

$$\nabla^2\phi = \frac{\rho}{\Delta t}\nabla\cdot\vec{V}^* \tag{81}$$

The Poisson equation in (81) must be solved to convergence during the time level k by imposing homogeneous conditions for the pressure correction field variable, ϕ. That is, Neumann homogeneous conditions are applied everywhere, $\partial\phi/\partial n = 0$, except at boundaries where the pressure is imposed, such as outlets, where homogeneous Dirichlet conditions are applied, $\phi = 0$. The right-hand side of Eq. (81) is explicitly known from evaluating the momentum equation in (78).

The particular step of solving a Poisson equation to convergence during each time level is precisely what makes the solution of incompressible fluid flows a slow process. This Poisson equation can be solved on the same Meshless point distribution defined for the transport problem by a number of approaches:

- Artificial time-stepping: Transform the Poisson equation into a transient diffusion equation and advance in artificial time, τ, until an asymptotic solution is reached.

$$\phi^{\tau+1} = \phi^{\tau} + \Delta\tau\left(\nabla^2\phi^{\tau} - \frac{\rho}{\Delta t}\nabla\cdot\vec{V}^*\right) \tag{82}$$

- Gauss-Seidel iteration: Introduce a relaxation parameter, θ, to iterate explicitly until a converged solution is reached. Here, L_c is the data center component of the Laplace interpolation vector, $\{L\}$.

$$\phi_c^{m+1} = \phi_c^{m} - \frac{\theta}{L_c}\left(\nabla^2\phi^{m} - \frac{\rho}{\Delta t}\nabla\cdot\vec{V}^*\right) \tag{83}$$

- Generalized Minimization of Residuals (GMRes) global iteration: Implicitly build a global algebraic system relating the field variable values at all data centers and solve using a preconditioned GMRes [67]. Here, each row of the global coefficient matrix, $[A]$, correspond to a data center and contains the influence from the localized topology of that particular data center. Therefore, most of the entries on a particular row are zeroes and hence the matrix can be stored in a vectorized form with an index of non-zero entries, making the approach feasible as no additional memory allocation is required.

$$[A]\{\phi\} = \{b\} \tag{84}$$

The GMRes global solution approach is the methodology adopted for the solution of the Poisson equation that results at every time level, k. The vectorized global coefficient matrix, $[A]$, and right-hand side vector, $\{b\}$, are preconditioned using a simple diagonal scaling such that:

$$\hat{a}_{ij} = \frac{a_{ij}}{d_{ii}}, \hat{b}_i = \frac{b_i}{d_{ii}} \therefore d_{ii} = a_{ii};\ i,j = 1\ldots N \tag{85}$$

This simple preconditioning is sufficient to ensure the convergence direction of the system in Eq. (84) due to its high degree of sparsity and near diagonal dominance. The GMRes global solution approach is very efficient and ideal for this application as its convergence rate is proven to be in the order of $N\ln(N)$ where N is the global size of the system.

Once the Poisson equation (81) is solved to convergence, the velocity field at the new time level, $k + 1$, can be computed as:

$$\vec{V}^{k+1} = \vec{V}^* - \frac{\Delta t}{\rho}\nabla\phi \tag{86}$$

which by definition is guaranteed to satisfy the continuity equation (79).

The pressure field at the new time level, $k + 1$, can be computed directly from Eq. (75). However, a more stable approach results from computing this new pressure field by deriving an additional equation for it. This is accomplished by taking the divergence of the momentum equation (74) resulting in the following additional Poisson equation:

$$\nabla^2 p^{k+1} = -\rho\nabla\cdot\left[\left(\vec{V}^{k+1}\cdot\nabla\right)\vec{V}^{k+1}\right] \tag{87}$$

Notice that the convective derivative of the velocity on the right-hand side is evaluated at the new time level, $k + 1$. This is not consistent with the momentum equation (74) but it ensures that the pressure field solution is staggered in time with respect to the velocity field solution, which helps with the stability of the solution scheme. Furthermore, the boundary conditions for this new Poisson equation (87) for the pressure field are also estimated from the momentum equation (74) by scalarly multiplying it by the boundary unit normal vector, $\hat{n}$, yielding a set of Neumann boundary conditions as:

$$\frac{\partial p}{\partial n}^{k+1} = \left[\mu \nabla^2 \vec{V}^{k+1} - \rho\left(\vec{V}^{k+1} \cdot \nabla\right)\vec{V}^{k+1}\right] \cdot \hat{n} \tag{88}$$

This Neumann boundary condition is applied everywhere on the boundary except on sections where a pressure value is imposed such as outlets. Notice again that the right-hand side of Eq. (88) is evaluated at the new time level, $k + 1$, for consistency with the governing equation (87).

The Poisson equation (87) for the pressure field is solved using the same GMRes global iteration scheme employed to solve the Poisson equation for the velocity correction field, ϕ, as well as the same preconditioning approach. However, it should be noted that the solution for the pressure field in Eq. (87) requires significantly fewer iterations because the pressure correction expression in Eq. (75) provides an initial guess for the iteration process that is very close to the converged solution.

Once the velocity and pressure fields are updated at the new time level, $k + 1$, sub-level iterations can be performed by evaluating these fields on the right-hand side (term in brackets) of the explicit time-marching equation (78) and estimate a new intermediate velocity, $\vec{V}^*$. Although sub-level iteration is not a requirement for the stability of the time-marching scheme it adds accuracy especially when large time steps are used.

The final step of the time-marching algorithm is to update the temperature field by explicitly marching the energy equation (72) in as:

$$T^{k+1} = T^k + \Delta t\left[\frac{k}{\rho c}\nabla^2 T^k - \left(\vec{V}^{k+1} \cdot \nabla\right)T^k\right] \tag{89}$$

Again, the upwinding scheme discussed in the previous section is implemented to evaluate the convective derivative on the right-hand side of Eq. (89). In addition, sub-level iterations can be performed by evaluating the term in brackets of Eq. (89) using the updated temperature field, T.

The boundary conditions for the thermal problem can be of the first kind (Dirichlet: imposed temperature), second kind (Neumann: imposed heat flux), or third kind (Robin/Convective: imposed heat flux to temperature ratio). However, special care must be taken at outlet boundaries where no information is known about the thermal conditions. In this case, an appropriate approach is to impose a boundary condition that guarantees non-reflection of the thermal field. This condition is known as a Sommerfeld non-reflective condition and it is formulated as:

$$\frac{\partial T}{\partial t} + V_n \frac{\partial T}{\partial n} = 0 \tag{90}$$

The boundary condition in Eq. (90) can be imposed as a first kind condition by expressing it in the form of an explicit time-marching scheme as:

$$T^{k+1} = T^k - \Delta t V_n \frac{\partial T}{\partial n}^k \tag{91}$$

A non-reflective boundary condition scheme guarantees that any artificial oscillations generated in the domain due to numerical noise is carried out of the domain through outlet boundaries without reflecting back into the domain helping mitigate the promotion of instabilities.

9.1.1 Decaying Vortex Flow

The decaying vortex problem is used to test the spatial accuracy of the LCMM on a 2D domain bounded by $x \in (-0.5, 0.5)$ and $y \in (-0.5, 0.5)$. This test problem for incompressible flows have been used by many [64,65] as a standard for spatial error analysis. The velocity and pressure exact solutions of the decaying vortex problem are provided in terms of space and time as shown in Equation (92) below:

$$
\begin{aligned}
u(x,y,t) &= -\cos(\pi x)\sin(\pi y)e^{-2\pi^2 t/\mathrm{Re}} \\
v(x,y,t) &= \sin(\pi x)\cos(\pi y)e^{-2\pi^2 t/\mathrm{Re}} \\
p(x,y,t) &= -\frac{1}{4}\left[\cos(2\pi x)+\cos(2\pi y)\right]e^{-4\pi^2 t/\mathrm{Re}}
\end{aligned} \tag{92}
$$

Here, $u(x,y,t)$ and $v(x,y,t)$ are the x and y components of the velocity vector, $\vec{V}$. The term Re is a reference Reynolds number that is equal to $\rho UL/\mu$, where U is the maximum velocity at $t = 0$ and L is the maximum vortex size. Given the current square domain, both U and L will be equal to 1. Then by choosing $\rho = 1$ and $\mu = 0.1$, Re will be equal to 10. In order to determine the spatial accuracy, the square domain will be discretized by four different point distribution densities such that: $N_1 = 444$, $N_2 = 968$, $N_3 = 1740$, and $N_4 = 2664$. For all the point distribution density cases, a constant time step $\Delta t = 10^{-4}\,s$ will be used to render the current error analysis time-step-independent. As a benchmark, Fig. 27 shows the velocity and pressure contours of the decaying vortex at $t = 0.2\,s$. Then, Fig. 28 reveals the velocity contours and vectors of the decaying vortex at $t = 0.2\,s$ for the four point distribution density cases as determined by the LCMM. An excellent agreement between the analytical and LCMM x-velocity solutions is indeed demonstrated by the results. Subsequently, Fig. 29 unveils the pressure contours featuring the LCMM point distributions at $t = 0.2\,s$ for all the point density cases. At a glance, the pressure solution trends for all the cases follow the analytical solution trend. Quantitatively, the pressure drops for the four cases are: $\Delta p = 0.39$ for the $N_4 = 2664$ case, $\Delta p = 0.38$ for the $N_3 = 1740$ case, $\Delta p = 0.37$ for the $N_2 = 968$ case, and $\Delta p = 0.34$ for the $N_1 = 444$ case. Compared to the analytical case pressure drop of $\Delta p = 0.4$, the case with the most dense point distribution exhibits the best spatial accuracy. Nonetheless, the pressure drop is just a rough accuracy estimate; a more detailed analysis should be performed to exactly evaluate the order of spatial accuracy for the LCMM. The spatial accuracy of the LCMM upwinding scheme can be determined by examining either the x-velocity or y-velocity errors while the spatial accuracy of the LCMM velocity-correction scheme can

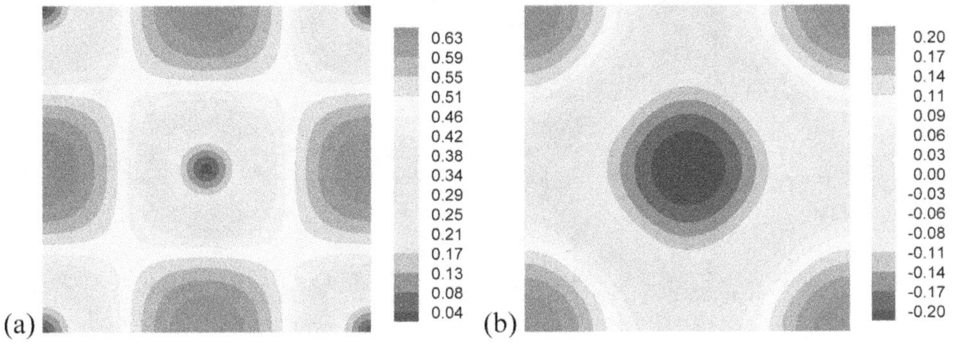

Figure 27. The decaying vortex (a) velocity and (b) pressure analytical solutions (from [45]).

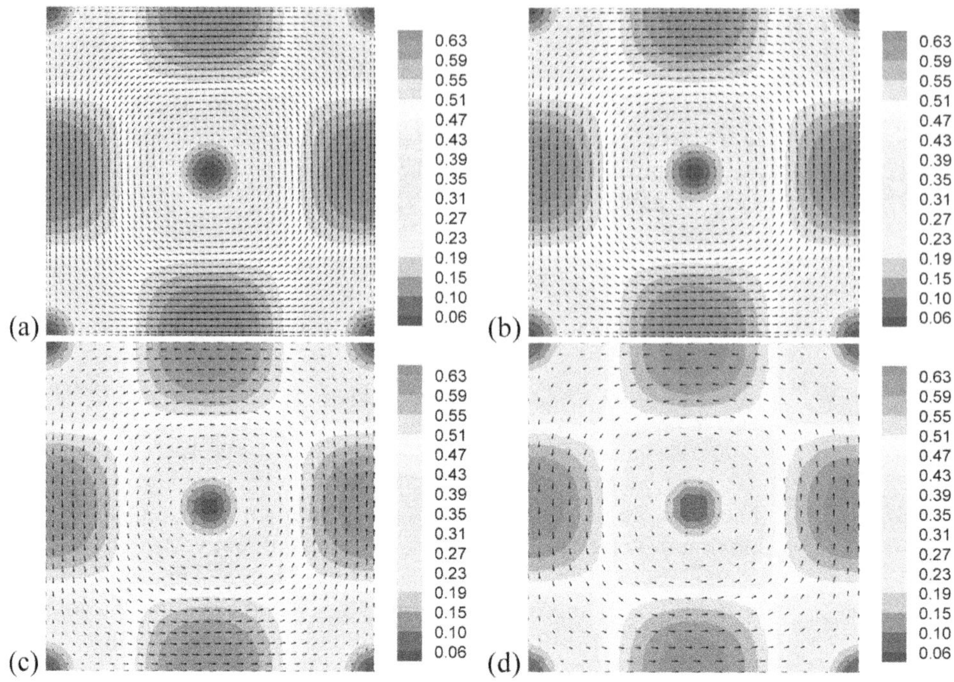

Figure 28. The decaying vortex velocity LCMM solutions with (a) 2664, (b) 1740, (c) 968, and (d) 444 point distribution density cases (from [45]).

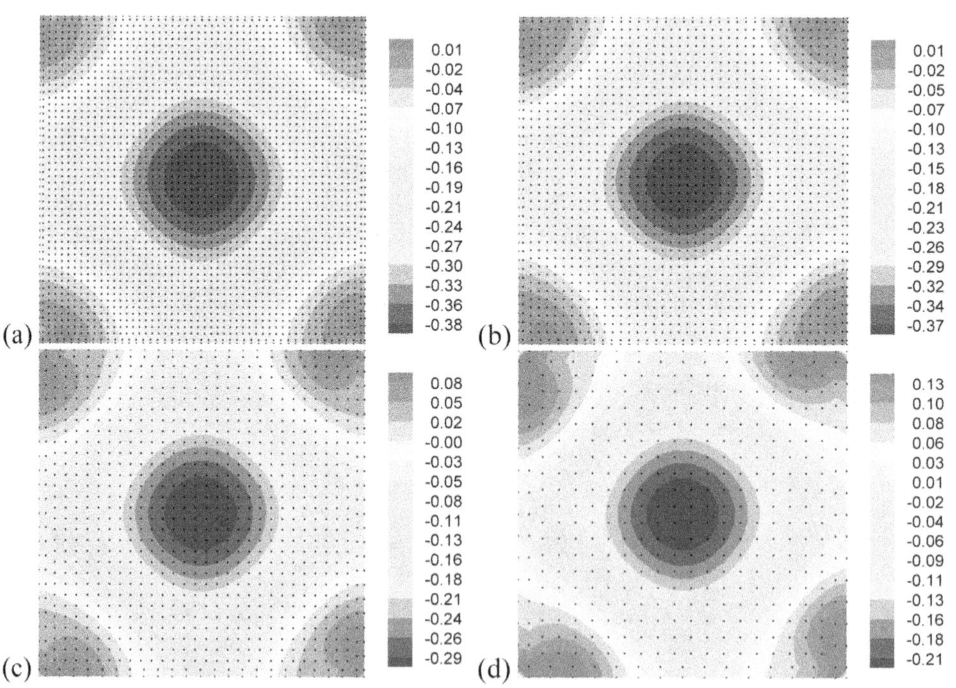

Figure 29. The decaying vortex pressure LCMM solutions with (a) 2664, (b) 1740, (c) 968, and (d) 444 point distribution density cases (from [45]).

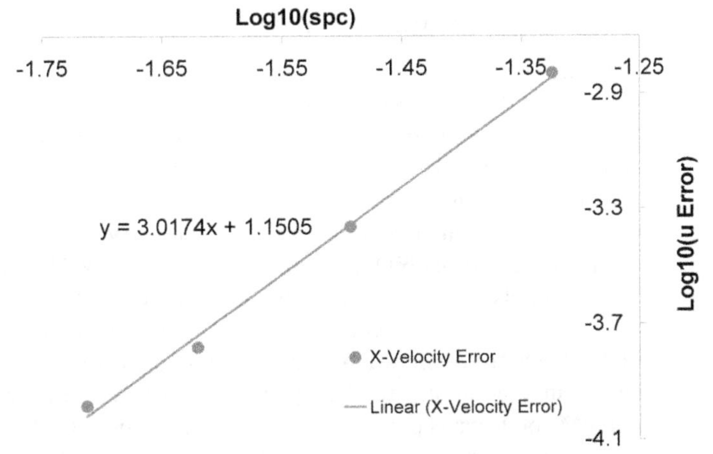

Figure 30. The linear curve-fit of the x-velocity error (from [45]).

be determined by examining the pressure errors. As such, a formal numerical error norm $\|e\|$ is defined as the average of the absolute errors between the exact and the LCMM solutions at each point in the domain

$$\|e\| = \frac{1}{N}\sum_{k=1}^{N}\left|\phi_{k,exact} - \phi_{k,LCMM}\right| \tag{93}$$

where ϕ represents the field variable being tested for error.

For a given point distribution density N over the square of area A, an average point spacing *spc* can be determined such that $spc = \sqrt{A/N}$. The spatial accuracy is determined by curve-fitting the Log_{10} of the error norm as a function of the Log_{10} of the average point spacing. Fig. 30 shows the distribution of the x-velocity numerical error as a function of the average point spacing. Similarly, Fig. 31 shows the distribution of the pressure numerical error. The distribution of the x-velocity error proves that the LCMM upwinding scheme is third-order accurate, whereas the pressure error distribution suggests that the velocity-correction scheme is second-order accurate.

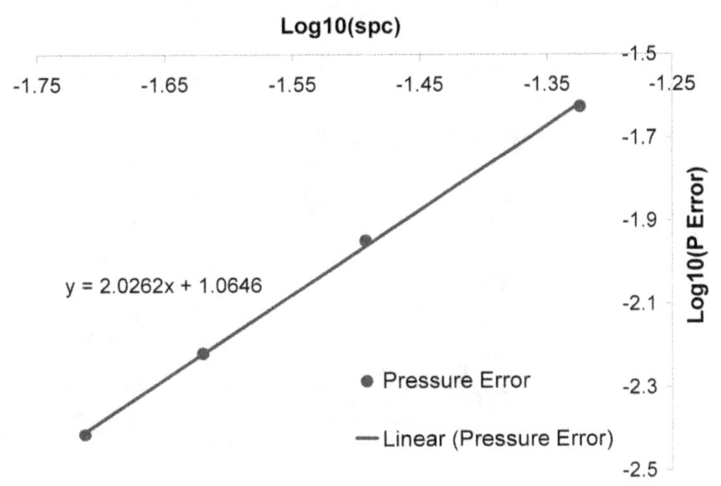

Figure 31. The linear curve-fit of the pressure error (from [45]).

9.1.2 Lid-Driven Flow in a Square Cavity

This numerical example validates both the accuracy and efficiency of the localized Meshless approach using the lid-driven square cavity problem as a benchmark compared to FVM solutions provided by a popular commercial CFD code and results reported by Ghia, Ghia, and Shin [68]. A square closed cavity is considered with the top wall moving at a constant speed. In addition, the bottom wall is kept at a temperature of 323 K while the top wall is kept at a temperature of 288 K. Initially, FVM and Meshless solutions are generated using three uniform point distributions (26×26, 51×51 as seen in Fig. 32, and 101×101). The Reynolds number based on the side of the cavity is $\mathrm{Re} = 68$ and the Prandtl number is $\mathrm{Pr} = 0.7$. Qualitative comparison of the results are shown in Fig. 33 where FVM and LCMM velocity vectors and magnitude contours are displayed, as well as in Fig. 34 where temperature contours are displayed. A measure of the efficiency of the LCCM compared to FVM solutions is presented in Table 1. Here both FVM and LCMM codes were implemented in the same platform (Single Xeon 64-bit, 3.2 GHz, 6 GB RAM) using the same grid and point distribution respectively. Also, the settings of the FVM solution scheme were set to match those of the LCMM solution scheme, i.e. explicit, coupled, unsteady with 1 sub-level iteration, and $\Delta t = 10^{-4}\,s$. The values in Table 1 reveal that the LCMM solution process is found to be more efficient than the FVM solution process under the same conditions and scheme. The difference in performance is more noticeable with the coarsest grid and point distribution which can be attributed to the amount of overhead involved in FVM computations. In addition to the average computation times over 1000 time steps, Table 1 also shows the LCMM pre-processing time for generating the topologies and

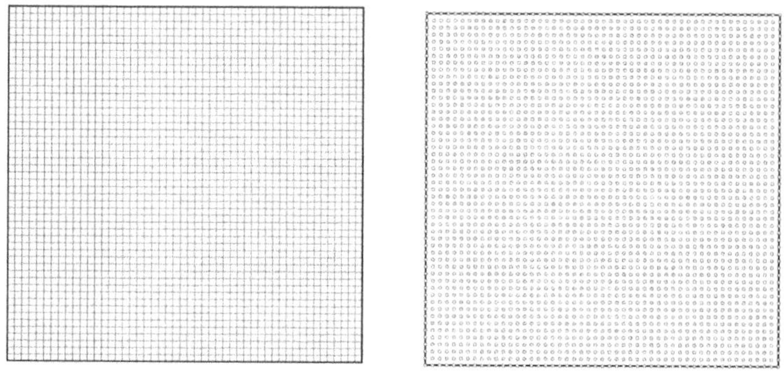

Figure 32. FVM mesh and LCMM point distribution for lid-driven flow in a square cavity (from [43]).

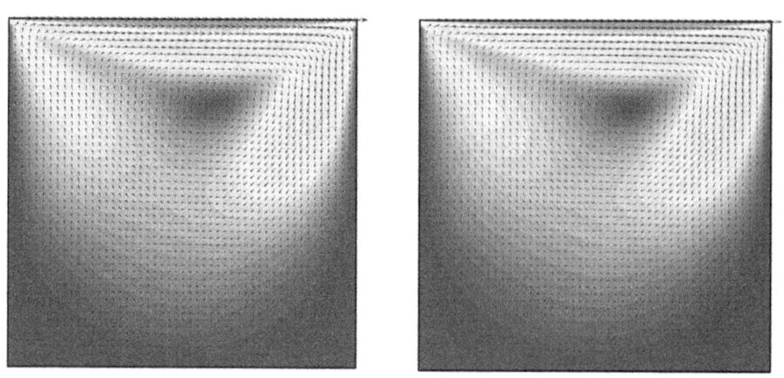

Figure 33. FVM and LCMM velocity vectors and magnitude contours for the lid-driven cavity (from [43]).

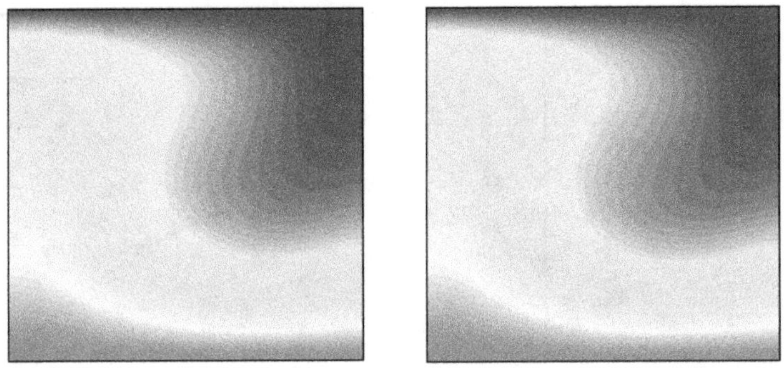

Figure 34. FVM and LCMM temperature contours for the lid-driven cavity (from [43]).

TABLE 1. Computation times for FVM and LCMM solutions for different point resolutions

	FVM	LCMM		
Mesh/Points	**Total**	**Topology**	**Collocation**	**Solution**
26 × 26	264s	<1s	<1s	5.19s
51 × 51	326s	<2s	<1s	27.86s
101 × 101	970s	<8s	<3s	164.14s

performing the RBF collocation and algebra, again revealing very little overhead at the startup of the process.

In addition to the qualitative comparisons, the LCMM results were compared along the geometrical center-lines with those reported by Ghia, Ghia, and Shin [68]. For this purpose, the Reynolds number was increased to 100 and 400 and the LCMM resolution was increased to include 129 × 129 uniformly distributed points in order to match the resolution used in [68]. Fig. 35 and Fig. 36 display the x and y velocity components along the vertical and horizontal geometrical center-lines respectively, revealing very close agreement between the LCMM and reported benchmark FVM results.

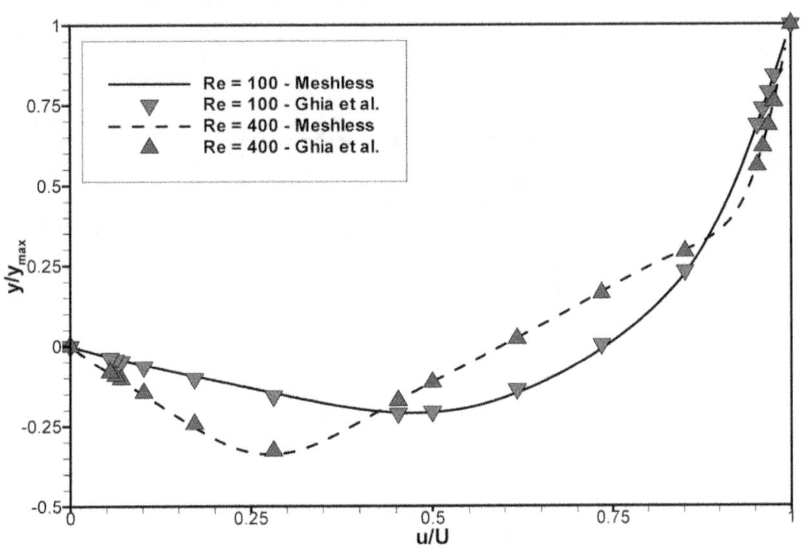

Figure 35. FVM and LCMM x-velocity component at the vertical center-line (from [43]).

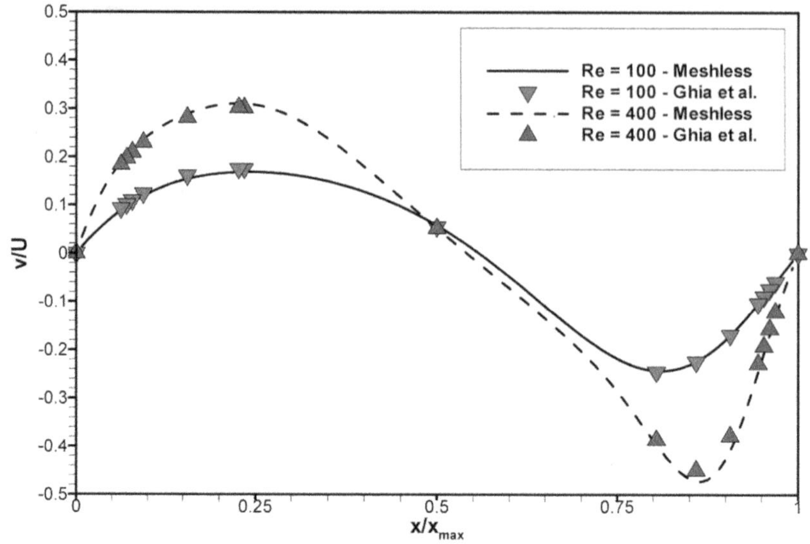

Figure 36. FVM and LCMM y-velocity component at the horizontal center-line (from [43]).

9.1.3 Air Jet into a Square Cavity

In this example a 1×1 m square cavity has a 20 cm opening centered on the left-hand wall through which air enters and a 20 cm outlet centered on the right-hand wall through which the air exits. The inlet velocity is imposed at 0.01 m/s so that the Reynolds number based on the size of the cavity is $\mathrm{Re}_L = 675$. The air density is assumed as $\rho = 1.22$ kg/m^3 while its viscosity as $\mu = 1.79 \cdot 10^{-5}$ $Pa \cdot s$. Here, the LCMM will be tested by collocating 1,681 (41×41) equally-spaced data centers in comparison with a commercial FVM code using a considerably more refined mesh with 13,685 nodes as seen in Fig. 37. The velocity contours for the FVM and LCMM solutions are shown in Fig. 38 revealing accurate qualitative results. The FVM and LCMM x-velocity profiles are shown in Fig. 39 at $x = 0$, 0.25, 0.5, 0.75, 1 m revealing very accurate quantitative results.

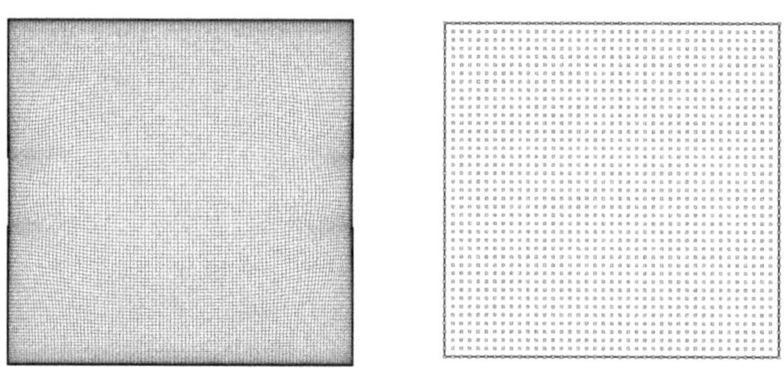

Figure 37. FVM mesh and LCMM point distribution for the air jet into a square cavity problem (from [41]).

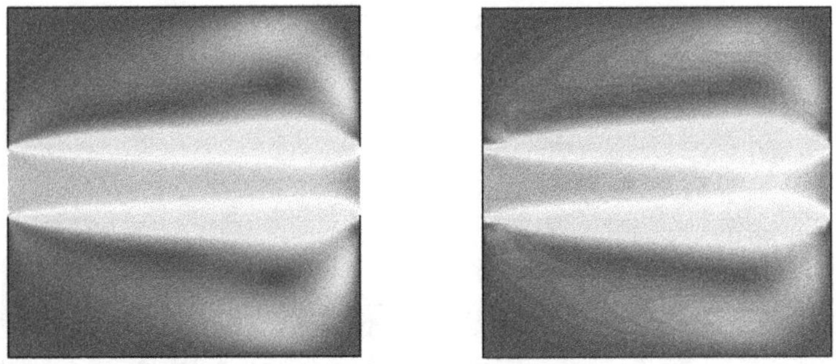

Figure 38. FVM and LCMM velocity contours for the air jet into a square cavity problem (from [41]).

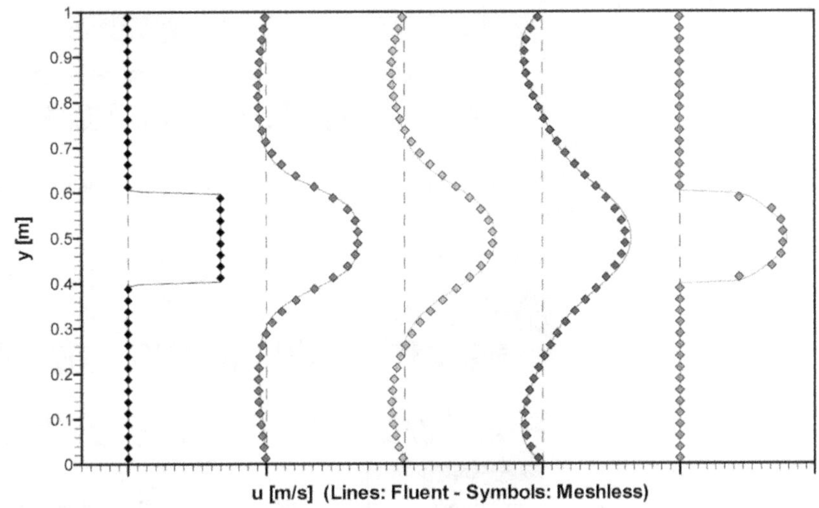

Figure 39. FVM and LCMM x-velocity profiles at $x = 0, 0.25, 0.5, 0.75, 1$ m for the air jet into a square cavity problem (from [41]).

9.1.4 Conjugate Heat Transfer between Parallel Plates

A conjugate heat transfer solution is evaluated for developing air flow entering at 0°C and 0.167 m/s between parallel steel plates. The plates are 10 cm long and 1 cm apart. The plates have a finite thickness of 4 mm and the top and bottom are maintained at 100°C to test the effects of conjugate heat transfer. The geometry and thermo-physical properties (in SI units) for this problem are shown in Fig. 40. The Prandtl number is taken as Pr = 0.7, and the Reynolds number based on the hydraulic diameter is Re = 225.

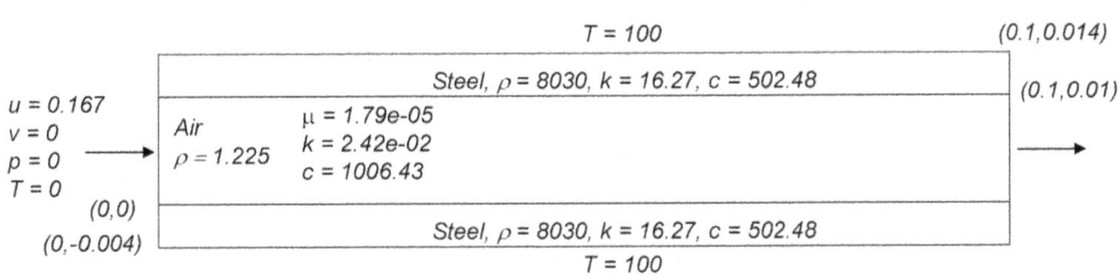

Figure 40. Geometry, conditions, and properties for conjugate heat transfer problem between parallel plates (from [41]).

The interfacing between the solid and fluid regions is accomplished by requiring continuity of temperatures and heat flux as:

$$T_i^s = T_i^f$$
$$-k^s \frac{\partial T_i^s}{\partial n} = k^f \frac{\partial T_i^f}{\partial n} \tag{94}$$

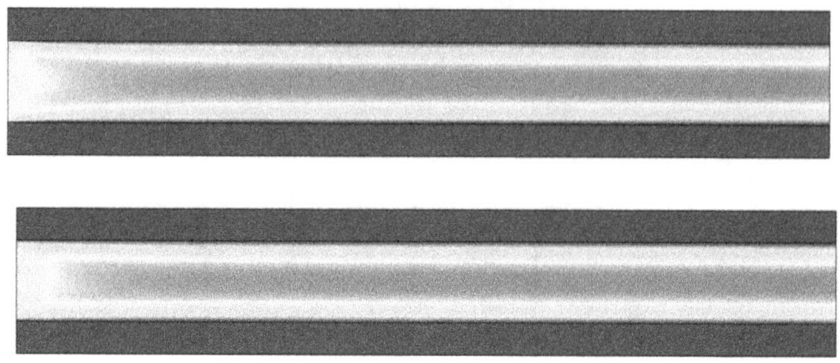

Figure 41. FVM and LCMM velocity contours for conjugate heat transfer problem between parallel plates (from [41]).

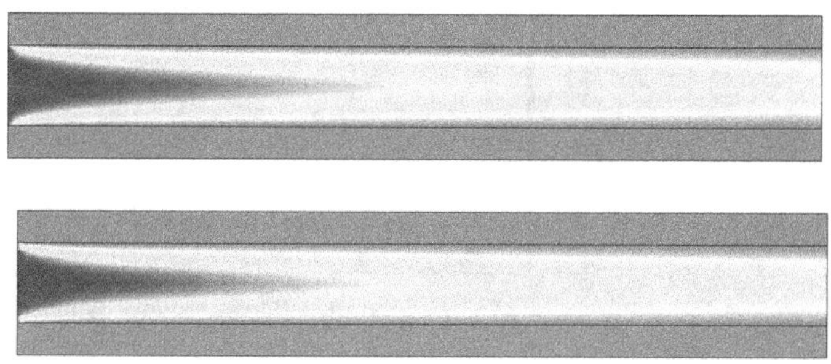

Figure 42. FVM and LCMM temperature contours for conjugate heat transfer problem between parallel plates (from [41]).

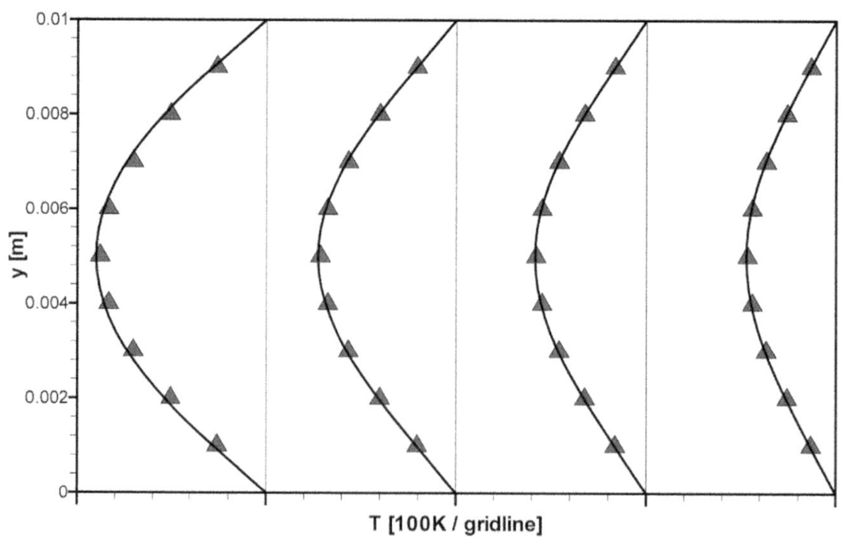

Figure 43. FVM and LCMM temperature profiles at 1/4, 1/2, 3/4, and 1/1 channel length (from [41]).

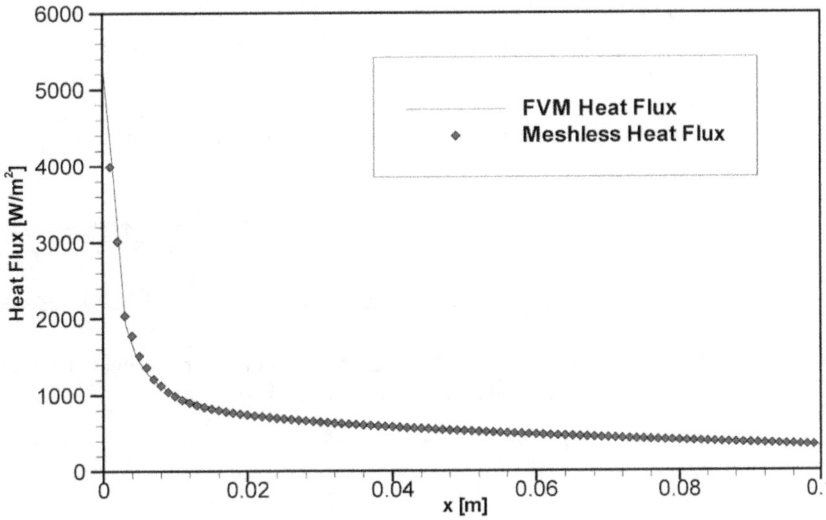

Figure 44. FVM and LCMM heat flux along interface (from [41]).

which can be ensured by performing a few sub-level iterations within each time level as long as the transient evolution of the problem starts from an equilibrium state such as uniform temperature in all regions.

The LCMM solution was obtained by collocating 1,919 (101 × 19) data centers while the FVM mesh contains 4,091 nodes. The FVM and LCMM velocity magnitude contours are shown in Fig. 41 while the temperature contours are shown in Fig. 42. A plot of the temperature profiles across the plates at 1/4, 1/2, 3/4, and 1 length of the plates is shown in Fig. 43 revealing high accuracy when compared to the FVM results. The heat flux along the interface wall between the air flow and the steel plates is shown in Fig. 44 again with very good agreement between the FVM and LCMM solutions.

9.1.5 Conjugate Heat Transfer Flow over a Rectangular Obstruction

This numerical example examines air flowing through a channel with a titanium rectangular obstruction, all parameters are provided in Fig. 45 using SI units.

The Prandtl number is taken as Pr = 0.7 and the Reynolds number based on the hydraulic diameter is Re = 338. Again the effects of conjugate heat transfer will be tested as the bottom of the obstruction is kept at 100°*C* while the flow enters at 0°*C* with a velocity of 0.25 *m/s*. The channel is 11 *cm* long by 1 *cm* high while the obstruction is 10 × 5 *mm* and is placed

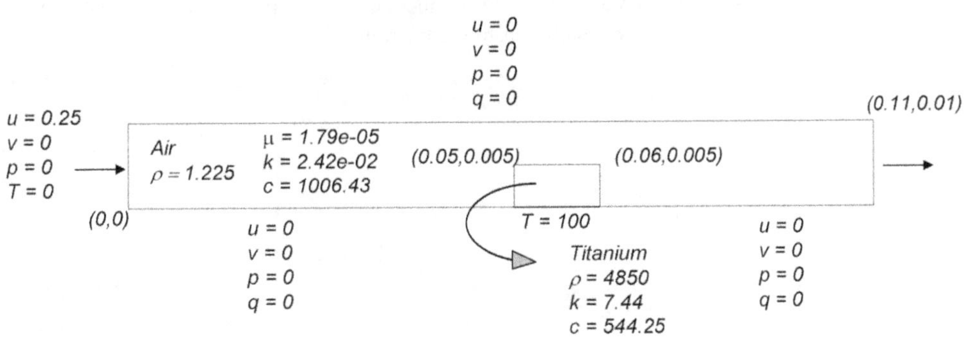

Figure 45. Geometry, conditions, and properties for conjugate heat transfer flow problem over a rectangular obstruction (from [41]).

5 *cm* after the inlet as shown in Fig. 45. The LCMM solution was obtained by collocation of 4,221 equally-spaced data centers while the FVM mesh contains 9,473 nodes. The FVM and LCMM velocity contours are shown in Fig. 46. A plot of the velocity profiles after the obstruction at $x = 7, 8, 9, 10, 11$ *cm* is shown in Fig. 47 revealing very good agreement between the FVM and LCMM solutions. Finally, the FVM and LCMM temperature contour plots are

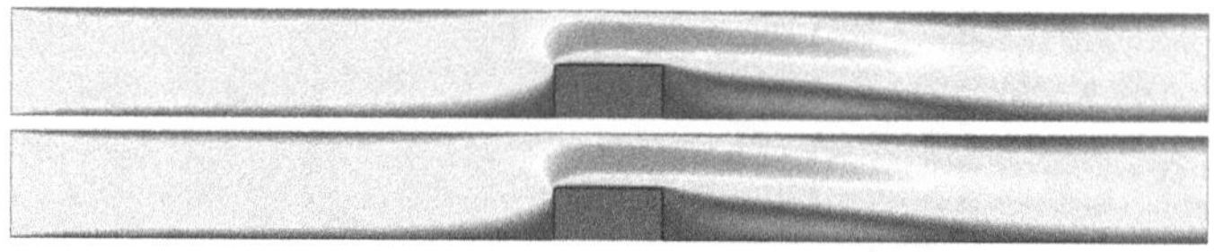

Figure 46. FVM and LCMM velocity contours for conjugate heat transfer flow problem over a rectangular obstruction (from [41]).

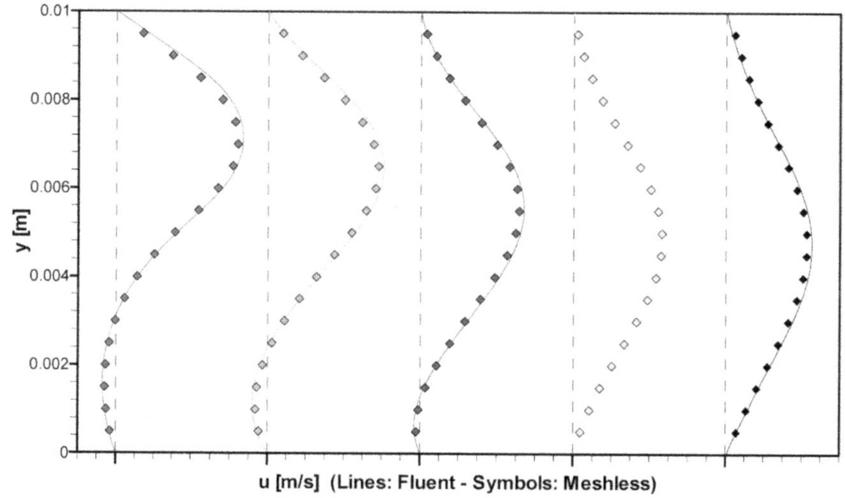

Figure 47. FVM and LCMM velocity profiles at $x = 7, 8, 9, 10, 11$ *cm* for conjugate heat transfer flow problem over a rectangular obstruction (from [41]).

Figure 48. FVM and LCMM temperature contours for conjugate heat transfer flow problem over a rectangular obstruction (from [41]).

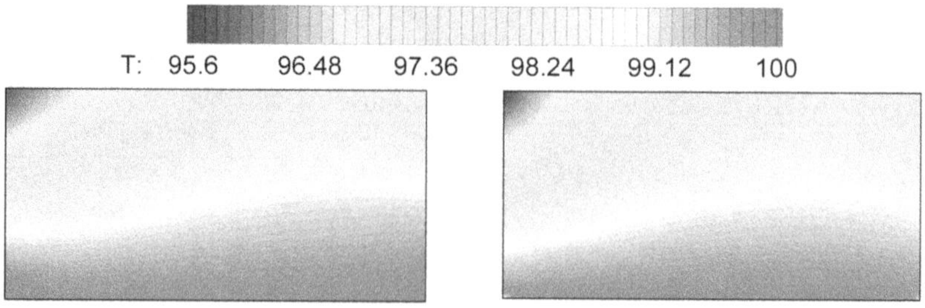

Figure 49. FVM and LCMM temperature contours on the rectangular titanium obstruction (from [41]).

shown in Fig. 48 as well as a zoom of the plot on the titanium obstruction in Fig. 49 clearly showing the multi-dimensional temperature gradients across the solid.

9.1.6 Conjugate Film-Cooling Heat Transfer

This example consists of a flow of colder air from a plenum through a stainless steel plate into a hotter incoming flow. The problem is conjugate and the boundary conditions are shown in Fig. 50 along with the LCMM and FVM discretizations consisting of 13,345 LCMM data points and 14,859 FVM nodes respectively. Temperature contours are compared in Fig. 51 while temperature profiles on the upper channel after the film-cooling hole are displayed in Fig. 52. The velocity contours are shown in Fig. 53 while the x-velocity profiles are shown in Fig. 54 again revealing very good qualitative and quantitative agreement between the LCCM and FVM solutions.

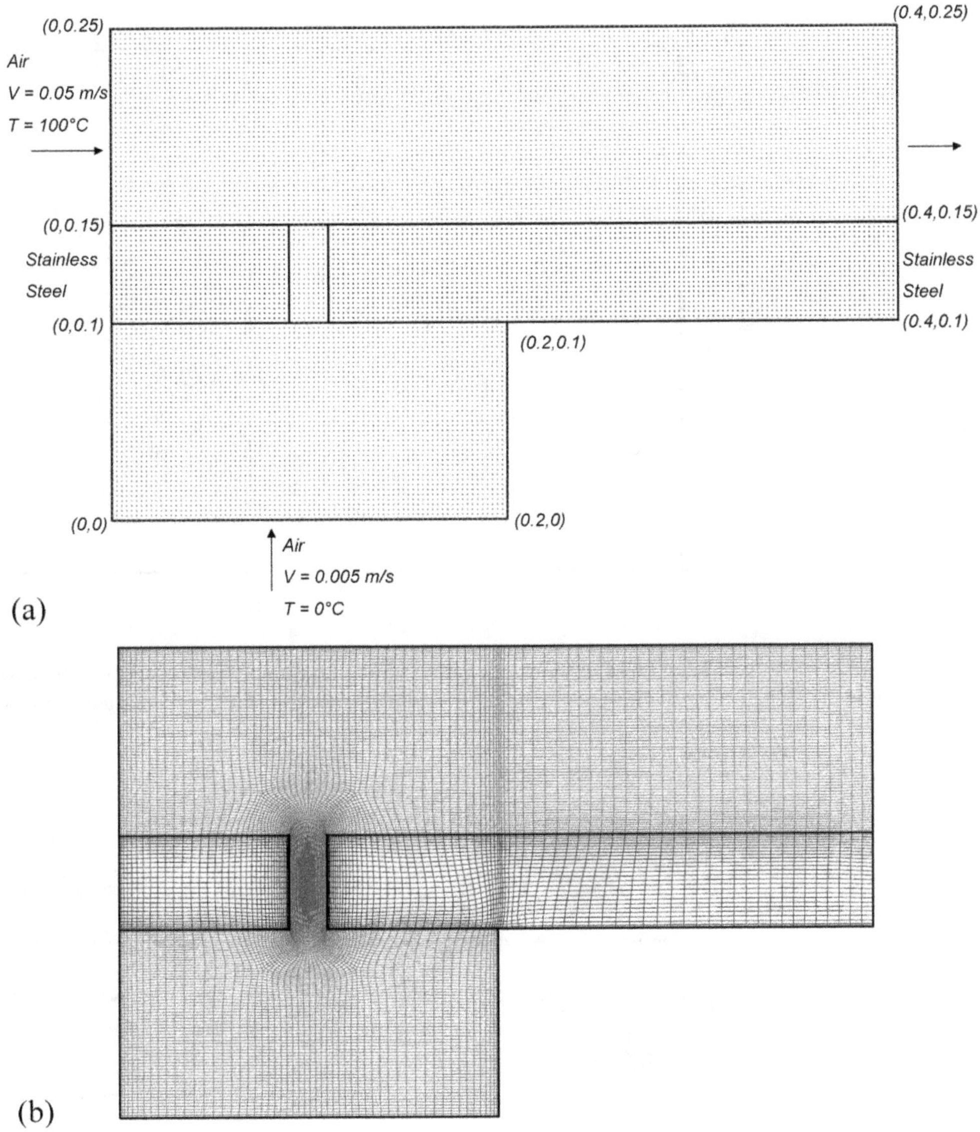

Figure 50. (a) LCMM point distribution including boundary conditions and (b) FVM mesh for cooling plenum, cooling hole, and main cooling channel (from [41]).

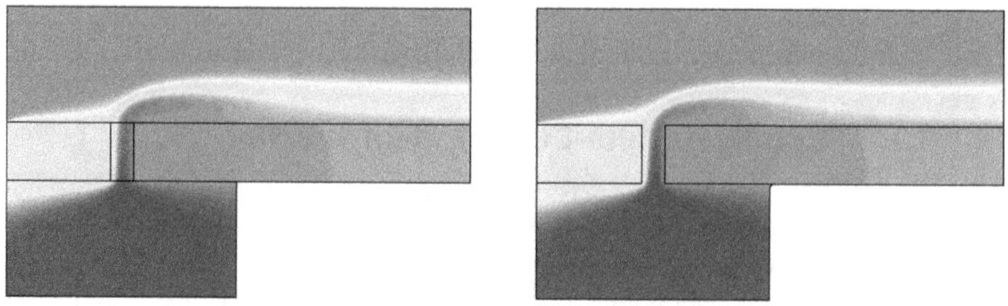

Figure 51. LCMM and FVM temperature contours for conjugate heat transfer film-cooling example (from [41]).

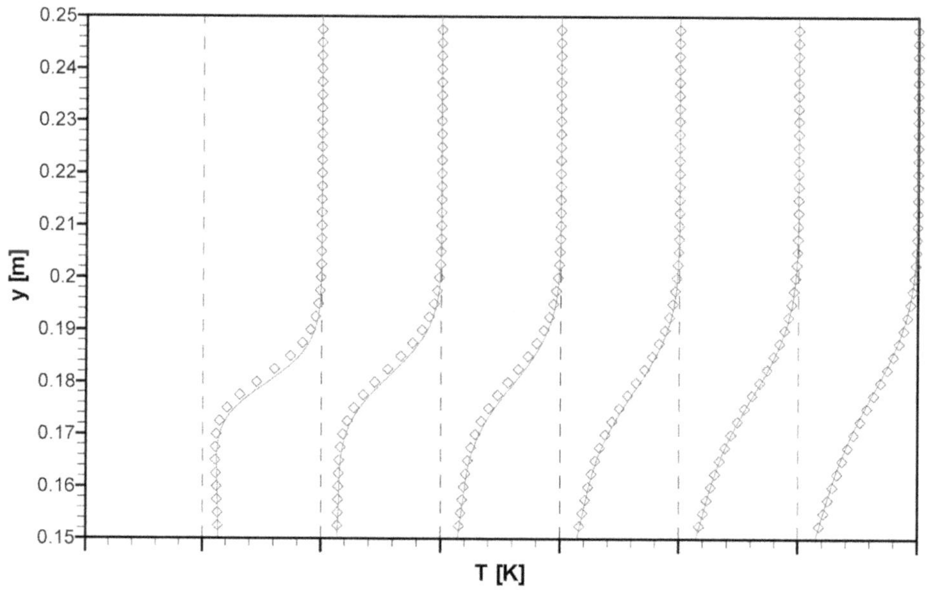

Figure 52. LCMM (symbols) FVM (solid lines) temperature profiles along upper channel after the film-cooling hole for conjugate heat transfer film-cooling example (from [41]).

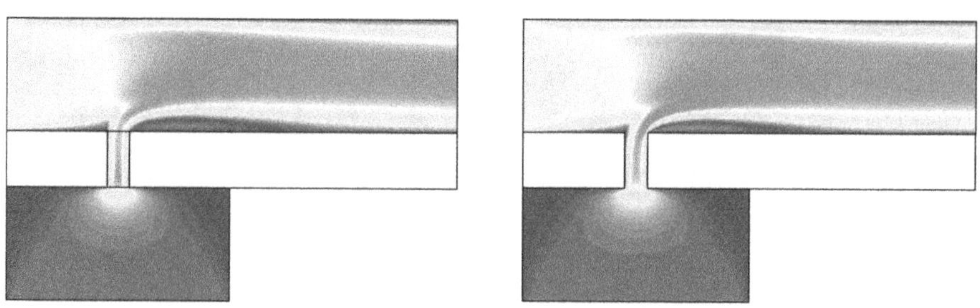

Figure 53. LCMM and FVM velocity contours for conjugate heat transfer film-cooling example (from [41]).

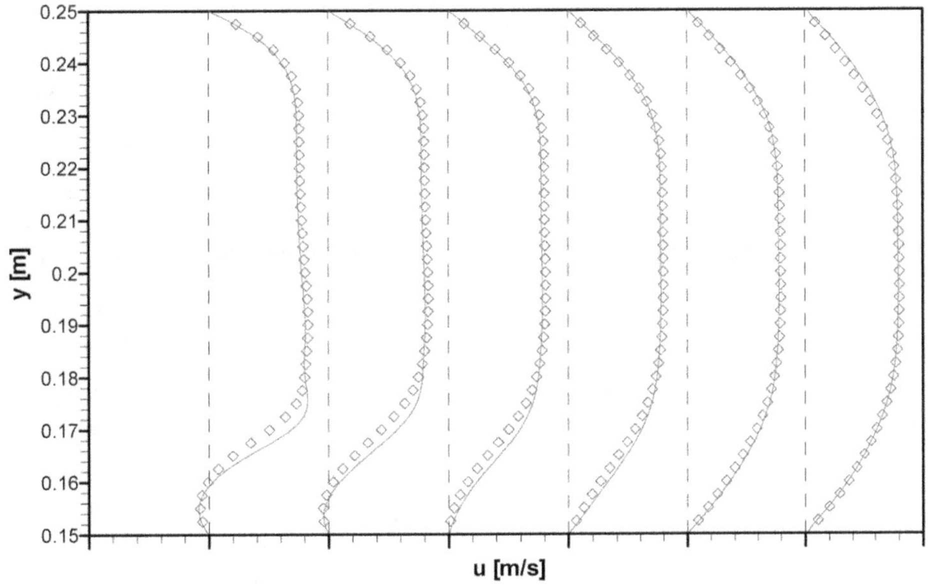

Figure 54. LCMM (symbols) FVM (solid lines) x-velocity profiles along upper channel after the film-cooling hole for conjugate heat transfer film-cooling example (from [41]).

9.1.7 Flow over a Cylinder

This numerical example presents results on a non-uniform data center distribution for LCMM predictions of flow over a cylinder centered in a rectangular channel. The cylinder of diameter d is enclosed between two parallel plates 3 d apart by 10 d long. The center of the cylinder is located at 1/3 of the length of the channel. A uniform flow field with an entrance Reynolds number Re = 10 based on the cylinder diameter enters the channel from the left boundary. The LCMM results are compared to grid-converged FVM results. The LCMM results are obtained using a non-uniform distribution of 8,989 points while the FVM results are obtained using a clustered mesh with 7,872 quadrilateral cells adapted to 31,488 quadrilateral cells to verify grid convergence. Fig. 55 displays a close up of the FVM grid and LCMM point distribution around the cylinder clearly showing the clustering of the grid and the non-uniformity of the point distribution. Fig. 56 shows the FVM and LCMM contours of the velocity magnitude revealing close qualitative agreement. A quantitative comparison is presented in Fig. 57 and Fig. 58 where the FVM, adapted FVM, and LCMM pressure coefficients and viscous stress coefficients (normalized magnitude of the traction vector) are displayed around the top of the cylinder from leading edge to

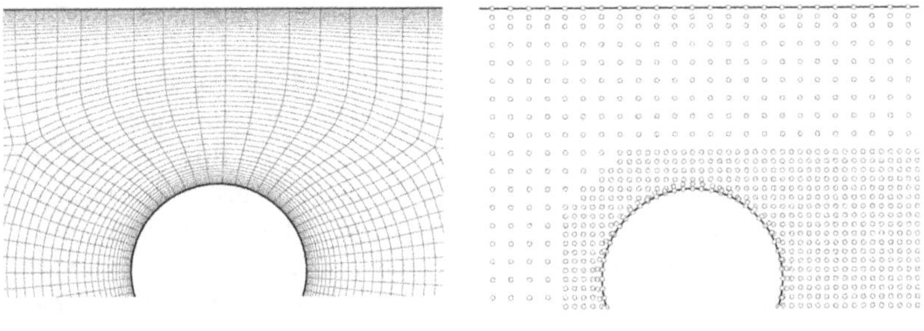

Figure 55. FVM grid and LCMM point distribution around the cylinder (from [41]).

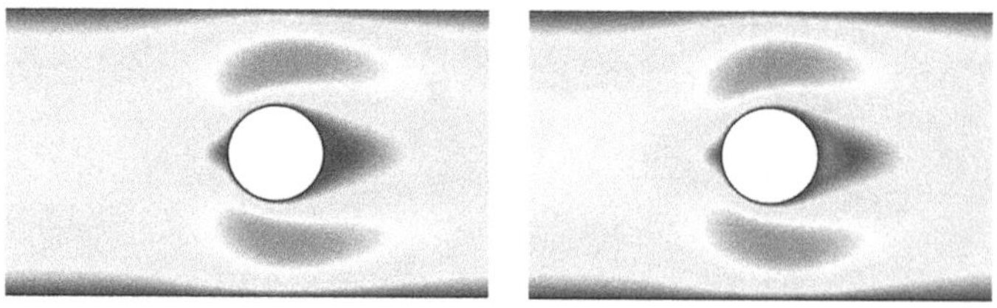

Figure 56. FVM and LCMM velocity magnitude contours around the cylinder (from [41]).

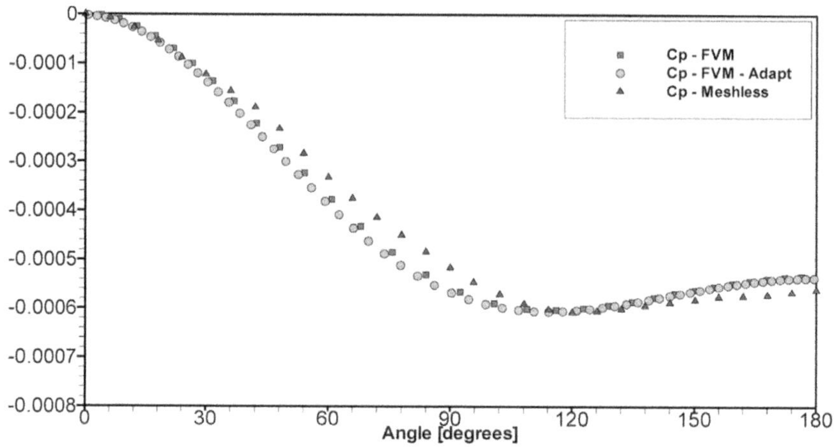

Figure 57. FVM and LCMM pressure coefficients around the cylinder (from [41]).

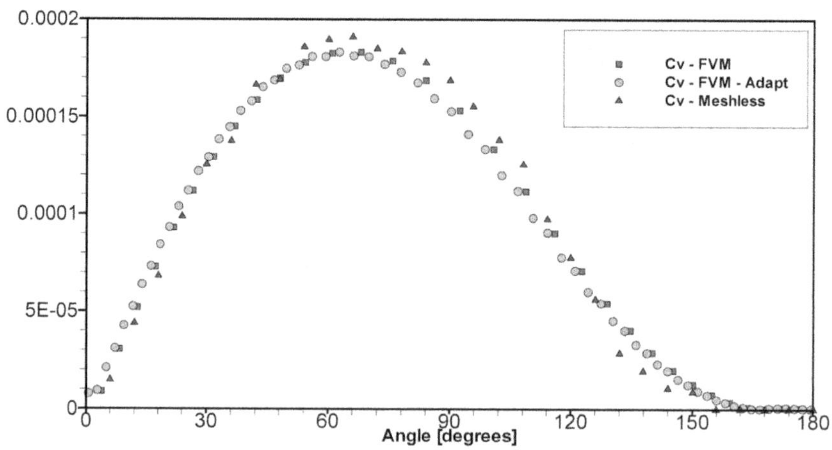

Figure 58. FVM and LCMM viscous stress coefficient around the cylinder (from [41]).

trailing edge, again revealing very good agreement between grid-converged FVM results and LCMM results even at very low Reynolds numbers.

Additional testing is performed for the flow over a cylinder case by increasing the Reynolds number to Re = 1000 which is expected to promote separation and vortex shedding. In this case the diameter of the cylinder was increased from 1/3 to 1/2 of the width of the channel. The velocity magnitude contour plots for this case are displayed at five different (progressive) time levels in Fig. 59 clearly revealing flow separation and periodic sustained vortex shedding.

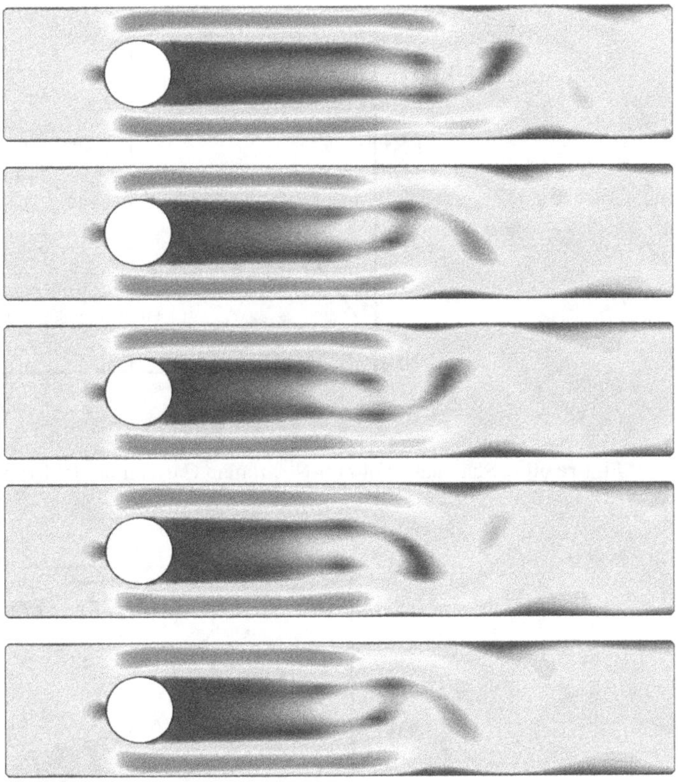

Figure 59. Separation and vortex shedding for flow over a cylinder at five times levels.

9.1.8 Steady Blood Flow through a Femoral Bypass

Blood flow is incompressible and pulsatile by nature, however, special attention should be given to the non-Newtonian rheology of blood. In this case, the Carreau non-Newtonian model is adopted where the governing momentum equation may be expressed as:

$$\rho \frac{\partial \vec{V}}{\partial t} + \rho \left(\vec{V} - \frac{\nabla \mu}{\rho} \right) \cdot \nabla \vec{V} = -\nabla p + \mu \nabla^2 \vec{V} \tag{95}$$

Note that the μ term appearing in Eq. (95) implies the spatial variability of the dynamic viscosity due to the blood non-Newtonain rheology. Actually, the space dependence of μ results from its dependence on the shear strain rate $\dot{\gamma}$ especially at low values ($\dot{\gamma} < 100 s^{-1}$). The expression for μ is given by the Carreau model as:

$$\mu = \mu_\infty + (\mu_0 - \mu_\infty)\left[1 + (\lambda \dot{\gamma})^2\right]^{\frac{n-1}{2}} \tag{96}$$

Here μ_∞ is the infinite shear viscosity, μ_0 is the zero shear viscosity, λ is the time constant, and n is the power law index. The Carreau model is experimentally validated as seen in Fig. 60 and Fig. 61 with the following parameter values for blood: $\mu_\infty = 0.00345\ Pa \cdot s$, $\mu_0 = 0.056\ Pa \cdot s$, $\lambda = 3.31 s$, and $n = 0.357$, see [69].

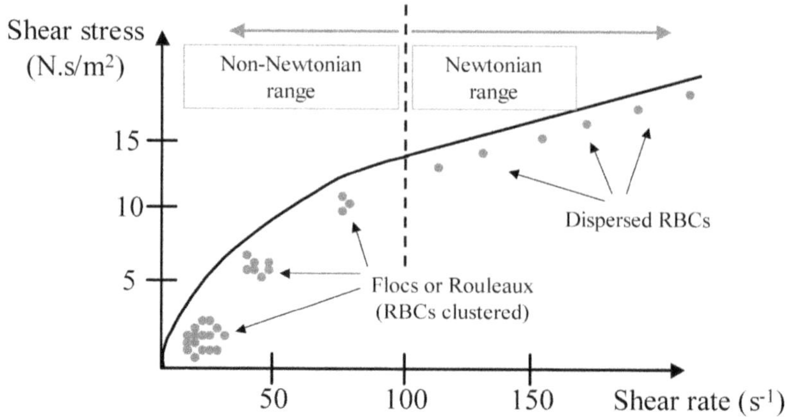

Figure 60. Schematics of non-Newtonian Carreau model for blood.

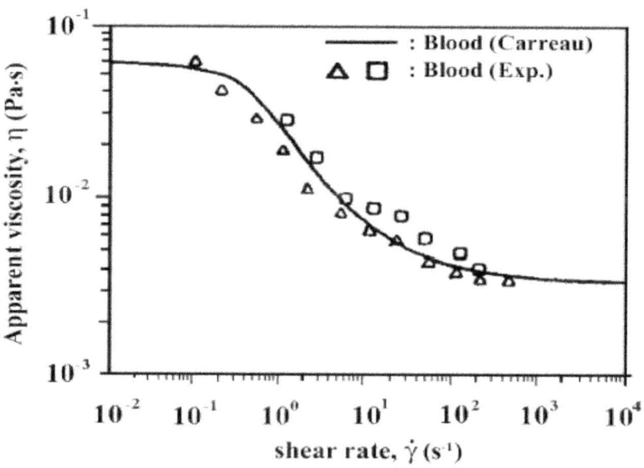

Figure 61. Experimental validation of Carreau model.

Three surgical junction geometries pertaining to the interconnection between a bypass graft and a host artery were selected for a steady flow validation. In the medical glossary, a bypass surgical junction is referred to as an end-to-side distal anastomosis (ETSDA). The selected ETSDA geometry models are the conventional model, the Miller Cuff model, and the hood model all shown in Fig. 62. The steady flow is simulated with a Reynolds number Re = 450 based on the bypass graft diameter and a density ρ = 1060 kg/m^3.

Fig. 63 shows the LCMM point distributions for the conventional, Miller Cuff, and hood ETSDA models respectively. The velocity magnitude comparison is provided in Fig. 64, Fig. 65, and Fig. 66 for the conventional, Miller Cuff, and hood ETSDA models respectively as obtained by both the LCMM and the FVM solvers. The velocity magnitude contours of the three ETSDA models demonstrate a very good qualitative agreement between the values predicted by the FVM and the LCMM solvers. Particularly, the LCMM is capable of capturing the flow recirculation at the floor of the host artery consistently with the FVM. To quantitatively validate the accuracy of the LCMM, the x-velocity profiles are compared with FVM results at five vertical sections (X1, X2, X3, X4, and X5) along the anastomoses of the three given ETSDA models. These locations for the three given ETSDA

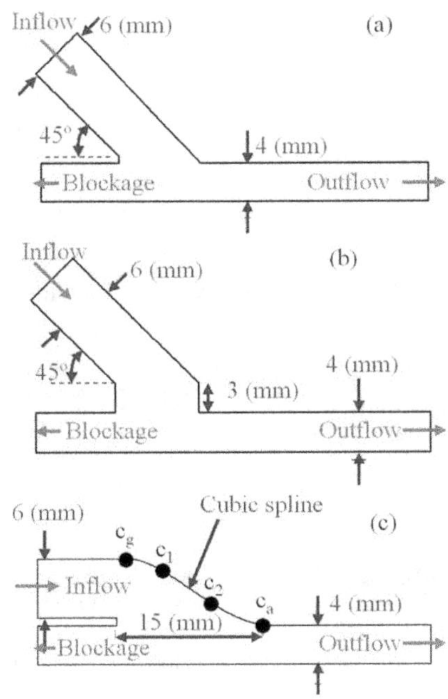

Figure 62. Schematics of the bypass grafts geometries: (a) conventional, (b) Miller Cuff, and (c) hood model (from [45]).

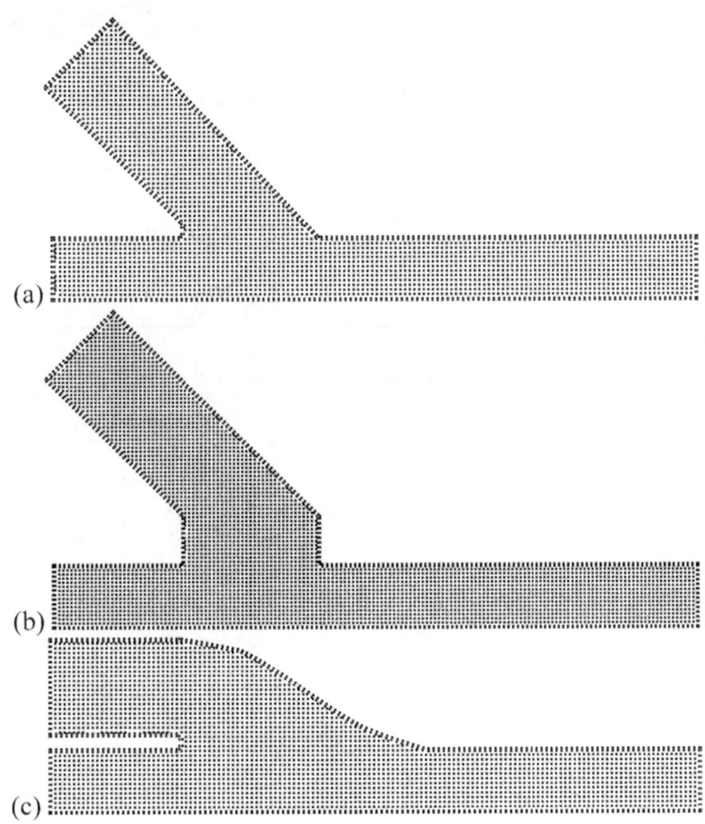

Figure 63. The LCMM point distribution for the (a) conventional, (b) Miller cuff, and (c) hood model geometries (from [45]).

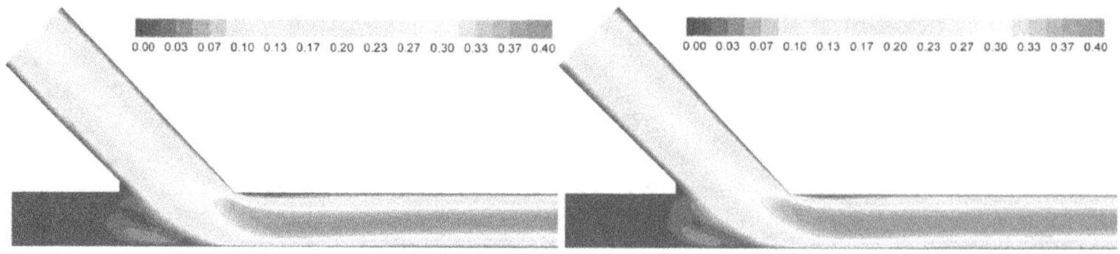

Figure 64. LCMM and FVM velocity contours for conventional ETSDA model (from [45]).

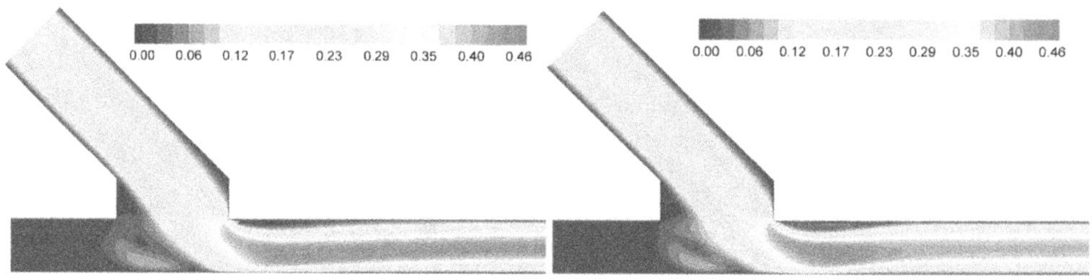

Figure 65. LCMM and FVM velocity contours for Miller Cuff ETSDA model (from [45]).

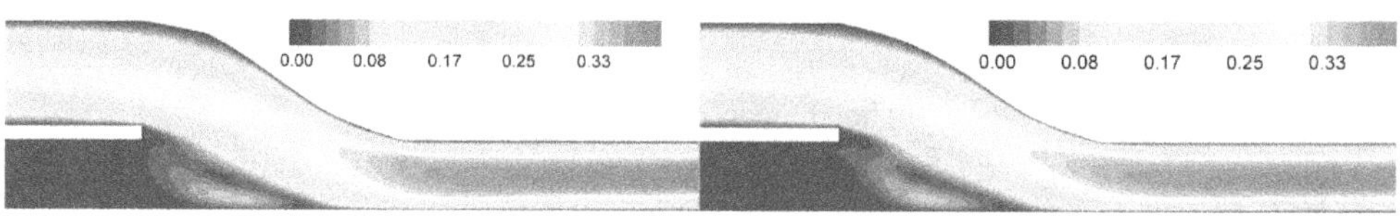

Figure 66. LCMM and FVM velocity contours for hood ETSDA model (from [45]).

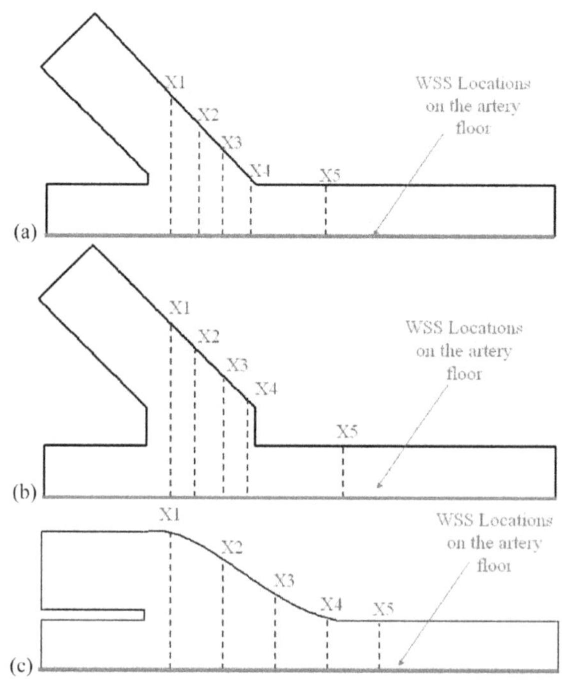

Figure 67. The x-velocity profile locations for the (a) conventional, (b) Miller cuff, and (c) hood models (from [45]).

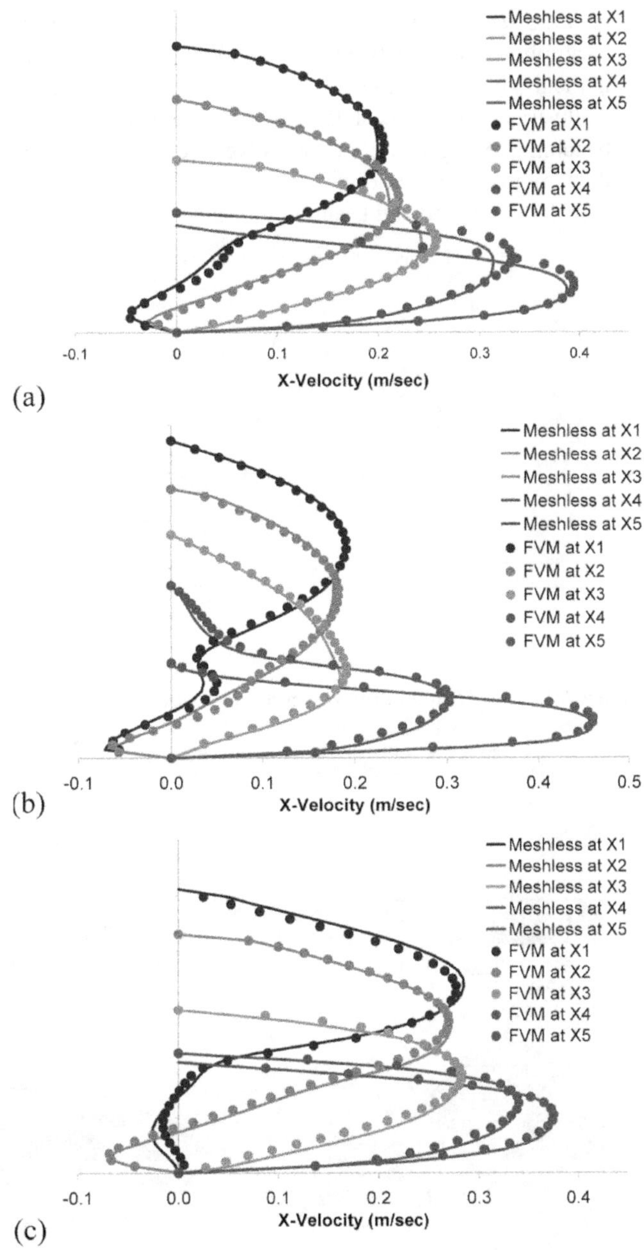

Figure 68. The x-velocity profiles for the (a) conventional, (b) Miller cuff, and (c) hood models (from [45]).

models are revealed in Fig. 67. The x-velocity profiles results for the conventional ETSDA model, Miller Cuff ETSDA model, and the hood ETSDA model are all displayed in Fig. 68. Although minor discrepancies are evident, the velocity profiles indicate a high level of quantitative agreement between the LCMM and the FVM.

9.1.9 Pulsatile Blood Flow through a Femoral Bypass

Following the validation of the LCMM steady-state results, the time-accurate capability of the LCMM is now tested through a case involving a pulsatile flow in the conventional ETSDA geometry shown in Fig. 62. This case consists of a femoral artery flow input waveform with an amplitude range between 0 and 480 *ml*/min and a period of 0.7 *s*; the femoral artery extends in the thigh between the pelvis and the knee levels. The plot of the femoral inflow waveform is illustrated in Fig. 69. The velocity contours of the LCMM femoral pulsatile flow results as benchmarked against the FVM solver are revealed in Fig. 70 at different time values. Then, the femoral pulsatile flow meshless and FVM x-velocity profiles at sections X1, X2, X3, X4, and X5 are plotted at the same time values within the cycle Fig. 71.

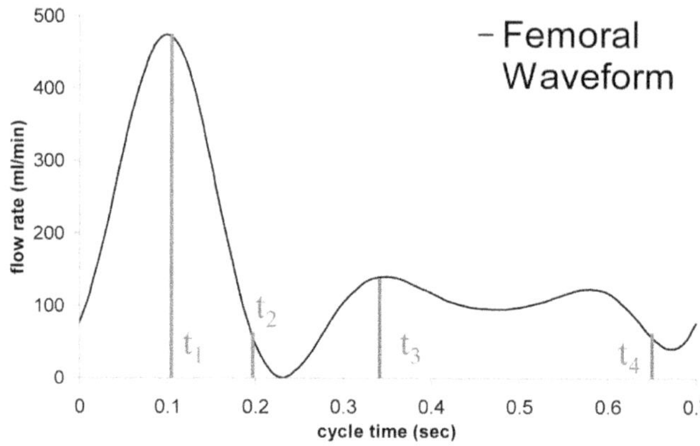

Figure 69. The femoral flow waveform at the conventional ETSDA geometry inlet (from [45]).

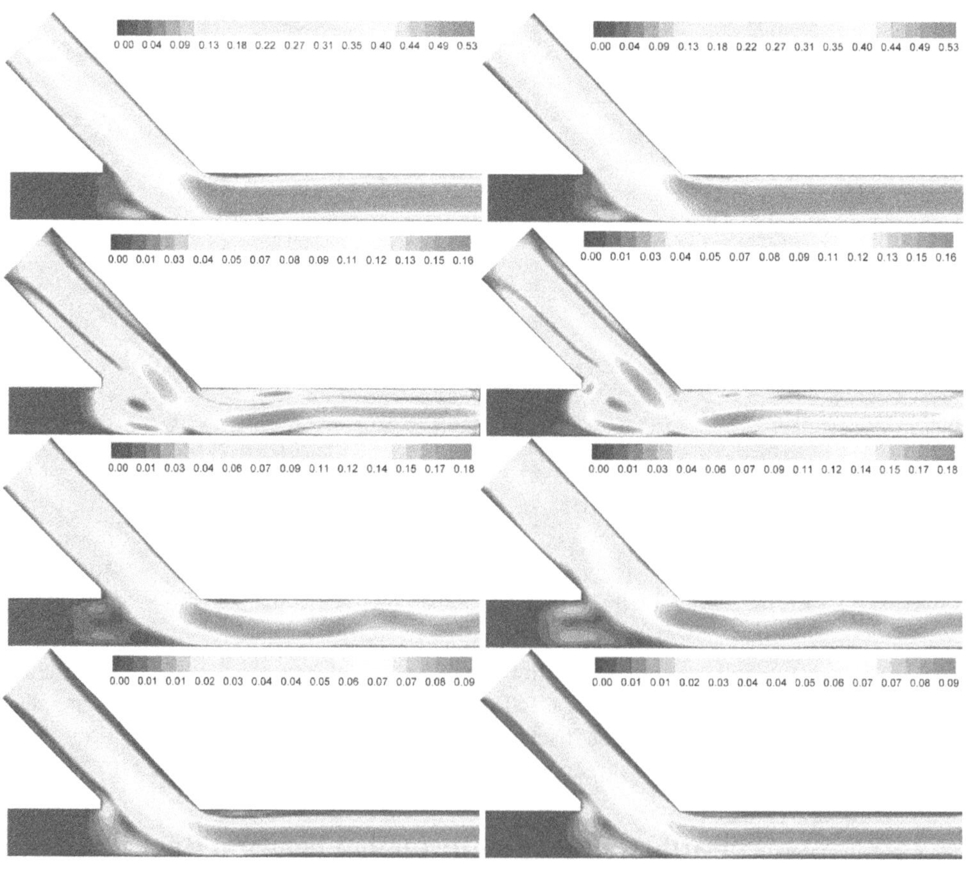

Figure 70. LCMM and FVM velocity contours at t_1, t_2, t_3, t_4 for the conventional ETSDA model (from [45]).

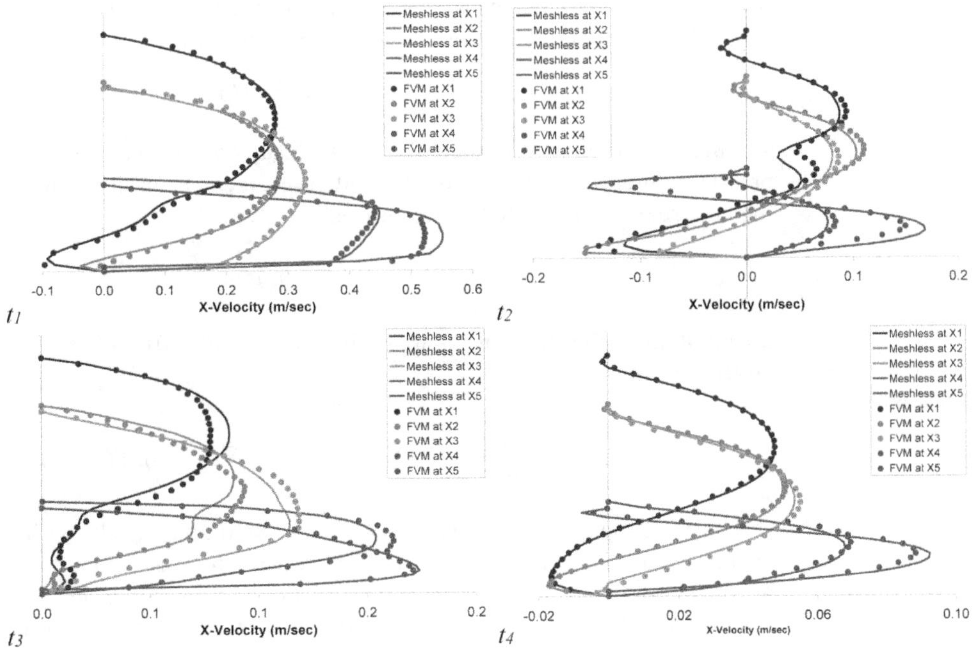

Figure 71. LCMM and FVM *x*-velocity profiles at t_1, t_2, t_3, t_4 for the conventional ETSDA model (from [45]).

9.2 NATURAL CONVECTION

In natural convection problems where buoyancy forces generated by temperature gradients drive the flow field, a special approximation is made that allows, under certain conditions, solving the problem as a coupled incompressible fluid flow and heat transfer problem. Initially, the specific body force in the momentum equation (72) is replaced by the acceleration of gravity and the pressure field, p, is replaced by the total pressure field, p_t, as:

$$\rho\frac{\partial\vec{V}}{\partial t}+\rho\left(\vec{V}\cdot\nabla\right)\vec{V}=\mu\nabla^2\vec{V}-\nabla p_t+\rho\vec{g} \tag{97}$$

The total pressure field, p_t, is a combination of the hydrostatic pressure, p_h, and the motion pressure, p, as:

$$p_t=p_h+p \tag{98}$$

And the hydrostatic pressure, hydrostatic pressure, p_h, can be expressed in terms of the acceleration of gravity and the reference density, ρ_o, as:

$$\nabla p_h=\rho_o\vec{g} \tag{99}$$

Substitution of equations (98) and (99) into the momentum equation (97) leads to:

$$\rho\frac{\partial\vec{V}}{\partial t}+\rho\left(\vec{V}\cdot\nabla\right)\vec{V}=\mu\nabla^2\vec{V}-\nabla p+\vec{g}\left(\rho-\rho_o\right) \tag{100}$$

The gravitational term may now be replaced using the Boussinesq approximation as:

$$\left(\rho-\rho_o\right)=\beta\rho\left(T_o-T\right) \tag{101}$$

where β is the thermal expansion coefficient of the fluid and T_o is a reference temperature. The Boussinesq approximation is applicable as long as:

$$|T_o - T| << 1/\beta \tag{102}$$

The thermal expansion coefficient β can be assumed to be a small constant in most incompressible fluids while in the case of ideal gases, it can be easily shown that it is defined as the inverse of the absolute temperature as:

$$\beta = 1/T \tag{103}$$

Substitution of the Boussinesq approximation in Eq. (101) into the momentum equation (100) leads to:

$$\rho \frac{\partial \vec{V}}{\partial t} + \rho\left(\vec{V} \cdot \nabla\right)\vec{V} = \mu \nabla^2 \vec{V} - \nabla p + \rho \vec{g} \beta \left(T_o - T\right) \tag{104}$$

The momentum equation must be solved simultaneously with the continuity and energy equations to account for the time variation of the temperature field. The same pressure-velocity correction explicit time-marching scheme described in the previous section is followed for the implementation of natural convection problems shown as follows.

9.2.1 Buoyancy-Driven Flow in a Square Cavity

This example tests the effects of pure natural convection by analyzing buoyancy-driven flow of air assumed as an ideal gas in a 1 × 1 *cm* square cavity with the right-hand side wall at 283 *K*, the left-hand side wall at 273 *K*, and both top and bottom walls insulated for a Raleigh number Ra = 1147.8. The LCMM and FVM solutions were obtained using the same distribution of nodes of the lid-driven flow example as seen in Fig. 32. The FVM and LCMM velocity vectors and magnitude contours are shown in Fig. 72 revealing excellent agreement. Also, the FVM and LCMM x and y velocity components profiles are shown in Fig. 73 at x = 0, 0.2, 0.4, 0.6, 0.8, 1 *cm*, where the horizontal distance between each grid-line in the plots represents 0.01 *m/s*.

Finally, the FVM and LCMM temperature contours are displayed in Fig. 74 and the temperature profiles are displayed in Fig. 75 showing very close agreement in the distribution of temperatures. The horizontal distance between each grid-line in the plot represents 10 *K*. All the results shown for this example are converged at steady-state.

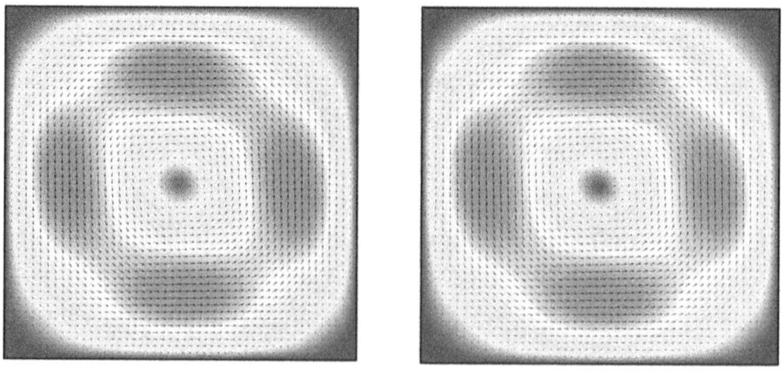

Figure 72. FVM and LCMM velocity vectors and magnitude contours for buoyancy-driven flow in a square cavity (from [43]).

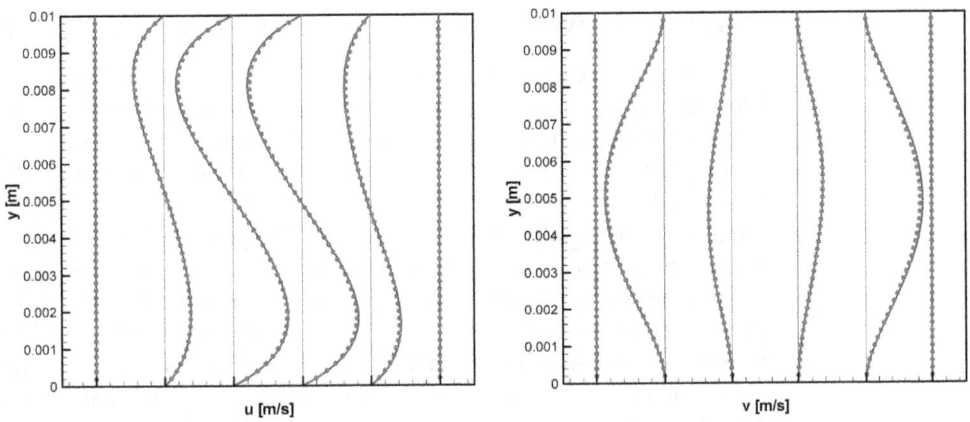

Figure 73. FVM and LCMM x and y velocity components profiles at $x = 0, 0.2, 0.4, 0.6, 0.8, 1$ *cm* (from [43]).

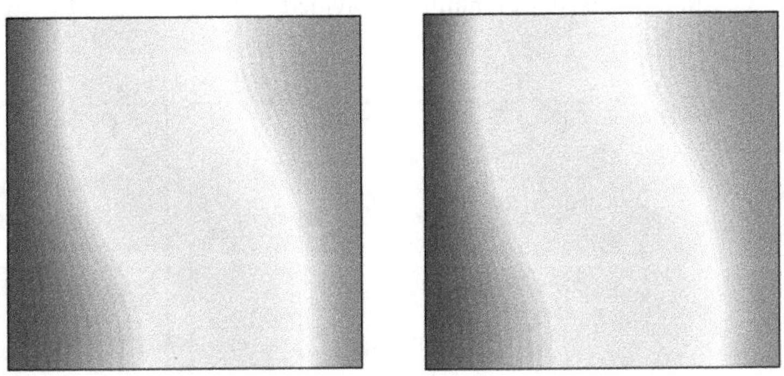

Figure 74. FVM and LCMM temperature contours for buoyancy-driven flow in a square cavity (from [43]).

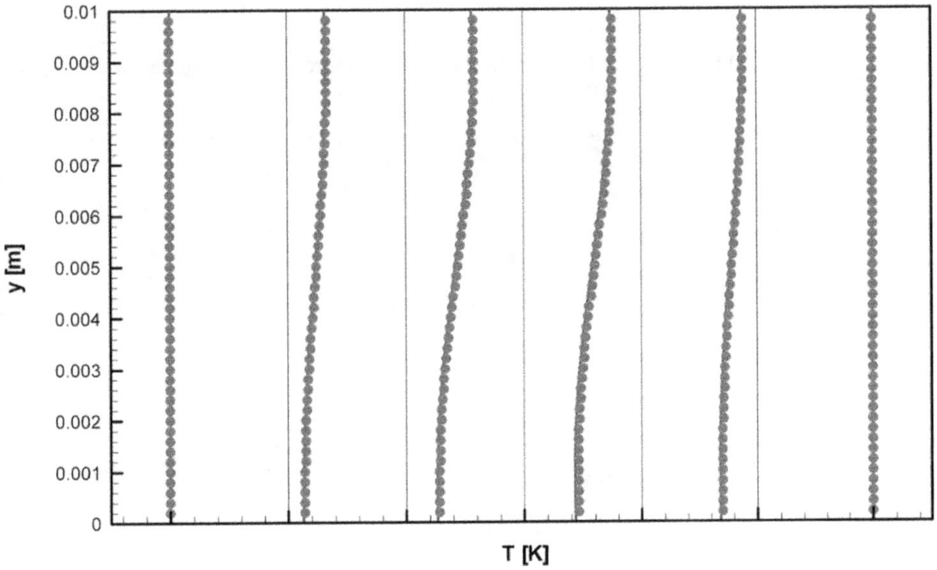

Figure 75. FVM and LCMM temperature profiles at $x = 0, 0.2, 0.4, 0.6, 0.8, 1$ *cm* (from [43]).

9.2.2 Buoyancy-Driven Flow of Liquid Aluminum in a Rectangular Cavity

This numerical example presents a buoyancy-driven flow of liquid aluminum in a 1.25 × 5 *cm* rectangular cavity. The left-hand side wall is kept at 960 *K* and the right-hand side wall is kept at 920 *K*, while the top and bottom walls are kept insulated. The liquid aluminum has a constant thermal expansion coefficient $\beta = 1.17 \times 10^{-4} K^{-1}$ yielding a Raleigh number $Ra =$ 4394 based on the width of the cavity. The LCMM solution was obtained using a distribution of 41 × 161 equally-spaced data centers. The problem is initialized with a uniform temperature field of 960 *K* and advanced in time using a time-step size $\Delta t = 10^{-4}\, s$. In this case, the solution does not settle to steady-state as it maintains a periodic sustained behavior. The LCMM velocity vectors colored by magnitude are shown if Fig. 76 while the temperature contours are shown in Fig. 77 at several time levels: 5 *s*, 10 *s*, 15 *s*, 20 *s*, 25 *s*, 30 *s*. The Nusselt number time-evolution over the first 15 *s* on the left-hand side and the right-hand side walls is presented in Fig. 78 comparing the LCMM solution to the solution provided by a cell-centered research FVM code. The FVM code uses third order accurate time stepping and second order upwinding and results are computed using a mesh consisting of 101 × 401 nodes and a time-step size $\Delta t = 10^{-3}\, s$. It is noted that the magnitudes and transient evolution of the Nusselt number provided by the FVM and LCMM codes coincide very closely.

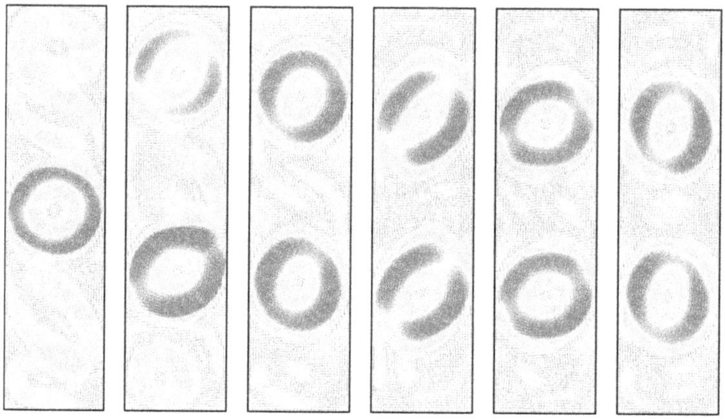

Figure 76. LCMM velocity vectors colored by magnitude for the liquid aluminum cavity at 5 *s*, 10 *s*, 15 *s*, 20 *s*, 25 *s*, 30 *s* (from [43]).

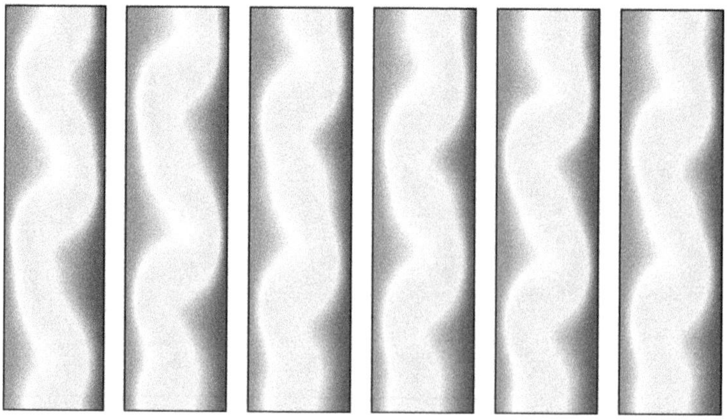

Figure 77. LCMM temperature contours for the liquid aluminum cavity at 5 *s*, 10 *s*, 15 *s*, 20 *s*, 25 *s*, 30 *s* (from [43]).

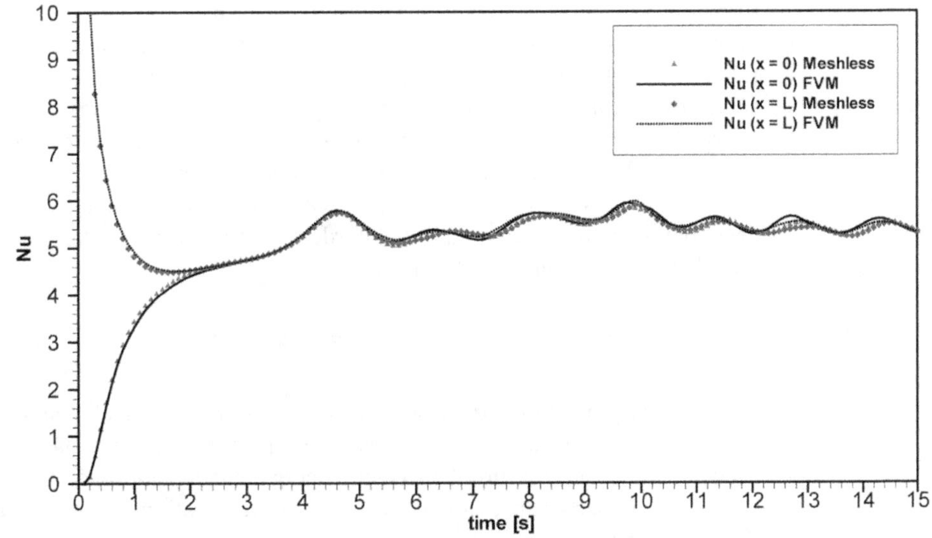

Figure 78. FVM and LCMM Nusselt number evolution in time on the left-hand and right-hand side walls (from [43]).

9.3 TURBULENT FLUID FLOWS

Standard turbulence modeling techniques have been developed and applied to Finite Volume, Finite Element, and Finite Differencing CFD routines for years. Some of these turbulence modeling schemes are found to be directly applicable to the LCMM approach. Since the foundation of the turbulence modeling approach is well known, the reader is referred to classic texts, such as that of Wilcox [70] for a full description.

A Reynolds Averaged Navier-Stokes (RANS) approach is used, which reduces the 2D compressible Navier-Stokes equations to the following:

$$\frac{\partial \bar{\rho}}{\partial t} + \frac{\partial \overline{\rho u}}{\partial x} + \frac{\partial \overline{\rho v}}{\partial y} = 0$$

$$\frac{\partial \bar{u}}{\partial t} + \bar{u}\frac{\partial \bar{u}}{\partial x} + \bar{v}\frac{\partial \bar{u}}{\partial y} = -\frac{1}{\bar{\rho}}\frac{\partial \bar{p}}{\partial x} + \nu\nabla^2\bar{u} + \frac{\nu}{3}\left[\frac{\partial}{\partial x}\left(\frac{\partial \bar{u}}{\partial x} + \frac{\partial \bar{v}}{\partial y}\right)\right] - \frac{1}{\bar{\rho}}\frac{\partial}{\partial x}\left(\rho\overline{u'^2}\right) - \frac{1}{\bar{\rho}}\frac{\partial}{\partial y}\left(\rho\overline{u'v'}\right) \quad (105)$$

$$\frac{\partial \bar{v}}{\partial t} + \bar{u}\frac{\partial \bar{v}}{\partial x} + \bar{v}\frac{\partial \bar{v}}{\partial y} = -\frac{1}{\bar{\rho}}\frac{\partial \bar{p}}{\partial y} + \nu\nabla^2\bar{v} + \frac{\nu}{3}\left[\frac{\partial}{\partial y}\left(\frac{\partial \bar{u}}{\partial x} + \frac{\partial \bar{v}}{\partial y}\right)\right] - \frac{1}{\bar{\rho}}\frac{\partial}{\partial x}\left(\rho\overline{u'v'}\right) - \frac{1}{\bar{\rho}}\frac{\partial}{\partial y}\left(\rho\overline{v'^2}\right)$$

The energy equation has been purposely omitted as its changes are analogous to those which will be shown for momentum. The prime (') symbol is used to denote a fluctuating (turbulent) component and the overbar denotes an average or mean value. These equations are now solved in hopes of properly capturing the flow field. The difficulty arises in modeling the fluctuating components, or Reynolds stresses. All turbulence modeling approaches attempt to approximate these terms using quantities which are determined from the mean flow variables; a process called closing the model. Various methods have been proposed to close the model, the simplest of which use algebraic relations only. Such methods are referred to as zero-order or zero-equation models as no additional differential equations are used to model the fluctuations. Such models suffer significantly from localization, as turbulence levels are dictated solely by local properties (often local velocity gradients). However, these methods often employ enough empirical closure constants that they can be tuned to specifically handle certain classes of flow and have been shown to perform well in such cases. More complex models employ turbulence parameters, such as turbulent kinetic energy, and utilize differential transport equations for these parameters to better capture the non-local effects of turbulence. The most popular of these types are one-equation and two-equation models, as higher order models offer little benefit in terms of increased performance, but do increase computational burden. One-equation models most commonly use the turbulent kinetic energy, k, defined as:

$$k = \frac{1}{2}\left(u'^2 + v'^2\right) \tag{106}$$

The necessary transport equation is developed semi-empirically, and will contain convective, diffusive, dissipative, and generation terms. Finally, two-equation models will employ transport equations for turbulent kinetic energy and an additional turbulence property. This additional property is usually either the rate of dissipation of kinetic energy or the specific dissipation rate.

The Boussinesq assumption is then employed as the first step in the approximation such that:

$$-\overline{u_i' u_j'} = \mu_T \left(\frac{\partial u_i}{\partial x_j} + \frac{\partial u_j}{\partial x_i} \right) \tag{107}$$

This assumption allows the standard Navier-Stokes equations to be used with a simple modification to the standard (laminar) stresses τ_{ij} such that:

$$\tau_{ij} = \left[\mu \left(\frac{\partial u_i}{\partial x_j} + \frac{\partial u_j}{\partial x_i} \right) - \delta_{ij} \frac{2}{3} \mu \frac{\partial u_k}{\partial x_k} \right] + \left[\mu_T \left(\frac{\partial u_i}{\partial x_j} + \frac{\partial u_j}{\partial x_i} - \frac{2}{3} \delta_{ij} \frac{\partial u_k}{\partial x_k} \right) - \frac{2}{3} \rho k \delta_{ij} \right] \tag{108}$$

where k is the turbulent kinetic energy and μ_T is the turbulent or eddy viscosity. The first term in Eq. (108) is considered the laminar shear stress, while the second is the so-called turbulent shear stress (Reynolds Stress).

The most simplistic turbulence models are the zero-equation models where algebraic expressions are used to define the turbulent viscosity based on the local mean flow properties, typically as:

$$\mu_T = \rho l^2 \left[\left(\frac{\partial u}{\partial y} \right)^2 + \left(\frac{\partial v}{\partial x} \right)^2 \right]^{1/2} \tag{109}$$

Numerous specific models exist to define l such as the Cebeci-Smith as well as the Baldwin-Lomax models.

Two-equation turbulence models are the lowest order, physically-based models that obtain full closure of the Reynolds Stress terms without the need for algebraic specification of turbulence parameters. Two-equation models have come to be the standard CFD turbulence techniques for FVM and FDM and it is expected that this will be the case for Meshless methods as well. These models employ detailed transport equations which reasonably account for the convection, diffusion, production, and dissipation of various turbulence parameters. A combined $k-\omega/k-\varepsilon$ model is presented herein and implemented in the LCMM framework. This turbulence model attempts to bring the desirable features of each model together by incorporation of a single transport equation for ε/ω, see Hoffmann [71]. The transport equations and relationships amongst the various turbulence parameters are given by:

$$\rho\left(\frac{\partial k}{\partial t}+u\frac{\partial k}{\partial x}+v\frac{\partial k}{\partial y}\right)=\frac{\partial}{\partial x_j}\left[\left(\mu+\frac{\mu_T}{\sigma_k}\right)\frac{\partial k}{\partial x_j}\right]+P_k-\rho\varepsilon$$

$$\rho\left(\frac{\partial \omega}{\partial t}+u\frac{\partial \omega}{\partial x}+v\frac{\partial \omega}{\partial y}\right)=\frac{\partial}{\partial x_j}\left[\left(\mu+\frac{\mu_T}{\sigma_\varepsilon}\right)\frac{\partial \omega}{\partial x_j}\right]+\alpha P_k\frac{\omega}{k}-\beta\rho\omega^2 \tag{110}$$

These transport equations along with the following relations are used to define the two-equation model:

$$\mu_T=\rho C_\mu\frac{k^2}{\varepsilon}\quad\text{or}\quad \mu_T=\rho\frac{k}{\omega}\quad\text{and}\quad \varepsilon=k\omega\beta^* \tag{111}$$

9.3.1 Turbulent Flow over a Flat Plate

Developing flow over an infinitely thin flat plate is a classic test case for turbulence model verification. This simple test case is useful as the desired turbulent solutions are well known and understood. This case consists of a rectangular flow domain with uniform fixed inlet conditions, an initial short length of zero shear wall, followed by a no-slip flat plate and standard outflow boundary, with inlet conditions of $P_0 = 100{,}000\ Pa$ and $T_0 = 300\ K$. Since this problem focuses on the leading edge of the plate, the Reynolds number is quite small, but nonetheless turbulent results are sought.

Traditionally, RANS type turbulence models themselves are incapable of predicting the transition from laminar to turbulent flow and are simply an on/off type of model. Therefore, allowing the model to remain on at the leading edge is equivalent to enforcing a turbulent boundary layer from the onset of the plate surface. Therefore the results of this study should correlate well to empirical power-law solutions assuming turbulent flow from the leading edge, and indeed we have found such correlations to be quite good. Fig. 79(a) shows the normalized velocity profiles at a location of $x = 40\ mm$ from the leading edge for two distinct turbulence models (empirical boundary layer profiles for laminar and 1/7th law turbulent flows are included for comparison purposes). There is reasonable agreement between the computed and theoretical turbulent profiles with deviations mainly present in the viscous sublayer where it is well known the power-law profiles do not hold well. Fig. 79(b) shows the overall boundary layer growth using the $k-\omega$ model, including a comparison to the classic power law results. Fig. 79 shows that the BL thickness is predicted within 5% of the theoretical distance at all locations along the wall using a definition 99.5% of the free-stream velocity for the BL edge. Fig. 80 shows the wall shear stress for the computed and theoretical results, where again the laminar values are shown for comparison purposes. The matching here is again reasonable and the asymptotic behavior as the flow becomes removed from the leading edge is captured quite well.

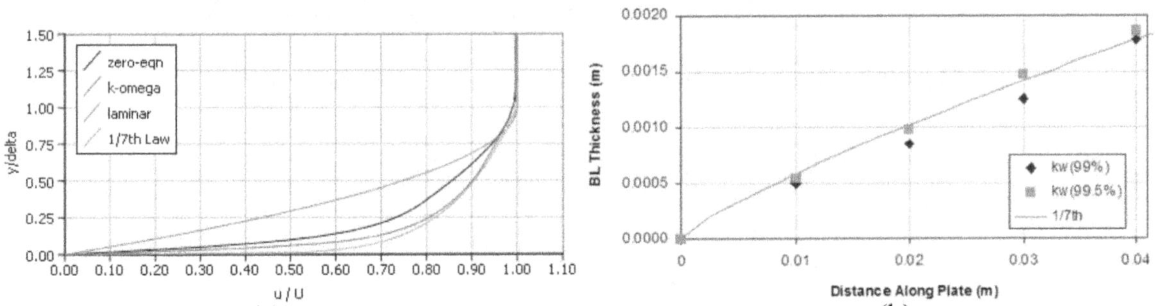

(a) (b)

Figure 79. Turbulent flat plate boundary layer (from [47]).

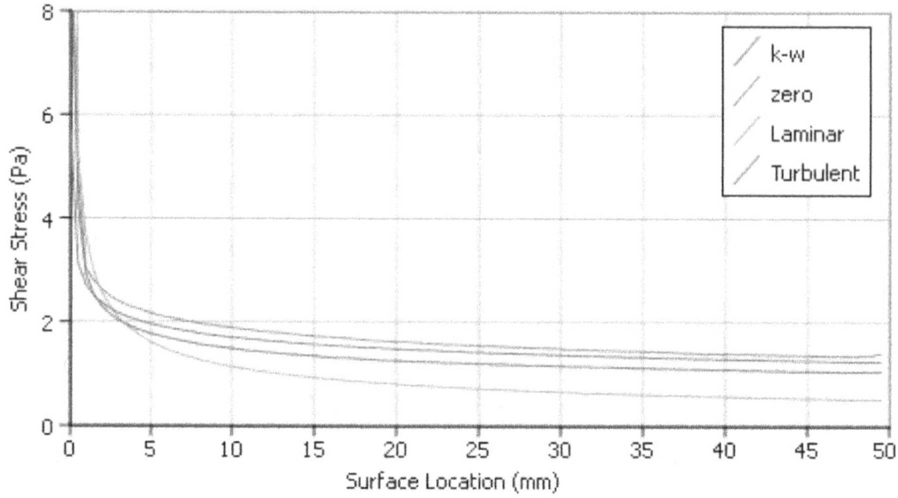

Figure 80. Wall shear stress distribution (from [47]).

9.3.2 Turbulent Flow over a Backward-Facing Step

The following test case is widely used throughout the literature as a benchmark for turbulent solution quality and involves the flow over a step or sudden-expansion in an internal channel flow (classically referred to as the backward facing step). Fig. 81 shows the specific geometric and flow conditions utilized for this test case (this problem has a Reynolds number of approximately 250,000). This problem is widely used because it tests turbulence model performance for both near wall and separated or wake-like flows. Many publications, such as the DNS simulations of Le et al. [72], show that for a wide range of Reynolds number the flow reattachment location should fall near to 6.28 step heights beyond the expansion. Our Meshless based solutions utilizing the combined $k-\varepsilon/k-\omega$ model match this value quite well, as evidenced by the results provided below which show 6.5 step heights as the attachment length. Fig. 82 shows the x-velocity flow contours, which are typical for this type of flow. Additionally, the profiles of velocity are shown for several locations following the step in Fig. 83 along with those given by Kim [73] from his experimental work. Good agreement

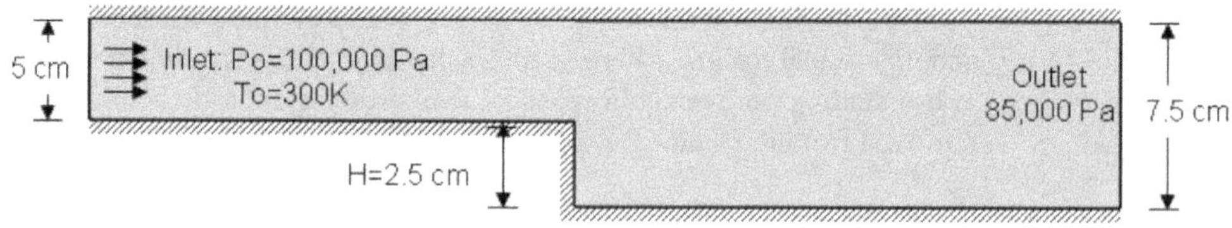

Figure 81. Geometry and boundary conditions for backward-facing step (from [47]).

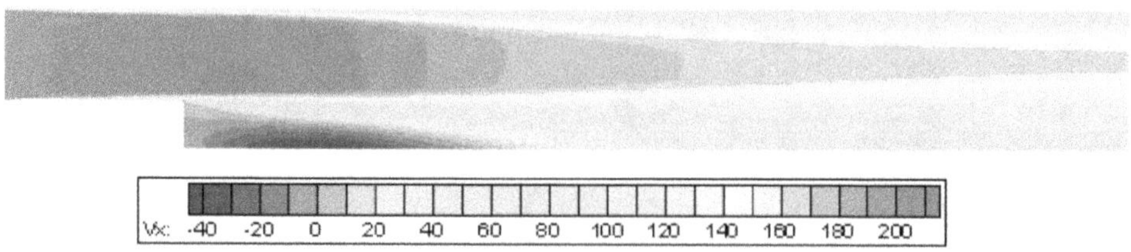

Figure 82. X-direction velocity contours behind the step ($k-\varepsilon/k-\omega$ model) (from [47]).

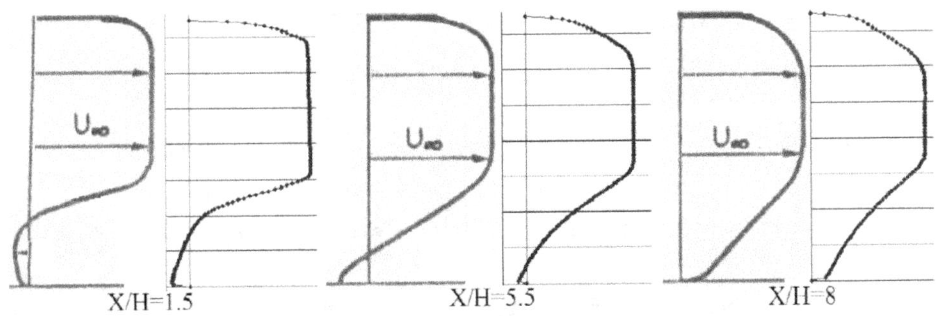

Figure 83. Downstream velocity profiles (Kim [73] vs. Meshless) (from [47]).

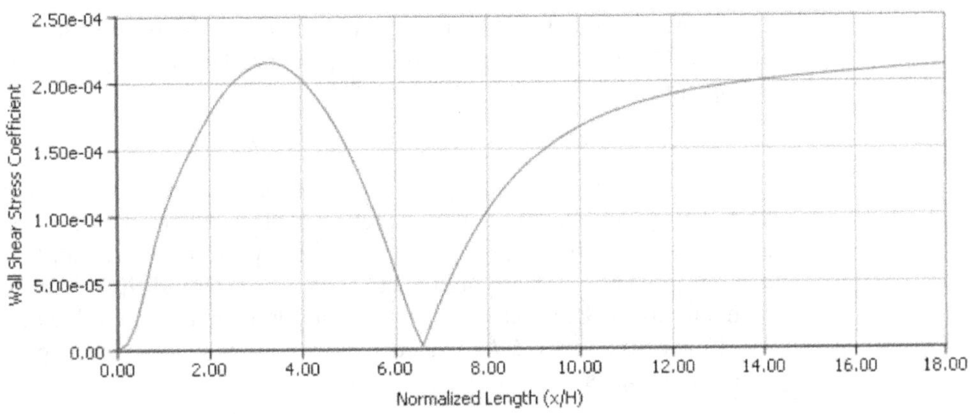

Figure 84. Wall shear stress distribution along the bottom plate after the step (from [47]).

is visible between the two sets of data, demonstrating the well captured recirculation zone. Finally, Fig. 84 shows the shear stress coefficient along the lower wall, which allows for precise determination of the location of the reattachment point ($x/H = 6.5$), which is easily within the acceptable range based on published experimental data.

9.4 COMPRESSIBLE FLUID FLOWS

The specific LCMM formulation for compressible flows starts from the three-dimensional Navier-Stokes (N-S) equations in conservative vector form as:

$$\frac{\partial Q}{\partial t}+\frac{\partial E}{\partial x}+\frac{\partial F}{\partial y}+\frac{\partial G}{\partial z}=\frac{\partial E_v}{\partial x}+\frac{\partial F_v}{\partial y}+\frac{\partial G_v}{\partial z} \tag{112}$$

where:

$$Q=\begin{bmatrix}\rho\\ \rho u\\ \rho v\\ \rho w\\ \rho e_t\end{bmatrix},\ E=\begin{bmatrix}\rho u\\ \rho u^2+p\\ \rho uv\\ \rho uw\\ (\rho e_t+p)u\end{bmatrix},\ F=\begin{bmatrix}\rho v\\ \rho vu\\ \rho v^2+p\\ \rho vw\\ (\rho e_t+p)v\end{bmatrix},\ G=\begin{bmatrix}\rho w\\ \rho wu\\ \rho wv\\ \rho w^2+p\\ (\rho e_t+p)w\end{bmatrix} \tag{113}$$

The viscous terms are:

$$E_v = \begin{bmatrix} 0 \\ \tau_{xx} \\ \tau_{xy} \\ \tau_{xz} \\ u\tau_{xx} + v\tau_{xy} + w\tau_{xz} - q_x \end{bmatrix}, \; F_v = \begin{bmatrix} 0 \\ \tau_{yx} \\ \tau_{yy} \\ \tau_{yz} \\ u\tau_{yx} + v\tau_{yy} + w\tau_{yz} - q_y \end{bmatrix}, \; G_v = \begin{bmatrix} 0 \\ \tau_{zx} \\ \tau_{zy} \\ \tau_{zz} \\ u\tau_{zx} + v\tau_{zy} + w\tau_{zz} - q_z \end{bmatrix} \tag{114}$$

where the shear stress τ_{ij} is:

$$\tau_{ij} = \mu\left(\frac{\partial u_i}{\partial x_j} + \frac{\partial u_j}{\partial x_i}\right) - \delta_{ij}\frac{2}{3}\mu\frac{\partial u_k}{\partial x_k} \tag{115}$$

Equations (112), (113), (114) and (115) represent the full Navier-Stokes equations. In addition, to arrive at a complete set of equations, the ideal gas equation of state must also be imposed as:

$$p = \rho RT \tag{116}$$

In order to solve this set of equations the Advection Upstream Splitting Method (AUSM) proposed by Liou and Steffen [74] is followed. This method has been shown to combine the accuracy of the Roe splitting method with the speed of more simplified splitting methods and has proven to be both highly stable as well as non-diffusive. It should be noted that both the Navier-Stokes and turbulence transport equations are solved using this technique.

Convective terms (E,F,G) are computed following the AUSM approach, which begins with a splitting as:

$$E = \begin{bmatrix} \rho \\ \rho u \\ \rho v \\ \rho w \\ (\rho e_t + p) \end{bmatrix} u + \begin{bmatrix} 0 \\ p \\ 0 \\ 0 \\ 0 \end{bmatrix} = E_i^{(c)} + \begin{bmatrix} 0 \\ 0 \\ 0 \\ 0 \\ up \end{bmatrix} + \begin{bmatrix} 0 \\ p \\ 0 \\ 0 \\ 0 \end{bmatrix} \tag{117}$$

where $E_i^{(c)}$ refers to the "true" convective terms which travel at the local flow speed, u, and p is the pressure term which travels at the local acoustic wave speed. The "true" convective terms are manipulated by factoring u, then multiplying and dividing by the local sound speed, a, as:

$$E_i^{(c)} = u\begin{bmatrix} \rho \\ \rho u \\ \rho v \\ \rho w \\ \rho e_t \end{bmatrix} = M\begin{bmatrix} \rho a \\ \rho a u \\ \rho a v \\ \rho a w \\ \rho a e_t \end{bmatrix} \tag{118}$$

where M is the signed directional Mach number computed as $M = u/a$. These terms are computed at the half nodes following the guidance of Liou and Steffen [74] as:

$$E_{1/2}^{(c)} = M_{1/2}\begin{bmatrix} \rho a \\ \rho a u \\ \rho a v \\ \rho a w \\ \rho a e_t \end{bmatrix}_{L/R} \tag{119}$$

where the terms in the brackets are evaluated at the left-hand node ($i-1$) for $M_{1/2} > 0$ or at the right-hand node (i) otherwise. The term $M_{1/2}$ is computed as:

$$M_{1/2} = M_L^- + M_R^+$$

$$M^{\pm} = \begin{cases} \pm\frac{1}{4}(M \pm 1)^2 & \text{if } |M| \le 1 \\ \pm\frac{1}{2}(M \pm |M|) & \text{if } |M| > 1 \end{cases} \quad (120)$$

Finally the convective derivative at node i is found by central differencing at the half nodes as:

$$\left.\frac{\partial E}{\partial x}\right|_i = \frac{E_{i+1/2} - E_{i-1/2}}{\Delta x} \quad (121)$$

9.4.1 Subsonic and Supersonic Smooth Expanding Diffuser

This case illustrates subsonic flow through a simple smooth expanding nozzle, of which the geometry is shown in Fig. 85 (all dimensions given in meters and z depth is 0.05 m). The inlet Mach number $M = 0.4$ with a corresponding stagnation pressure and temperature of $P_0 = 100{,}000\ Pa$ and $T_0 = 300\ K$ were imposed. Additionally, all walls (other than inlet and outlet) are assumed to have no friction (slip walls). An initial point distribution consisting of approximately 45,000 nodes was automatically generated for this geometry, and after comparing to two-dimensional results generated by FLUENT (structured grid, approximately 70,000 two dimensional cells), no refinement was deemed necessary. The resulting Mach number contours on the central plane along with those generated by FLUENT are shown in Fig. 86 revealing very good qualitative agreement. Furthermore, the comparison of the Meshless and FLUENT pressure and Mach number distributions along the center line are shown in Fig. 87 and Fig. 88 respectively, revealing excellent quantitative agreement.

The smooth expanding diffuser found in Fig. 85 was tested with supersonic flow with an inlet Mach number $M = 2$ and the same stagnation pressure and temperature as in the subsonic case. The Meshless solution began with an initial distribution of 45,000 nodes and was allowed 3 levels of refinement at intervals of 4,000 iterations, resulting in a final distribution consisting of approximately 160,000 nodes. The pressure and Mach number distribution along the centerline are shown in Fig. 89 and Fig. 90 respectively, clearly showing that the refinement process improves the solution as it tends to match the solution provided by FLUENT.

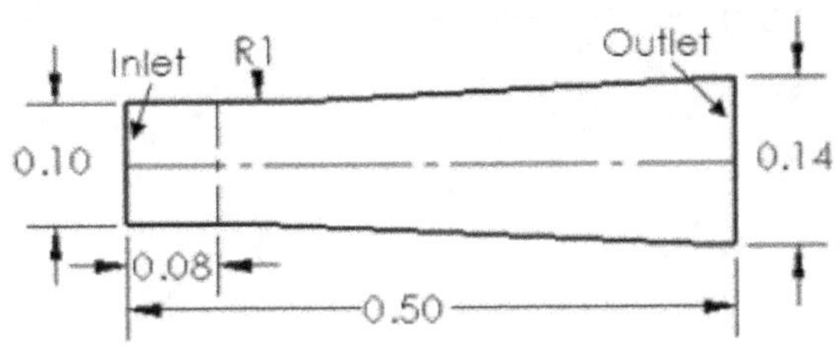

Figure 85. Smooth expanding diffuser geometry (from [47]).

(a) (b)

Figure 86. Central pane Mach number contours: (a) Meshless, (b) FLUENT (from [47]).

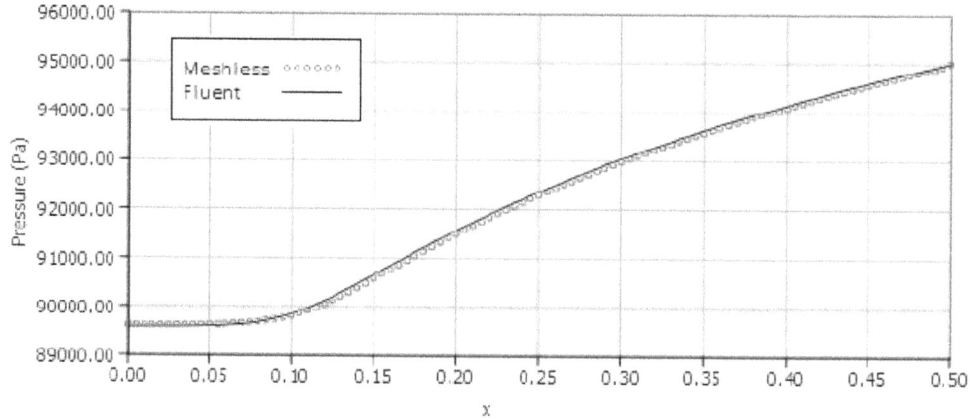

Figure 87. Center-line pressure distribution.

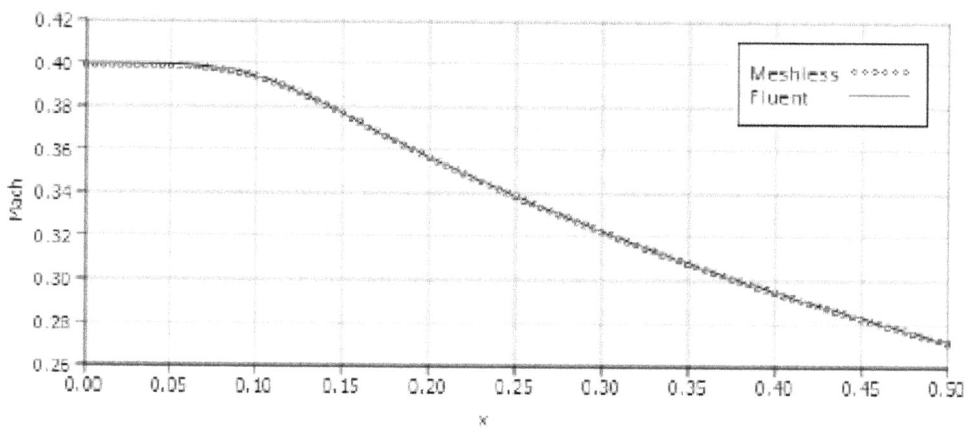

Figure 88. Center-line Mach number distribution.

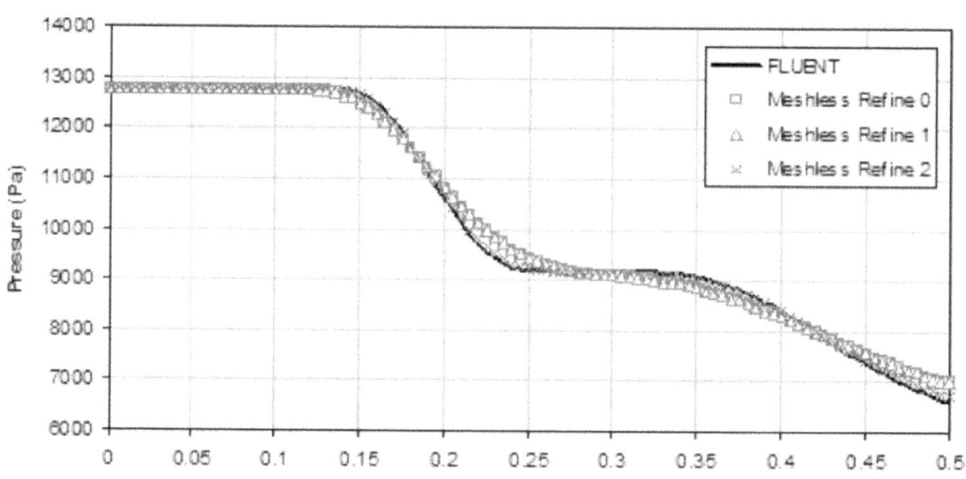

Figure 89. Center-line pressure distribution throughout refinement process (from [47]).

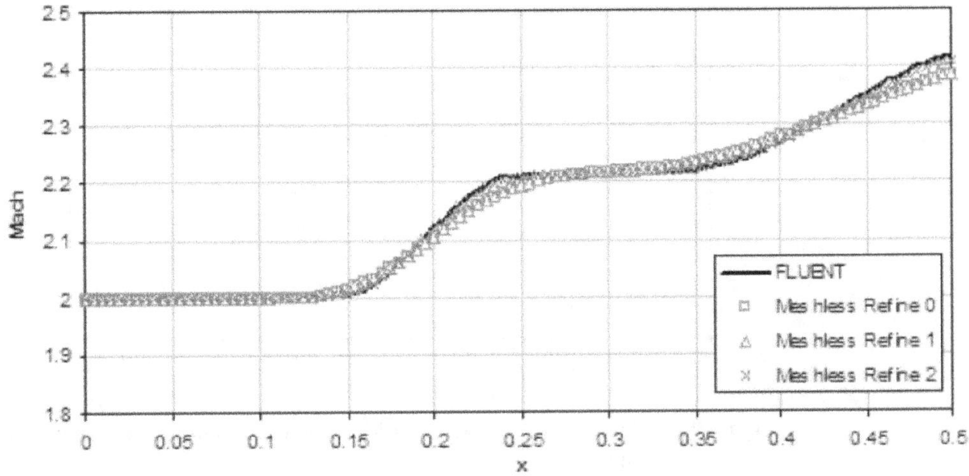

Figure 90. Center-line Mach number distribution throughout refinement process.

9.4.2 Characteristic Nozzle Flow

A characteristic nozzle shape obtained from Hoffman and Chiang [71] was tested assuming inlet conditions of $M = 1.5$, $P_0 = 100{,}000\ Pa$, $T_0 = 300\ K$, and $P = 27{,}240.31\ Pa$. The shock location was directly prescribed at $x = 5\ m$ by assuming an outlet pressure $P_{out} = 66{,}809.64\ Pa$ (found through utilization of 1D isentropic and normal shock equations). As usual, the solution was started on an initially coarse point distribution consisting of roughly 60,000 nodes and was allowed to adaptively refine 3 levels, resulting in a final distribution of just over 175,000 nodes. Note that for this problem the gradient window on which to adaptively refine was reduced in an attempt to isolate the shock location. The contours of Mach number over the refinement detail are shown in Fig. 91. Also, the contours of Mach number over the refinement detail are shown in Fig. 92 revealing high qualitative agreement with the expected shock

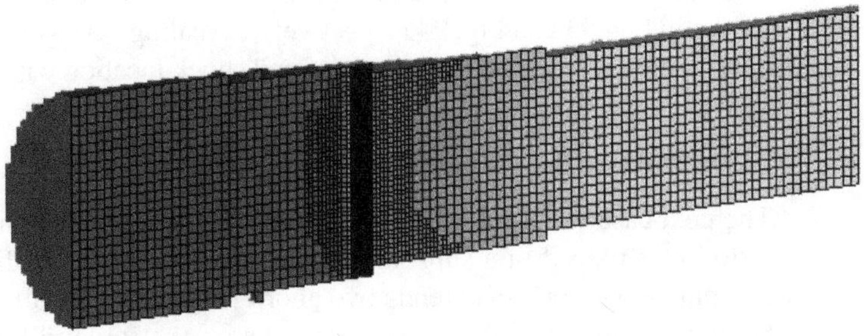

Figure 91. Mach number contours and refinement detail (from [47]).

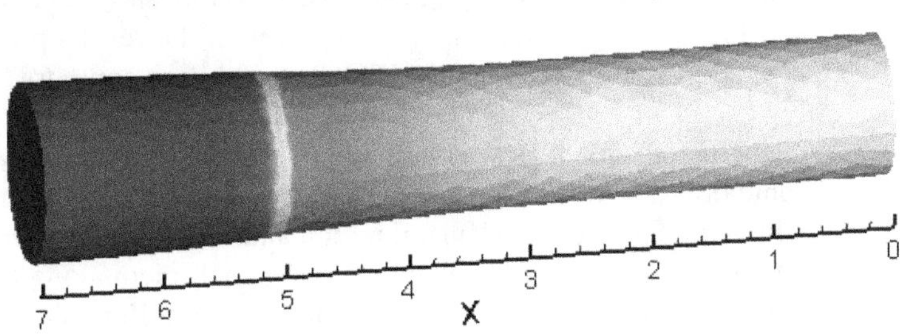

Figure 92. High qualitative agreement with expected shock location (from [47]).

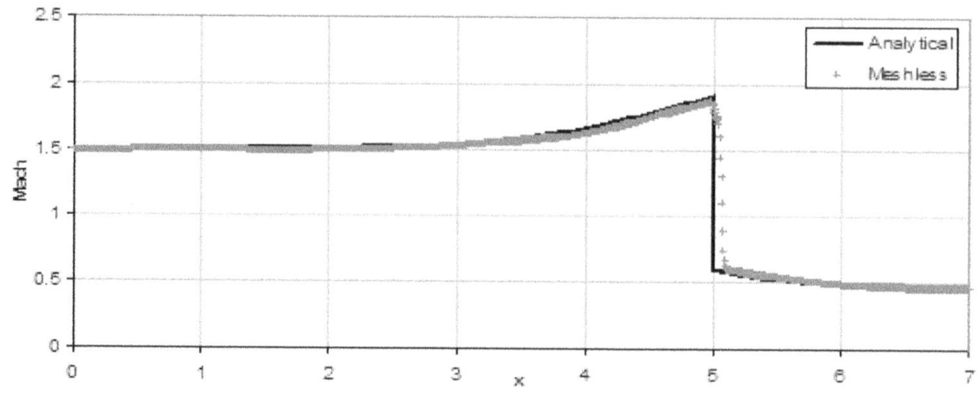

Figure 93. Qualitative comparison of Mach number along center-line of nozzle to analytical 1D solution.

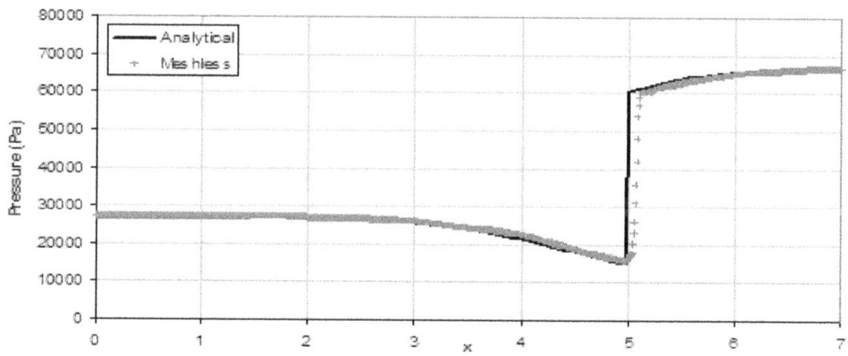

Figure 94. Qualitative comparison of pressure along center-line of nozzle to analytical 1D solution.

location. The Mach number and pressure distributions along the center-line of the Nozzle are shown in Fig. 93 and Fig. 94 respectively, revealing very good agreement with the analytical 1D solution as well as a very well defined shock location with very little spread.

9.4.3 Subsonic and Supersonic Flow Past an Airfoil

The next case was performed on a NACA-0012 airfoil at an angle of attack $\alpha = 10°$. The airfoil was placed in a subsonic flow with freestream conditions $M = 0.8$. The Meshless computational domain extends two chord lengths in front of the airfoil, eight chord lengths behind the airfoil, and six chord lengths on the top and bottom of the airfoil. Adaptive refinement was also performed on this problem. The resulting pressure coefficient and the skin friction coefficient along the top and bottom surfaces of the airfoil are shown in Fig. 95 and Fig. 96 respectively in comparison with the results provided in Marshall and Ruffin [75] and Casalini and Dadone [76]. The resulting contours of Mach number are shown in Fig. 97.

The freestream flow conditions of the airfoil problem were modified to model supersonic flow at a Mach number $M = 2.0$. A similar adaptive refinement process was performed with more focus on local airfoil region and less adaptation in bulk flow field. The resulting pressure coefficient and the skin friction coefficient along the top and bottom surfaces of the supersonic flow over the airfoil are shown in Fig. 98 and Fig. 99 respectively in comparison with the results provided in Marshall and Ruffin [75]. The resulting contours of Mach number for the supersonic case are shown in Fig. 100.

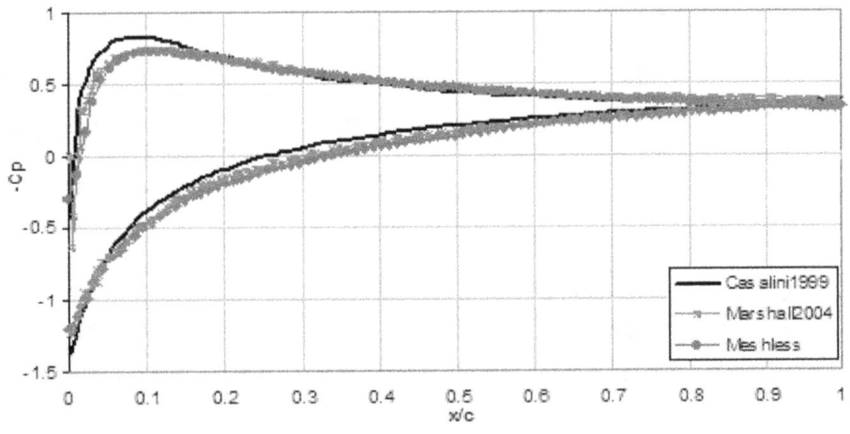

Figure 95. Pressure coefficient along the top and bottom surfaces of the subsonic flow airfoil (from [47]).

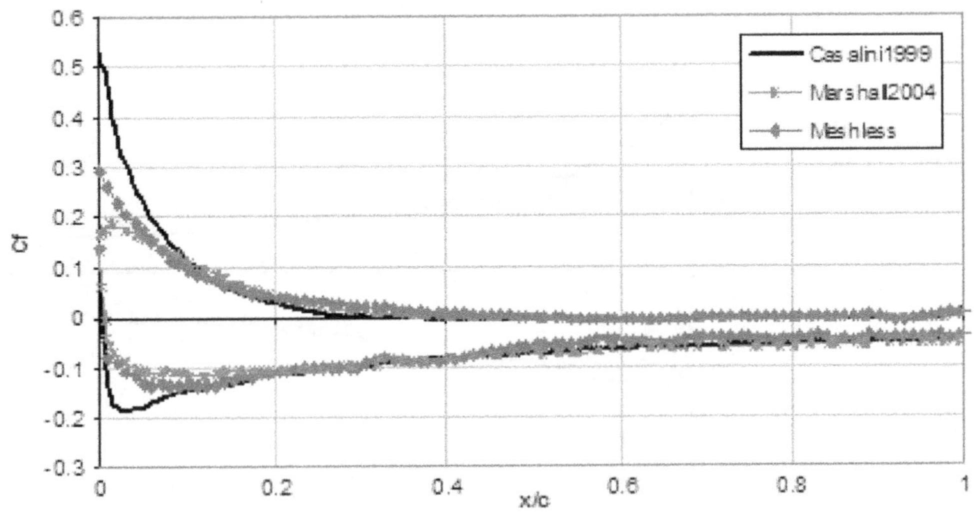

Figure 96. Skin friction coefficient along the top and bottom surfaces of the subsonic flow airfoil (from [47]).

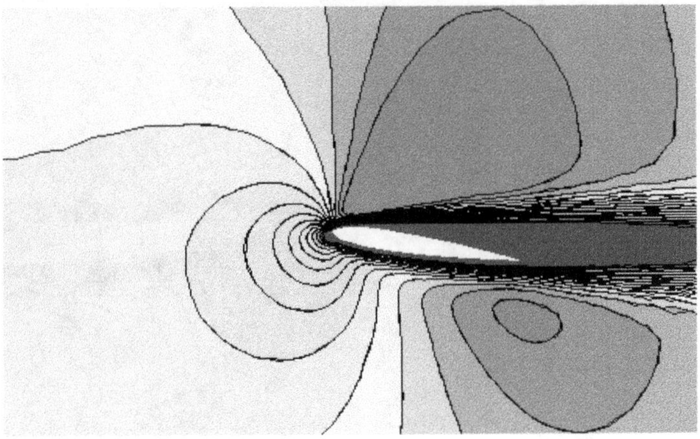

Figure 97. Contours of Mach number for subsonic flow over an airfoil (from [47]).

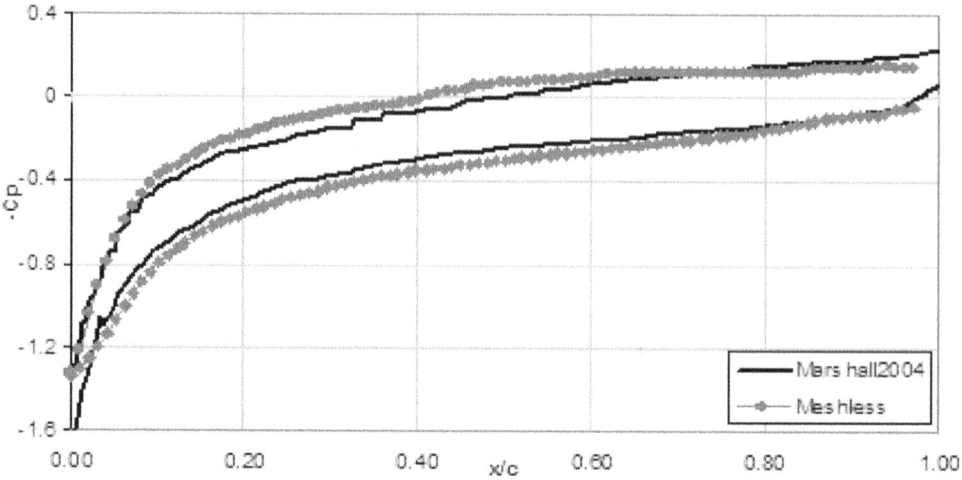

Figure 98. Pressure coefficient along the top and bottom surfaces of the supersonic flow airfoil (from [47]).

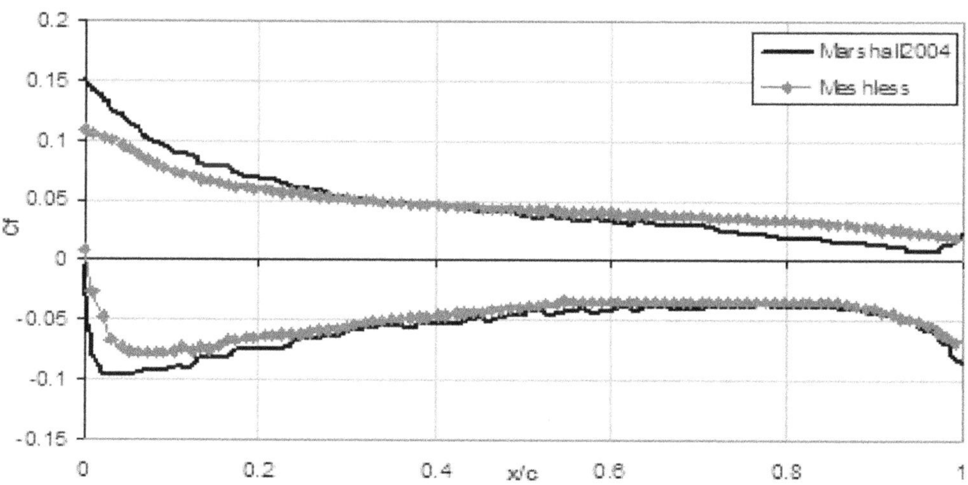

Figure 99. Skin friction coefficient along the top and bottom surfaces of the supersonic flow airfoil (from [47]).

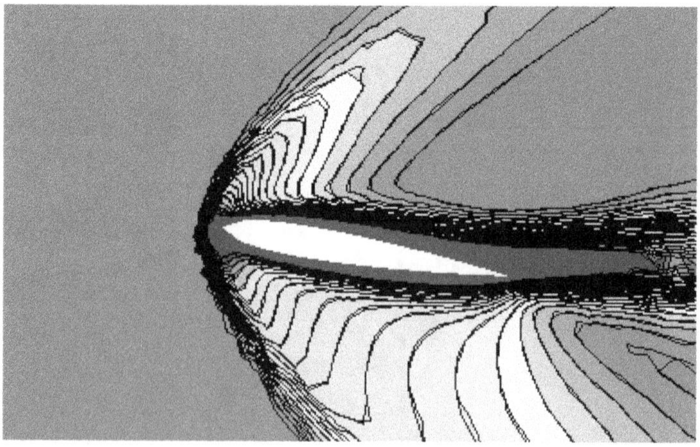

Figure 100. Contours of Mach number for supersonic flow over an airfoil (from [47]).

9.4.4 Turbulent Wake Flow

The flow in the wake behind a projectile traveling at a subsonic speed ($M = 0.6$) was analyzed. Again, the k–ω turbulence model was implemented in the Meshless code. The geometry and flow conditions are shown in Fig. 101. The resulting contours of Mach number are shown in Fig. 102, while the contours of turbulent viscosity, turbulent kinetic energy, and turbulent specific dissipation rate are shown in Fig. 103, Fig. 104, and Fig. 105 respectively.

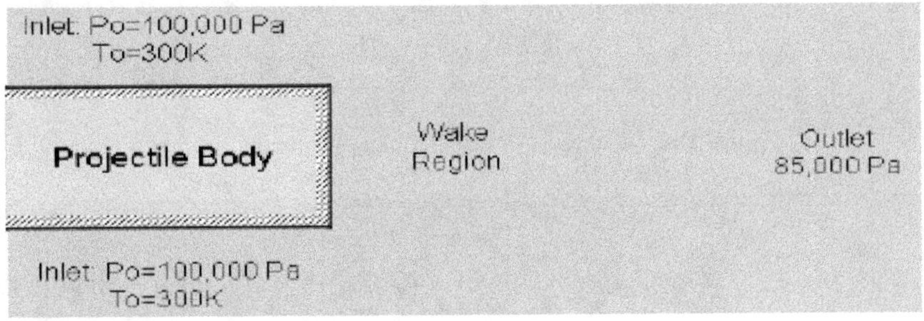

Figure 101. Projectile geometry, flow domain, and conditions.

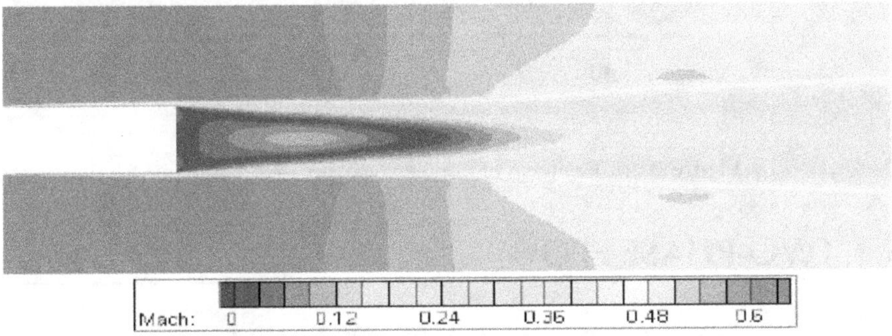

Figure 102. Contours of Mach number for flow in the wake of a projectile.

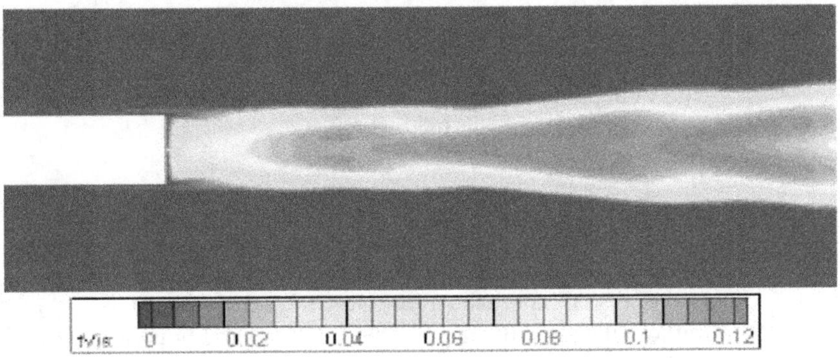

Figure 103. Contours of turbulent viscosity for flow in the wake of a projectile.

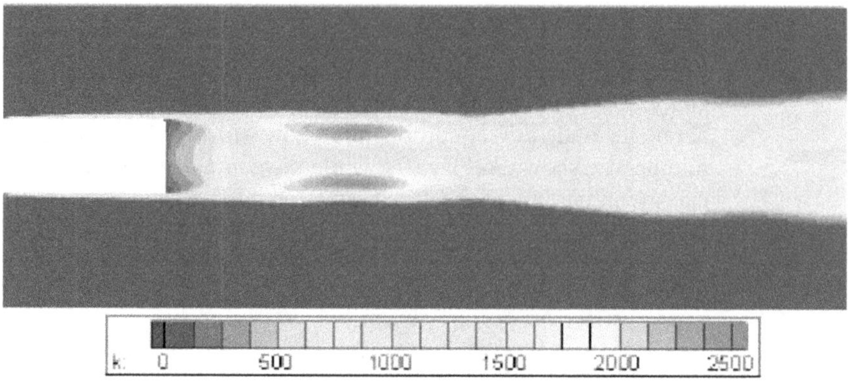

Figure 104. Contours of turbulent kinetic energy for flow in the wake of a projectile.

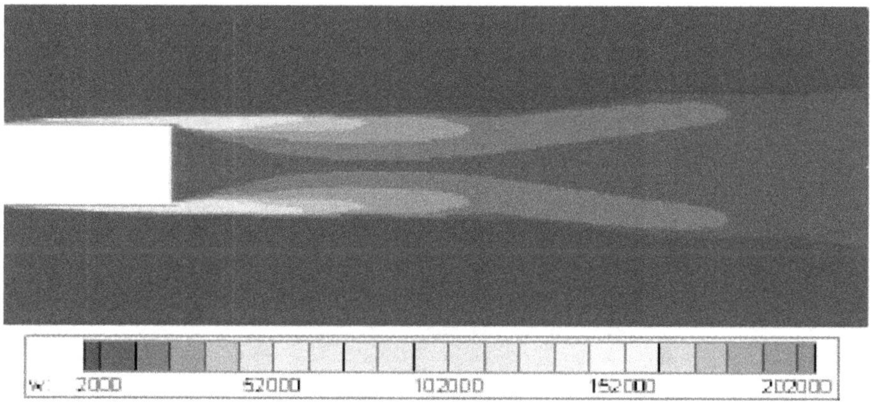

Figure 105. Contours of turbulent specific dissipation rate for flow in the wake of a projectile.

9.5 TWO-PHASE FLOW

Free surface flow problems as well as two-phase non-mixing problems can be approximated by the implementation of a Volume-of-Fluid (VOF) approach, see [77], to trace the interface at the free surface or between the two dissimilar fluids. In this approach a variable s is transported within the domain to quantify the absolute content of one of the fluids ($s = 1$) or the absolute absence of it ($s = 0$) as:

$$\frac{\partial s}{\partial t} + \left(\vec{V} \cdot \nabla\right) s = 0 \tag{122}$$

The VOF parameter, s, is then used as a weighting factor for the thermo-physical properties of the two fluids as, for instance:

$$\mu = \left(1 - s\right) \mu^1 + s\mu^2 \tag{123}$$

where μ^1 and μ^2 are the absolute viscosities of the two fluids.

The VOF approach allows to analyze two-phase, two-fluid, or free-surface problems without the need of tracking the interface and braking the problem up into two dynamic regions. Instead, the two-phase problem can be conceived as a single-region problem where an extra variable s is used to post-determine the location of the interface between the two phases, ($s = 0.5$). To this end, the solution approach followed in Section 9.1 for Incompressible Fluid Flow problems, for example, can be easily extended to account for two non-mixing fluids within the same domain by simply adding the explicit solution of the VOF parameter, s, at every time level as:

$$s^{k+1} = s^k - \Delta t\left(\vec{V}^k \cdot \nabla\right)s^k \tag{124}$$

And by expanding the VOF parameter, s, in the same localized Meshless topologies used for the rest of the field variables, leading to:

$$s_c^{k+1} = s_c^k - \Delta t\left(V_x^k\{\partial x\}^T\{s^k\} + V_y^k\{\partial y\}^T\{s^k\} + V_z^k\{\partial z\}^T\{s^k\}\right) \tag{125}$$

It is important to mention that because the VOF parameter, s, is being purely transported within the field, any artificial numerical diffusion of it will cause 'blurring' of the interface between the two fluids. Therefore, it is imperative that a well-tuned high-order upwinding approach is implemented as that described in detail in Chapter 6.

9.5.1 Dam-Breaking Test of Two-Phase Flow Formulation

The classical benchmark problem of the instantaneous dam-breaking in a 'sloshing' tank is tested using the LCMM VOF approach. Here, the tank is initially filled with air ($s = 0$) everywhere except on the bottom left corner where a block of water ($s = 1$) is trapped by a

Figure 106. Schematics of the initial condition and LCMM point distribution of the dam-breaking problem.

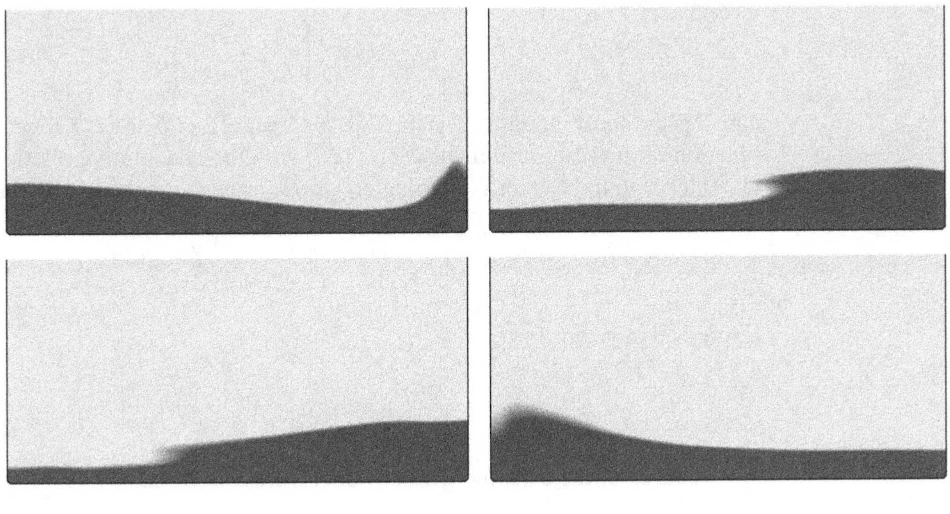

Figure 107. Time-progression of the dam-breaking problem at $t = 3, 6, 9, 12$ s.

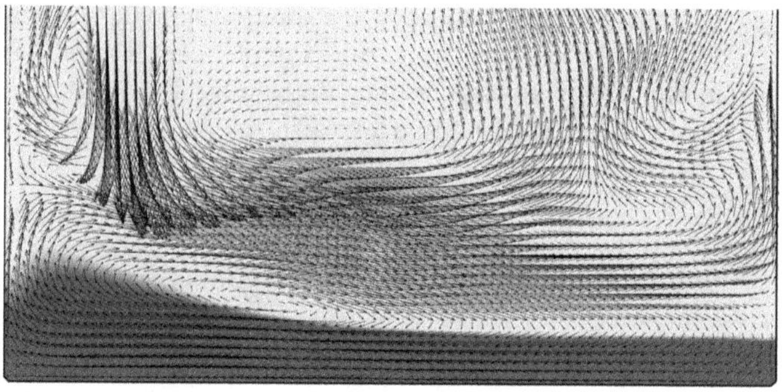

Figure 108. Velocity vector plot of the dam-breaking problem solution at $t = 12\ s$.

dam as seen in Fig. 106. The tank is 20 m wide by 10 m high and the initial water content is 8 m wide by 4 m high. The LCMM point distribution is also shown in Fig. 106.

The dam is instantaneously removed at $t = 0$ and the water is allowed to slosh towards the right-hand side of the tank. Plots of the time-progression of the water sloshing are shown in Fig. 107 for time values of $t = 3, 6, 9, 12\ s$. A plot of the velocity vectors is shown in Fig. 108 revealing the motion of the air ($s = 0$) induced by the water ($s = 1$) sloshing.

9.6 SOLID MECHANICS AND THERMO-ELASTICITY

The steady-state of stresses in a solid medium is governed by the equilibrium equation as follows:

$$\frac{\partial \sigma_{ij}}{\partial x_j} = -b_i \tag{126}$$

where σ_{ij} is the stress tensor, and b_i is the body force vector. In a Hookean solid under small deformations the stress tensor is linearly related to the strain tensor as:

$$\sigma_{ij} = \frac{2G\nu}{(1-\nu)} \delta_{ij} e_{ii} + G e_{ij} \tag{127}$$

where the strain tensor, e_{ij}, is defined in terms of the deformation vector, u_i, as:

$$e_{ij} = \frac{1}{2}\left(\frac{\partial u_i}{\partial x_j} + \frac{\partial u_j}{\partial x_i} \right) \tag{128}$$

and G is the shear modulus, ν is the Poisson's ratio, and δ_{ij} is the Kronecker delta. The combination of the equilibrium equation (126) with the constitutive relations in Eq. (127) and (128) leads to the Navier governing equation for the deformation field as:

$$G \frac{\partial^2 u_i}{\partial x_j \partial x_j} + \frac{G}{(1-2\nu)} \frac{\partial^2 u_j}{\partial x_i \partial x_j} = -b_i \tag{129}$$

Expressed in vector form:

$$G \nabla^2 \vec{u} + \frac{G}{(1-2\nu)} \nabla (\nabla \cdot \vec{u}) = -\vec{b} \tag{103}$$

In the case where thermo-elasticity effects are considered, the field stresses are defined as:

$$\sigma_{ij} = \sigma_{ij}^{e} + \sigma_{ij}^{T} \tag{131}$$

where the term σ_{ij}^{e} represents the contribution to the stress components due to the actual straining of the material, while the term σ_{ij}^{T} represents the thermal expansion effect, see [32,67,78], given by:

$$\sigma_{ij}^{T} = -m\delta_{ij}\left(T - T_{ref}\right) \tag{132}$$

where the thermo-elastic constant, m, is defined as:

$$m = \frac{2G\alpha(1+\nu)}{(1-2\nu)} \tag{133}$$

and α is the thermal expansion coefficient. Introducing the thermal stress field of Eq. (132) in the Navier equation (130) leads to:

$$G\nabla^2\vec{u} + \frac{G}{(1-2\nu)}\nabla\left(\nabla\cdot\vec{u}\right) = -m\nabla T \tag{134}$$

The thermo-elastic Navier equation in (134) is then coupled with the steady-state heat conduction equation:

$$k\nabla^2 T + \dot{u}_g''' = 0 \tag{135}$$

Notice that the thermo-physical properties, G, k, m, need not be assumed to be constant parameters. In the case these properties are considered to be space-dependent, generalized forms of the governing equations (134) and (135) can easily be derived.

The thermo-elasticity governing equations (134) and (135) are loosely coupled (one-way) unless the thermal conductivity, k, or the heat generation, $\dot{u}_g'''$, are affected by the deformation field, $\vec{u}$. Therefore, these steady-state equations may be solved independently using the LCMM and the Generalized Minimization of Residuals (GMRes) global iteration scheme devised for the solution of the Poisson equations that arise at every time step of the solution process for Incompressible Fluid Flow problems, see Eq. (81) and Eq. (87). That is, the thermal problem in Eq. (135) along with a complete set of boundary conditions may be algebraically expressed as:

$$[A_T]\{T\} = \{b_T\} \tag{136}$$

where each row of the system in equation (136) contains the localized influence over each data center defined by the LCCM expansion on the corresponding topology. For instance, an internal data center topology of influence will result in:

$$\{L\}^T\{T\} = -\frac{\dot{u}_g'''}{k} \tag{137}$$

and similarly for the boundary data centers where boundary conditions are applied.

Once the thermal problem has been solved, the Navier equation along with a complete set of boundary conditions for the elastic problem may be algebraically expressed as:

$$[A_u]\{u\} = \{b_u\} \tag{138}$$

In this case, each row of the system in equation (138) contains the localized influence over each data center defined by the LCCM expansion on the corresponding topology as:

$$\{L_u\}^T\{u\} = -\frac{m}{\mu}\{\partial\}^T\{T\} \tag{139}$$

Notice however that the size of the system equation (138) and hence each of its corresponding rows in equation (139) is multiplied by the dimensionality of the problem (2 for 2D and

3 for 3D) because the Navier equation in (134) is a vector equation and all the components of the deformation vector, $\vec{u}$, must be solved simultaneously. The boundary conditions for the elastic problem may be a combination of specified deformation, $\hat{u}$, and specified traction, $\hat{t}$. The boundary tractions are defined as:

$$t_i = \sigma_{ij} n_j \tag{140}$$

where n_j is the outward-drawn unit normal vector. This can be easily expressed in terms of the deformation vector by using the constitutive relations in Eqs. (127) and (128), as well as the corresponding traction due to thermal stresses from Eq. (132), leading to:

$$t_i = \frac{2G\nu}{(1-\nu)}\left(\frac{\partial u_k}{\partial x_k}\right)n_i + \frac{G}{2}\left(\frac{\partial u_i}{\partial x_j} + \frac{\partial u_j}{\partial x_i}\right)n_j - m\left(T - T_{ref}\right)n_i \tag{141}$$

The expression in Eq. (141) may then be solved for the deformation field, u_i, in cases when the boundary traction is imposed, leading to an algebraic expression of the form:

$$\{L_t\}^T \{u\} = f(\hat{t}, T) \tag{142}$$

where the LCMM operator $\{L_t\}$ contains the localized topology influence of the differential operators in Eq. (141).

9.6.1 Cantilever Beam under Constant Distributed Load

The first is the benchmark example of a simple 5 × 1 *m* cantilever beam under a constant distributed traction on the top wall of 10^4 *Pa*. The beam properties are assumed to be those of a steel with density $\rho = 7749.84$ kg/m^3, shear modulus $G = 79.61$ *GPa*, and Poisson ratio $\nu = 0.291$. The boundary is distributed with 120 equally-spaced points and the interior with 500 equally-spaced points for the LCMM analysis. The solution is shown in Fig. 109 where the Von Mises stress field is displayed in the undeformed geometry and the LCCM point distribution is displayed and colored with the corresponding deformation field magnified 10000 times. A maximum deformation of 5.09×10^{-5} m is obtained by the LCMM implementation which coincides with the exact solution.

9.6.2 Cortical Bone with Fixation Element under Bending Moment

A diaphysis section of a cortical bone model with a fixation element (Shanz screw) is presented. The cortical bone is considered a transversally isotropic material. The shear modulus $G = 11.9$ *GPa*, the Poisson ratio is set to $\nu = 0.31$, and the thermal conductivity is $k =$ 0.535 *W/mK*, see [79,80]. The thermal expansion coefficient is $\alpha = 9.15 \times 10^{-7} K^{-1}$, see [81]. The cortical bone geometry as seen in Fig. 110 was obtained from a tomography of an ac-

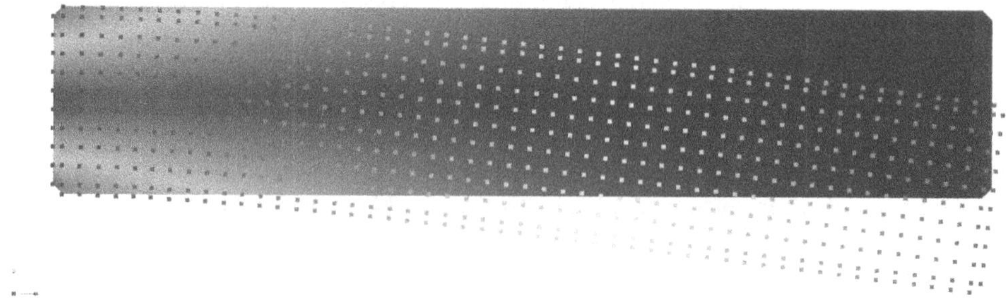

Figure 109. Von Mises stress field and magnified deformation on LCCM point distribution for the cantilever beam problem.

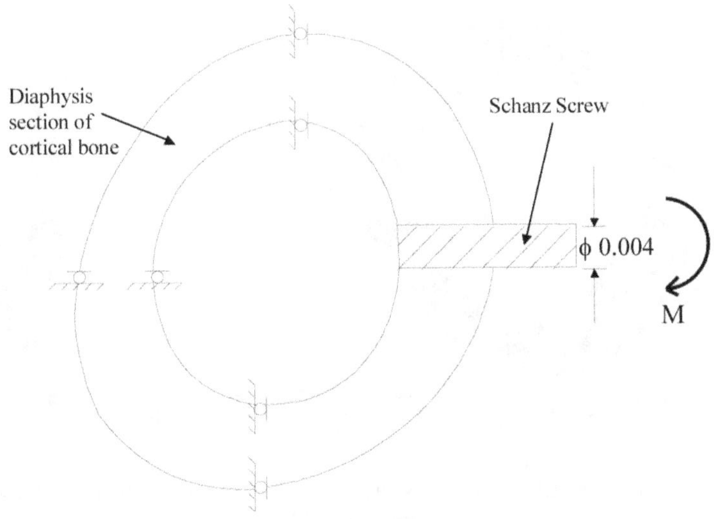

Figure 110. Cortical bone model geometry and boundary conditions.

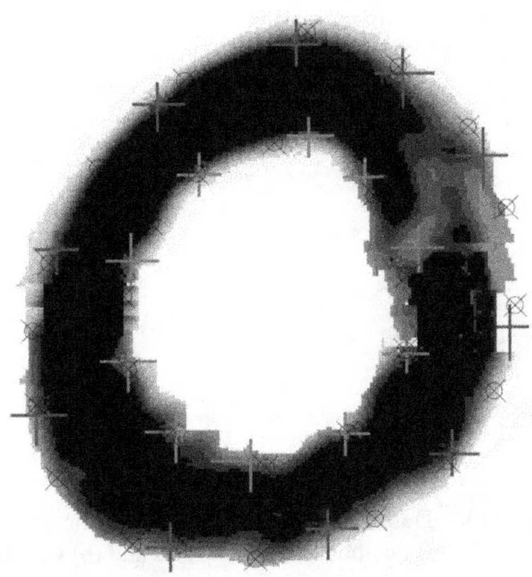

Figure 111. Cortical bone section tomography.

tual patient, see Fig. 111. The fixation element dimensions are $h = 4\ mm$ and $L = 20\ mm$. The bone is imposed with a temperature $T = 311\ K$ from its periphery while the Schanz screw is imposed with a lower temperature of $T = 293\ K$ on its end to simulate the effects of a colder fixation element being heated by the bone. A bending moment is applied to the end of the Schanz screw to simulate the fixation adjustment procedure. The thermal conductivity, thermal expansion coefficient, shear modulus, and Poisson ratio for the Schanz screw are $k = 14.9\ W/mK$, $\alpha = 13 \times 10^{-6} K^{-1}$, $G = 60\ GPa$, and $\nu = 0.3$, respectively.

The problem is solved as a two-domain problem with the bone sub-domain being solved independently from the fixation element sub-domain and requiring continuity of the deformation and traction fields in an Schwartz iteration process, see [32]. The automatic point distribution technique demonstrated in previous sections is employed to distribute points in the interior of the two domains maintaining a conformal point distribution along the interface between the two domains. The analysis was carried out with and without thermal conditions, and converged results to an iterative norm of 10^{-5} were obtained in just

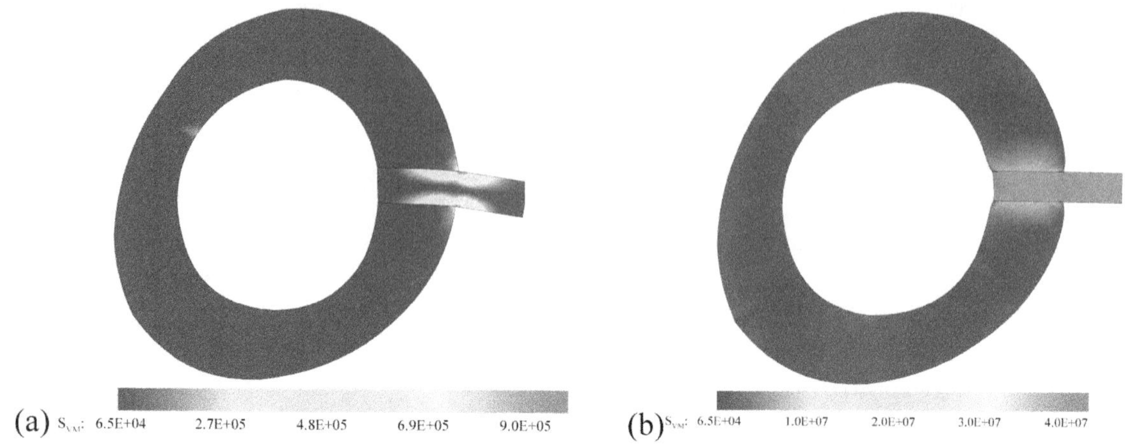

Figure 112. Von Mises stress field for the cortical bone model with fixation element: (a) without thermal effects and (b) with thermal effects.

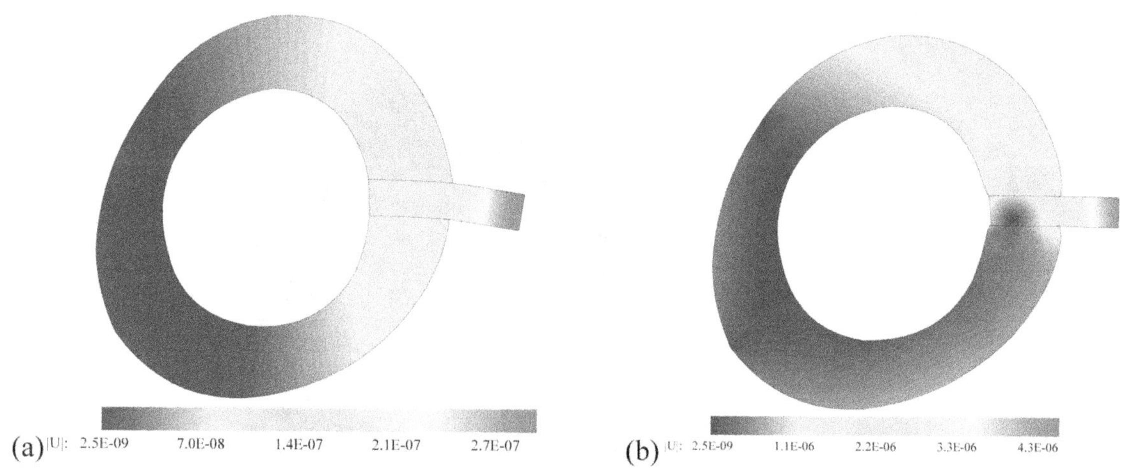

Figure 113. Deformation magnitude field for the cortical bone model with fixation element: (a) without thermal effects and (b) with thermal effects.

a few iterations. Plots of the Von Mises stress field for the elastic and thermo-elastic cases are shown in Fig. 112 while the deformation fields for the elastic and thermo-elastic cases are shown in Fig. 113. Note the strong influence of the thermal field over the elastic field when the coupled thermo-elastic model is analyzed.

9.7 POROUS MEDIA FLOW AND PORO-ELASTICITY

The equations that govern the flow of a fluid through a poro-elastic medium are the Navier equation coupled with the Richards equation [82,83] as:

$$\begin{aligned} G\nabla^2\vec{u} + \frac{G}{(1-2\nu)}\nabla(\nabla\cdot\vec{u}) &= \nabla p \\ \phi\beta_f\frac{\partial p}{\partial t} &= \nabla\cdot\left[\frac{\kappa}{\mu_f}(\nabla p + \rho_f\vec{g})\right] - \frac{\partial}{\partial t}(\nabla\cdot\vec{u}) \end{aligned} \tag{143}$$

Here, p is the pore pressure, ϕ is the porosity of the medium, κ is the permeability of the medium, and β_f, ρ_f, μ_f are respectively the compressibility, density, and viscosity of the fluid flowing through the porous medium. Notice that the two governing equation in (143) are strongly coupled (two-way) as the pore pressure appears on the right-hand side of the Navier equation while the deformation field dilatation, $(\nabla \cdot \vec{u})$, appears on the right-hand side of the Richards equation.

An important parameter in flow through porous media is the hydraulic conductivity which is defined as:

$$K_h = \frac{\kappa \rho_f \,|\, \vec{g} \,|}{\mu_f} \tag{144}$$

The Darcy velocity of the flow through the porous medium is defined as:

$$\vec{V} = -\frac{\kappa}{\mu_f}\left(\nabla p + \rho_f \vec{g}\right) \tag{145}$$

Therefore, the actual front velocity of the flow is given by:

$$\vec{V}_f = \frac{\partial \vec{u}}{\partial t} - \frac{\kappa}{\phi \mu_f}\left(\nabla p + \rho_f \vec{g}\right) \tag{146}$$

The front velocity of the flow, $\vec{V}_f$, is used to track the location of the saturation time within the porous medium by adopting the same VOF approach detailed in Section 9.5 for two-phase flows. Thus, the location of the saturation (phreatic) line, s, at any time-level can be traced by a transport equation as:

$$\frac{\partial s}{\partial t} + \left(\vec{V}_f \cdot \nabla\right) s = 0 \tag{147}$$

The VOF parameter, s, can be used to weight the value of the physical properties between the saturated and unsaturated values as:

$$G = (1 - s)G^u + sG^s \tag{148}$$

where, for instance, G^u is the shear modulus of the unsaturated medium while G^s is the shear modulus of the saturated medium.

Notice also that the Navier equation in (143) is assumed to be in steady-state while the Richards equation in (143) is assumed to be in transient mode. This can be implemented in a scheme where the Richards equation is explicitly evolved in time as:

$$p^{k+1} = p^k + \frac{\Delta t}{\phi \beta_f}\left\{\nabla \cdot \left[\frac{\kappa}{\mu_f}\left(\nabla p^k + \rho_f \vec{g}\right)\right] - \frac{\partial}{\partial t}\left(\nabla \cdot \vec{u}^k\right)\right\} \tag{149}$$

using the time-accurate LCMM scheme described in Section 9.1 while the deformation field, $\vec{u}$, is updated quasi-statically using the scheme described in Section 9.6. Updating the deformation field, $\vec{u}$, need not happen at every time-level but rather every few time-levels to accelerate the solution process. A simple systematic approach can be adopted to determine when the deformation field needs to be updated based on quantified changes of the pressure field.

Furthermore, it is important to note that there is a very significant difference in the time scale at which the pressure signal, p, propagates throughout the medium in comparison to the time scale at which the saturation front, s, moves within the medium. This is evident in Eq. (149) where it can be seen that fluid compressibility, β_f, appears in the denominator of the right-hand side. The fact that the fluid compressibility, β_f, is usually a very small number renders the propagation of the pressure signal, p, very fast in comparison with the velocity, $\vec{V}_f$, at which the saturation line, s, moves. Therefore, it is very important to take special care when progressing equations (147) and (149) as completely different time-steps, Δt, must be used. A simple solution to this issue is to solve the Richards equation as a steady-state equation using the global GMRes iteration scheme described in Section 9.1.

9.7.1 Rectangular Poro-Elastic Medium

The first example deals with a 10×20 *cm* rectangular poro-elastic medium with a centered 2 *cm* opening on the upper wall at which a fixed pore pressure value of $p = 1000$ *Pa* is imposed. The domain is constrained to have zero displacements at this opening and is allowed to freely deform elsewhere. The domain is assumed to have the approximate properties of a biological tissue such as the lungs with density $\rho = 1000\ kg/m^3$, shear modulus $G = 10^4\ Pa$, Poisson ratio $\nu = 0.3$, permeability $\kappa = 10^{-5}\ \mathrm{m}^2$, and porosity $\phi = 0.8$. The fluid flowing through the porous medium is assumed to be air with density $\rho = 1.22\ kg/m^3$, viscosity $\mu = 1.79 \times 10^{-5}\ Pa \cdot s$, and compressibility $\beta = 10^{-5}\ Pa^{-1}$. The geometry and LCMM point distribution for this problem is shown in Fig. 114.

An incremental small deformation approach is implemented to account for the fact that large deformations are attained. Through this approach, the small deformation elastostatics formulation of Section 9.6 is employed in a setting where the geometry is updated every few time levels once the threshold of small deformations is surpassed. This allows to refer the current stress-deformation state to the previous one rather than to an undeformed reference state and, thus, employ the small deformation formulation described herein. The time progression of the pore pressure field and the deformation field (magnified 10 times) for the first millisecond is shown in Fig. 115.

9.7.2 Air Flow Coupled with Poro-Elastic Balloon

In this example incompressible air is flown through a fixed pipe coupled with a porous balloon. Again, the balloon is assumed to have the approximate properties of lung tissue as used in the previous example problem so that this problem simulates the interaction between a single bronchial branch and a lung lobe assumed to behave as a poro-elastic medium. The geometry and LCMM point distribution for this case is shown in Fig. 116.

Figure 114. Geometry and LCMM point distribution of rectangular poro-elastic medium.

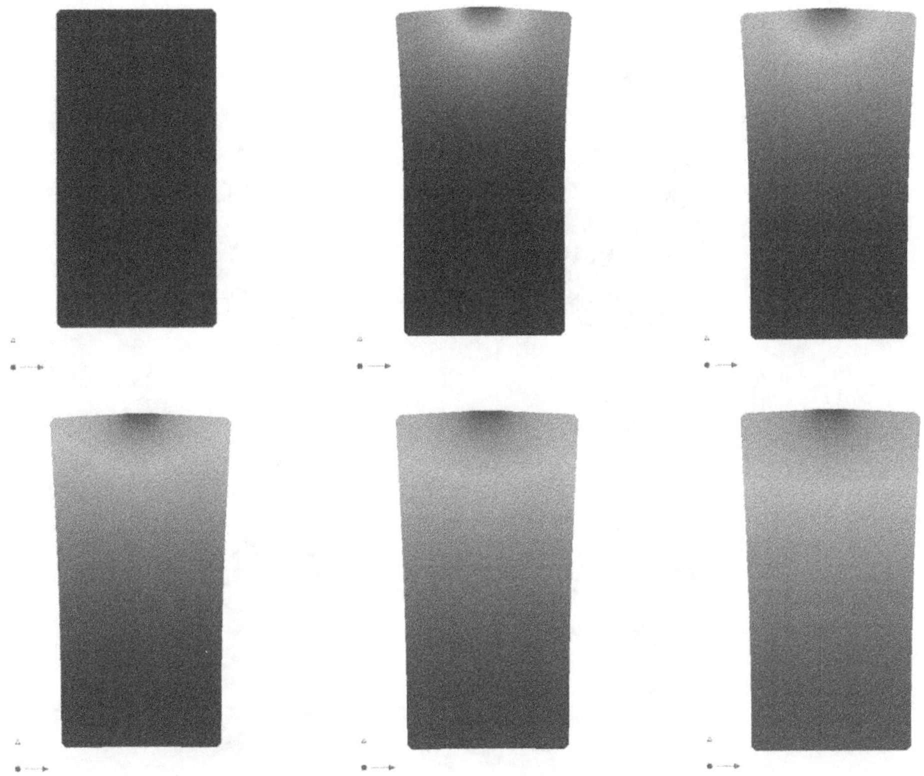

Figure 115. Time progression of the pressure and deformation for the rectangular poro-elastic medium.

The pipe is 2 × 30 *cm* while the balloon is 20 *cm* wide by 30 *cm* high. The inlet velocity of the air into the pipe is assumed to be 1 *m/s* while the inlet pressure is assumed to be 1000 *Pa*. Again, an incremental approach is followed to account for the large deformations that result from the poro-elastic analysis. The time progression of the deformation field (magnified 4 times) for the first few milliseconds is shown in Fig. 117.

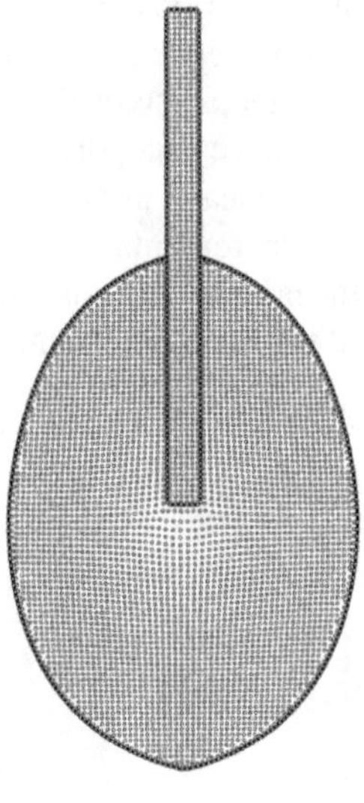

Figure 116. Geometry and LCMM point distribution for the coupled pipe poro-elastic balloon example.

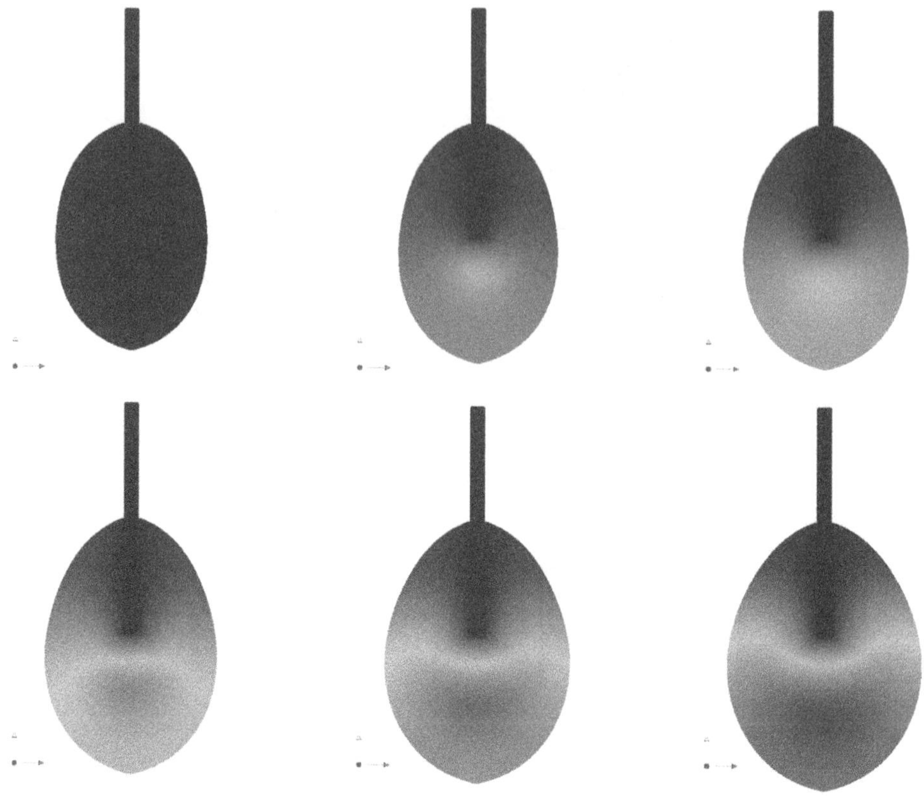

Figure 117. Time progression of the deformation field for the coupled pipe poro-elastic balloon example.

9.7.3 Coupled Tracheo-Bronchial Poro-Elastic Lung

In this example incompressible air is again flown through a system of pipes that simulate the trachea and the first bifurcations of the bronchi coupled with a porous medium that simulates the geometry of an antero-posterior slice of the lungs. The porous medium that simulates the lungs is assumed to have the approximate properties of lung tissue as used in the previous two examples. The geometry of the coupling between the tracheo-bronchial tract and the lungs assumed to behave as a poro-elastic medium is shown in Fig. 118. The inlet velocity of the air into the trachea is assumed to be 1 *m/s* while the inlet pressure is assumed to be 1000 *Pa*. Again, an incremental approach is followed to account for the large deformations that result from the poro-elastic analysis. The time progression of the velocity contours in the tracheo-bronchial tract and deformation field in the right lung is shown in Fig. 119.

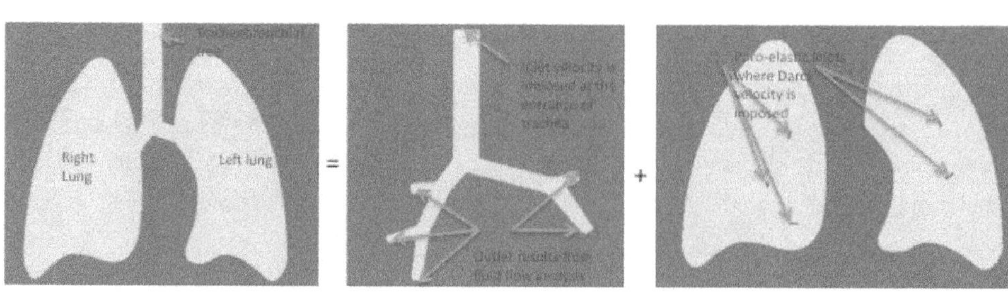

Figure 118. Gemoetry and coupling between the tracheo-bronchial tract and the lungs.

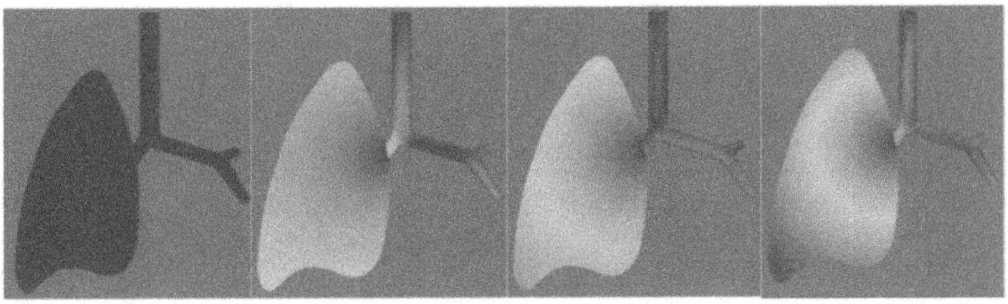

Figure 119. Time progression of the velocity contours in the tracheo-bronchial tract and deformation field in the right lung.

9.7.4 Groundwater Flow through a Poro-Elastic Levee

The final example of this section deals with the multi-disciplinary problem of water flow through a compacted sand poro-elastic levee. This is a problem of major relevance in the field of groundwater flow as it combines the effects of storm surges to the levee as well as the progressive piping (crack propagation) within the levee due to the pore pressure and stress field. The schematics of this problem are illustrated in Fig. 120. The trapezoidal portion of the levee is 40 *m* at the base and 3 *m* at the top with a height of 6 *m*. The water level on the left-hand side of the levee is 1 *m* from the top of the levee. A 3 *m*-long pipe (crack) inclined 45° is included 8 *m* from the left-hand corner of the trapezoidal section of the levee.

The levee hydraulic conductivity $K_h = 10^{-7}\ m/s$, the Poisson ratio $\nu = 0.35$, the modulus of elasticity $E = 120\ MPa$, the dry density $\rho = 1500\ kg/m^3$, the specific gravity of soil particles $SG = 2.65$, the porosity $\phi = 0.3$, the water density, viscosity, and compressibility

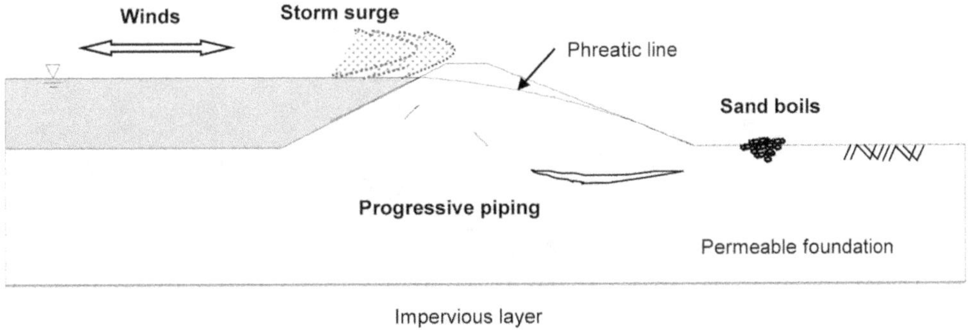

Figure 120. Schematics of the groundwater flow through a poro-elastic sand levee problem.

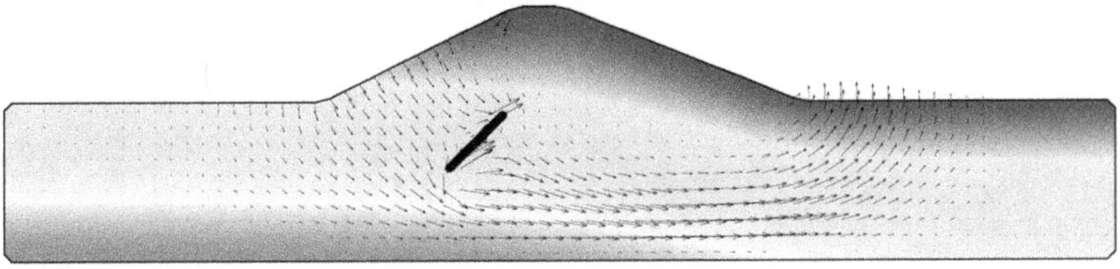

Figure 121. Contour plots of hydrostatic pressure levels and water flow vectors within the levee.

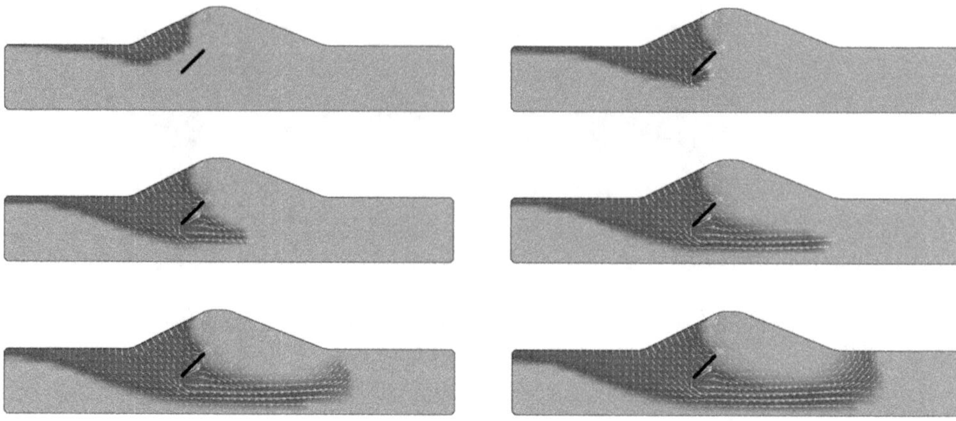

Figure 122. Time progression of the saturation front within the poro-elastic levee at $t = 20, 40, 60, 80, 100, 120$ days.

are $\rho_f = 1000\ kg/m^3$, $\mu_f = 10^{-3}\ Pa \cdot s$, and $\beta_f = 4.55 \cdot 10^{-10}\ m^2/N$ respectively. The contour plots of hydrostatic pressure levels and water flow vectors within the levee and around the pipe are shown in Fig. 121 after a steady-state solution is acheived. The time progression of the saturation front within the poro-elastic levee at $t = 20, 40, 60, 80, 100, 120$ days is shown in Fig. 122.

Chapter 10

CONCLUSIONS

The finite element method (FEM) and the boundary element method (BEM) have been developed to a mature stage such that they are now utilized routinely to model complex multi-physics problems. Meshless methods are a relative newcomer to the field of computational methods. The term "Meshless Methods" refers to the class of numerical techniques that rely on either global or localized interpolation on non-ordered spatial point distributions. As such, there has been much interest in the development of these techniques as they have the hope of reducing the effort devoted to model preparation. The approach finds its origin in classical spectral or pseudo-spectral methods that are based on global orthogonal functions such as Legendre or Chebyshev polynomials requiring a regular nodal point distribution. In contrast, Meshless methods use a nodal or point distribution that is not required to be uniform or regular due to the fact that most such techniques rely on global radial-basis functions (RBF). However, global RBF-based Meshless methods have some drawbacks including poor conditioning of the ensuing algebraic set of equations which can be addressed to some extent by domain decomposition and appropriate pre-conditioning. Moreover, care must be taken in the evaluation of derivatives in global RBF-based Meshless methods. Although, very promising, these techniques can also be computationally intensive. Recently, localized collocation Meshless methods have been proposed to address many of these issues.

An effective, efficient, and robust Meshless Method based on collocation of Hardy Multiquadrics RBF over scattered points on localized topologies was formulated, implemented, and applied to a variety of field problems ranging from laminar and turbulent, incompressible and compressible, subsonic and sonic, single and two-phase fluid flows and heat transfer, to elasticity and poro-elasticity. The Localized Collocation Meshless Method (LCMM) finds its efficiency through a formulation that allows the generation of central and upwinded derivative fields through simple inner products of small-order vectors while its robustness is found by blending the interpolation scheme with an effective moving least-squares smoothing scheme to mitigate the issues of oscillative derivative fields around large gradient areas. In addition, the highly automated nature of the point generation process has made it possible to implement an adaptive nodal distribution method which is capable of both boundary and interior refinement.

The results for the wide range of applications presented herein demonstrate the accuracy and stability of the LCMM in both qualitative and quantitative terms. In addition, the formulation illustrates a key advantage of this technique over more traditional methods; mainly that the robustness of the LCMM algorithms allows arrival at high quality solutions even when very coarse initial point distributions are used. Thus, by coupling the LCMM routines with an automatic refinement procedure results in an approach whereby the analyst need not have any prior understanding of the expected solution in order to arrive at accurate results. This provides confidence that important problem characteristics that are not accurately reproduced in the original point distribution will be automatically detected, properly refined, and well captured by the end of the solution analysis. Thereby, consolidating the LCMM as an industrially-relevant numerical solution tool.

REFERENCES

[1] Belytscho, T., Lu, Y.Y., and Gu, L., "Element-Free Galerkin Methods," *Int. J. Num. Methods*, Vol. 37, pp. 229–256, 1994.

[2] Atluri, S.N. and Shen, S., *The Meshless Method*, Tech. Science Press, Forsyth, 2002.

[3] Atluri, S.N. and Zhu, T., "A New Meshless Local Petrov-Galerkin (MLPG) Approach in Computational Mechanics," *Computational Mechanics*, Vol. 22, pp. 117–127, 1998.

[4] Onate, E., Idehlson, S., Zienkiewicz, O.C., Taylor, R.L., and Sacco, C., "A Stabilized Finite Point Method for Analysis of Fluid Mechanics Problems," *Computer Methods in Applied Mechanics and Engineering*, Vol. 139, pp. 315–346, 1996.

[5] Liu, G.R., *Mesh Free Methods: Moving Beyond the Finite Element Method*, CRC Press, Boca Raton, 2002.

[6] Melenk, J.M. and Babuska, I., "The Partition of Unity Finite Element Method: Basic Theory and Application," *Comp. Meth. Appl. Mechanics and Eng.*, Vol. 139, pp. 289–316, 1996.

[7] Gu, Y.T. and Liu, G.R., "Meshless Techniques for Convection Dominated Problems," *Computational Mechanics*, Vol. 38, pp. 171–182, 2006.

[8] Gottlieb, D. and Orzag, S.A., *Numerical Analysis of Spectral Methods: Theory and Applications*, Society for Industrial and Applied Mathematics, Bristol, England, 1977.

[9] Maday, Y. and Quateroni, A., "Spectral and Pseudo-Spectral Approximations of the Navier-Stokes Equations," *SIAM J. Numerical Analysis*, Vol. 19, No. 4, pp. 761–780, 1982.

[10] Patera, A., "A Spectral Element Method of Fluid Dynamics: Laminar Flow in a Channel Expansion," *J. of Computational Physics*, Vol. 54, pp. 468–488, 1984.

[11] Macaraeg, M. and Street, C.L., "Improvement in Spectral Collocation Discretization Through a Multiple Domain Technique," *Applied Numerical Mathematics*, Vol. 2, pp. 95–108, 1986.

[12] Hwar, C.K., Hirsch, R., Taylor, T., and Rosenberg, A.P., "A Pseudo-Spectral Matrix Element Method for Solution of Three Dimensional Incompressible Flows and Its Parallel Implementation," *J. of Computational Physics*, Vol. 83, pp. 260–291, 1989.

[13] Fasshauer, G., "RBF Collocation Methods as Pseudo-Spectral Methods," *Boundary Elements XVII*, Kassab, A., Brebbia, C.A., and Divo, E. (eds.), WIT Press, pp. 47–57, 2005.

[14] Powell, M.J.D., "The Theory of Radial Basis Function Approximation," in *Advances in Numerical Analysis, Vol. II*, Light, W., ed., Oxford Science Publications, Oxford, pp. 143–167, 1992.

[15] Buhmann, M.D., *Radial Basis Functions: Theory and Implementation*, Cambridge University Press, Cambridge, 2003.

[16] Dyn, N., Levin, D., and Rippa, S., "Numerical Procedures for Surface Fitting of Scattered Data by Radial Basis Functions," *SIAM J. of Sci. Stat. Computing*, Vol. 7, No. 2, pp. 639–659, 1986.

[17] Hardy, R.L., "Multiquadric Equations of Topography and Other Irregular Surfaces," *Journal of Geophysical Research*, Vol. 176, pp. 1905–1915.

[18] Kansa, E.J., "Multiquadrics – A Scattered Data Approximation Scheme with Applications to Computational Fluid Dynamics I-Surface Approximations and Partial Derivative Estimates," *Comp. Math. Appl.*, Vol. 19, pp. 127–145, 1990.

[19] Kansa, E.J., "Multiquadrics – A Scattered Data Approximation Scheme with Applications to Computational Fluid Dynamics II-Solutions to Parabolic, Hyperbolic and Elliptic Partial Differential Equations," *Comp. Math. Appl.*, Vol. 19, pp. 147–161, 1990.

[20] Kansa, E.J. and Hon, Y.C., "Circumventing the Ill-Conditioning Problem with Multiquadric Radial Basis Functions: Applications to Elliptic Partial Differential Equations," *Comp. Math. Appl.*, Vol. 39, pp. 123–137, 2000.

[21] Franke, R., "Scattered Data Interpolation: Test of Some Methods," *Math. Comput.*, Vol. 38, pp. 181–200, 1982.

[22] Mai-Duy, N. and Tran-Cong, T., "Mesh-Free Radial Basis Function Network Methods with Domain Decomposition for Approximation of Functions and Numerical Solution of Poisson's Equation," *Engineering Analysis with Boundary Elements*, Vol. 26, pp. 133–156, 2002.

[23] Cheng, A.H.-D., Golberg, M.A., Kansa, E.J., and Zammito, G., "Exponential Convergence and H-c Multiquadric Collocation Method for Partial Differential Equations," *Numerical Methods in P.D.E.*, Vol. 19, No. 5, pp. 571–594, 2003.

[24] Divo, E. and Kassab, A.J., "Boundary Element Method for Heat Conduction: With Applications in Non-Homogeneous Media," *Topics in Engineering Series Vol. 44*, WIT Press, Billerica, MA, 2002.

[25] Kassab, A.J., Wrobel, L.C., Bialecki, R.A., and Divo, E., "Boundary Elements in Heat Transfer," *Chapter 4 in Handbook of Numerical Heat Transfer*, Minkowycz, W. and Sparrow, E.M. (eds.), Vol. 1, 2nd Edition, John Wiley and Sons, New York, NY, pp. 125–166, 2005.

[26] Sarler, B. and Vertnik, R., "Local Explicit Radial Basis Function Collocation Method for Diffusion Problems," *Computers and Mathematics with Applications*, Vol. 51, No. 8, pp. 1269–1282, 2006.

[27] Vertnik, R. and Sarler, B., "Meshless Local Radial Basis Function Collocation Method for Convective-Diffusive Solid-Liquid Phase Change Problems," *International Journal of Numerical Methods for Heat and Fluid Flow*, Vol. 16, No. 5, pp. 617–640, 2006.

[28] Vertnik, R., Zaloznik, M., and Sarler, B., "Solution of Transient Direct-Chill Aluminium Billet Casting Problem with Simultaneous Material and Interphase Moving Boundaries by a Meshless Method," *Engineering Analysis with Boundary Elements*, Vol. 30, No. 10, pp. 847–855, 2006.

[29] Divo, E., Kassab, A.J., and Erhart, K., "Parallel Domain Decomposition BEM Techniques for Steady and Transient Heat Transfer," *Chapter 5 in Parallel BEM and Mesh Reduction Methods*, Popov, V., Power, H., and Skerget, L. (eds.), WIT Press, Billerica, MA, pp. 147–186, 2007.

[30] Divo, E.A., Kassab, A.J., and Rodriguez, F., "Parallel Domain Decomposition Approach for Large-Scale 3D Boundary Element Models in Linear and Non-Linear Heat Conduction," *Numerical Heat Transfer, Part B, Fundamentals*, Vol. 44, No. 5, pp. 417–437, 2003.

[31] Erhart, K., Divo, E., and Kassab, A.J., "A Parallel Domain Decomposition Boundary Element Method Technique for Large-Scale Transient Heat Conduction Problems," *Engineering Analysis with Boundary Elements*, Vol. 30, No. 7, pp. 553–563, 2006.

[32] Gamez, B., Ojeda, D., Divo, E., Kassab, A., and Cerrolaza, M., "Parallelized Iterative Domain Decomposition Boundary Element Method for Thermoelasticity," *Engineering Analysis with Boundary Elements*, Vol. 32, pp. 1061–1073, 2008.

[33] Divo, E., Kassab, A.J., Mitteff, E., and Quintana, L. "A Parallel Domain Decomposition Technique for Meshless Methods Applications to Large-Scale Heat Transfer Problems," *ASME Paper: HT-FED2004-56004*.

[34] Divo, E. and Kassab, A.J., "Effective Domain Decomposition Meshless Formulation of Fully-Viscous Incompressible Fluid Flows," *Boundary Elements XVII*, Kassab, A., Brebbia, C.A. and Divo, E. (eds.), WIT Press, pp. 67–77, 2005.

[35] Divo, E. and Kassab, A.J., "A Meshless Method for Conjugate Heat Transfer," *Proceedings of ECCOMAS Coupled Problems 2005*, Papadrakis, M., Oñate, E., and Schrefler, B. (Eds.), Santorini, Greece, 2005.

[36] Divo, E. and Kassab, A.J., "A Meshless Method for Conjugate Heat Transfer Problems," *Engineering Analysis*, Vol. 29, No. 2, pp. 136–149, 2005.

[37] Gerace, S.A., Divo, E., and Kassab, A.J., "A Localized Radial-Basis-Function Meshless Method Approach to Axisymmetric Thermo-Elasticity," *AIAA Paper 2006-3788*.

[38] Divo, E. and Kassab, A.J., "Modeling of Convective and Conjugate Heat Transfer by a Third Order Localized RBF Meshless Collocation Method," *Proc. of NHT-05 Eurotherm82*, Bialecki, R.A. and Nowak, A.J. (eds.), pp. 357–366, 2005.

[39] Divo, E., Kassab, A.J., and El Zahab, Z., "Parallel Domain Decomposition Meshless Modeling of Dilute Convection-Diffusion of Species," *Proceedings of BEM27/MRM*, Kassab, A.J., Brebbia, C.A., and Divo, E.A. (eds.), Computational Mechanics Publications, Southampton, U.K., pp. 79–90, 2005.

[40] Divo, E. and Kassab, A.J., "Iterative Domain Decomposition Meshless Method Modeling of Incompressible Flows and Conjugate Heat Transfer," *Engineering Analysis*, Vol. 30, No. 6, pp. 465–478, 2006.

[41] Divo, E. and Kassab, A.J., "An Efficient Localized RBF Meshless Method for Fluid Flow and Conjugate Heat Transfer," *ASME Journal of Heat Transfer*, Vol. 129, pp. 124–136, 2007.

[42] Divo, E. and Kassab, A., "Localized Collocation Meshless Method (LCMM) for Convectively Dominated Flows," *Proceedings of the 8th World Congress on Computational Mechanics and 5th European*

Congress on Computational Methods in Applied Sciences and Engineering, Venice, Italy, June 30–July 5, 2008.

[43] Divo, E. and Kassab, A.J., "Localized Meshless Modeling of Natural Convective Viscous Flows," *Numerical Heat Transfer, Part A: Fundamentals*, Vol. 53, Issue 6, pp. 487–509, June 2008.

[44] Mitteff, E., Divo, E., and Kassab, A.J., "Automated Point Distribution and Parallel Segmentation for Meshless Methods," *Proceedings of CIMENICS 2006, 8th International Congress of Numerical Methods in Engineering and Applied Sciences, held on Margarita Island, Venezuela, March 20–24, 2006*, Gamez, B., Ojeda, D., Larrazabal, G., and Cerrolaza, M. (eds.), Sociedad Venezolana de Methodos Numericos en Ingenieria, Valencia, Venezuela, pp. 93–100.

[45] El Zahab, Z., Divo, E., and Kassab, A.J., "A Localized Collocation Meshless Method (LCMM) for Incompressible Flows CFD Modeling with Applications to Transient Hemodynamics," *Engineering Analysis with Boundary Elements*, Vol. 33, No. 8–9, pp. 1045–1061, 2009.

[46] Divo, E., Kassab, A.J., Reddi, L., Kakuturu, S., and Hagen S., "RBF-FVM Numerical Solution of the Poro-Elasticity Levee Problem with Time-Varying Boundary Conditions," *Proceedings of ECCOMAS Couple Problems 2009*, Ischia Island, Italy, June 8–10, 2009.

[47] Gerace, S., Erhart, K., Divo, E., and Kassab, A.J., "Adaptively Refined Hybrid FDM-RBF Meshless Scheme with Applications to Laminar and Turbulent Viscous Fluid Flows," *Computer Modeling in Engineering and Science*, Vol. 81, No. 1, pp. 35–67, 2011.

[48] Gerace, S., Erhart, K., Kassab, A.J., and Divo, E., "A Model-Integrated Localized Collocation Meshless Method for Large Scale Three-Dimensional Heat Transfer Problems," *Computer Assisted Methods in Engineering and Science*, Vol. 20, pp. 207–225, 2013.

[49] Kelly, J., Divo, E., and Kassab, A.J., "Numerical Solution of the Two-Phase Incompressible Navier-Stokes Equations Using a GPU-Accelerated Meshless Method," *Engineering Analysis with Boundary Elements*, Vol. 40, pp. 36–49, 2014.

[50] Erhart, K., Kassab, A.J., and Divo, E., "An Inverse Localized Meshless Technique for the Determination of Non-Linear Heat Generation Rates in Living Tissues," *International Journal of Numerical Methods for Heat and Fluid Flow*, Vol. 18, No. 3/4, pp. 401–414, 2008.

[51] El Zahab, Z., Divo, E., and Kassab, A.J., "A Meshless CFD Approach for Evolutionary Shape Optimization of Bypass Grafts Anastomoses," *Journal Inverse Problems in Science and Engineering*, Vol. 17, No. 3, pp. 411–435, 2009.

[52] Leonard, B.P., "A Stable and Accurate Convective Modelling Procedure Based on Quadratic Interpolation," *Computer Methods in Applied Mechanics and Engineering*, Vol. 19, pp. 59–98, 1979.

[53] Borris, J.P. and Book, D.L., "Flux-Corrected Transport II: Generalizations of the Method," *Journal of Computational Physics*, Vol. 18, p. 248, 1983.

[54] Harten, A., "High Resolution Schemes for Hyperbolic Conservation Laws," *Journal of Computational Physics*, Vol. 49, pp. 357–393, 1983.

[55] Van Leer, B., "Towards the Ultimate Conservative Difference Scheme. V. A Second Order Sequel to Godunov's Method," *Journal of Computational Physics*, Vol. 32, pp. 101–136, 1979.

[56] Chakravarthy, S.R. and Osher, S., "High Resolution Applications of the OSHER Upwind Scheme for the Euler Equations," *AIAA Paper 83-1943*.

[57] Roe, P.L., "Some Contributions to the Modelling of Discontinuous Flows," *Lectures in Applied Mathematics*, Vol. 22, pp. 163–193, 1985.

[58] Gaskell, P.H. and Lau, A.K.C., "Curvature Compensated Convective Transport: SMART, a New Boundedness Preserving Transport Algorithm," *International Journal of Numerical Methods in Fluids*, Vol. 8, pp. 617–641, 1988.

[59] Leonard, B.P., "The ULTIMATE Conservative Difference Scheme Applied to Unsteady One-Dimensional Advection," *Computer Methods in Applied Mechanics and Engineering*, Vol. 88, pp. 17–74, 1991.

[60] Jasak, H., Weller, H.G., and Gosman, A.D., "High Resolution NVD Differencing Scheme for Arbitrary Unstructured Meshes," *International Journal of Numerical Methods in Fluids*, Vol. 31, pp. 431–449, 1999.

[61] Soni, B., Thompson, J., and Weathermill, N., *Handbook of Grid Generation*, CRC Press, 1999.

[62] Dirichlet, G.L., "Uber die Reduction der Positiven Quadratischen Formen Mit Drei Underestimmten Ganzen Zahlan," *Z. ReineAngew. Mathematics*, Vol. 40, No. 3, pp. 209–227, 1850.

[63] Harlow, F.H. and Welch, J.E., "Numerical Calculation of Time Dependent Viscous Incompressible Flow of Fluids with a Free Surface," *Physics of Fluids*, Vol. 8, pp. 2182–2189, 1965.

[64] Patankar, S.V., *Numerical Heat Transfer and Fluid Flow*, Hemisphere Press, 1980.

[65] Kim, D. and Choi, H., "A Second-Order Time-Accurate Finite Volume Method for Unsteady Incompressible Flow on Hybrid Unstructured Grids," *Journal of Computational Physics*, Vol. 162, pp. 411–428, 2000.

[66] Zhang, Y., Street, R.L., and Koseff, J.R., "A Non-Staggered Grid, Fractional Step Method for Time-Dependent Incompressible Navier-Stokes Equations in Curvilinear Coordinates," *Journal of Computational Physics*, Vol. 114, pp. 18–33, 1994.

[67] Kane, J., *Boundary Element Analysis in Engineering Continuum Mechanics*, Prentice-Hall, New Jersey, 1994.

[68] Ghia, U., Ghia, K.N., and Shin, C.T., "High-Resolutions for Incompressible Flow Using the Navier-Stokes Equations and a Multigrid Method," *Journal of Computational Physics*, Vol. 48, Issue 3, pp. 387–411, December 1982.

[69] Jung, H., Choi, J.W., and Park, C.G., "Asymmetric Flows of Non-Newtonian Fluids in Symmetric Stenosed Artery," *Korea-Australia Rheology Journal*, Vol. 16, pp. 101–108, 2004.

[70] Wilcox, D.C., *Tubulence Modeling for CFD*, DCW Industries, 2006.

[71] Hoffmann, K.A. and Chiang, S.T., *Computational Fluid Dynamics I-III*, vol. 2, Engineering Education System, 4th edn., 2004.

[72] Le, H., Moin, P., and Kim, J., "Direct Numerical Simulation of Turbulent Flow over a Backward-Facing Step," *Journal of Fluid Mechanics*, Vol. 330, pp. 349–374, 1997.

[73] Kim, J.J., Investigation of Separation and Reattachment of a Turbulent Shear Layer: Flow Over a Backward-Facing Step, Ph.D. thesis, Stanford University, 1978.

[74] Liou, M.S. and Steffen, C.J., "A New Flux Splitting Scheme," *Journal of Computational Physics*, Vol. 107, pp. 23–39, 1993.

[75] Marshall, D.D. and Ruffin, S.M., "An Embedded Boundary Cartesian Grid Scheme for Viscous Flows Using a New Viscous Wall Boundary Condition Treatment." Paper No. AIAA 2004-0581. In *42nd AIAA Aerospace Sciences Meeting and Exhibit*, Reno, NV, 2004.

[76] Casalini, F. and Dadone, A., "Computations of Viscous Flows Using a Multigrid Finite Volume Lambda Formulation." *Engineering Computations*, Vol. 16, No. 7, pp. 767–786, 1999.

[77] Hirt, C.W. and Nichols, B.D., "Volume of Fluid (VOF) Method for the Dynamics of Free Boundaries," *Journal of Computational Fluids*, Vol. 39, pp. 201–225, 1981.

[78] Biot, M.A., "Thermoelasticity and Irreversible Thermodynamics," *Journal of Applied Physics*, Vol. 23, No. 1, pp. 91–96, 1956.

[79] Relly, D.T. and Burstein, A.H., "The Elastic and Ultimate Properties of Compact Bone Tissue," *Journal of Biomechanics*, Vol. 8, pp. 393–405, 1975.

[80] Davidson, S. and James, D., "Measurement of Thermal Conductivity of Bovine Cortical Bone," *Medical Engineering & Physics*, Vol. 22, pp. 741, 747, 2000.

[81] Parker, J and Russell, K., *Thermal Expansion of Bone*. Department of Physics. College of Wooster, Wooster, Ohio.

[82] Bear, J., *Dynamics of Fluids in Porous Media*, Dover Publications, Inc., New York, 1988.

[83] Wang, H.F. and Anderson, M.P., *Introduction to Groundwater Modeling: Finite and Finite Element Methods*, W. H. Freeman and Co., San Francisco, 1982.

www.ingramcontent.com/pod-product-compliance
Lightning Source LLC
LaVergne TN
LVHW062115200926
848484LV00007B/1677